Kadimakara

EXTINCT VERTEBRATES OF AUSTRALIA

Kadimakara

EXTINCT VERTEBRATES OF AUSTRALIA

Editors
P.V. Rich
G.F. van Tets

Illustrator
Frank Knight

Princeton University Press
Princeton, New Jersey

Published by Princeton University Press
41 William Street, Princeton, New Jersey 08540

Library of Congress Cataloging-in-Publication Data

Kadimakara: extinct vertebrates of Australia/editors, P.V. Rich, G.F. van Tets; illustrator Frank Knight.
p. cm.
Originally published: Victoria, Australia: Pioneer Design Studio, 1985.
Includes bibliographical references and index.
ISBN 0-691-08733-4 (cloth:): $49.50
1. Vertebrates, Fossil –Australia. 2. Paleontology – Australia. 3. Geology – Australia. I. Rich, Pat Vickers. II. Van Tets, Gerard F. III. Knight, Frank, 1941- .
QE841.K33 1991
566'.0994 – dc20 90-8846
CIP

Printed in Singapore

Contents

Acknowledgements

Kadimakara would never have been possible were it not for a number of individuals, museums and granting bodies, in addition to all those who provided chapters for this book. Our thanks go especially to the many amateur palaeontologists, land owners, miners and quarrymen amongst others who have brought many important fossils to our attention over the years and provided hospitality and help on a number of occasions. Others have provided illustrative material or information used in *Kadimakara*: Alan Charig (British Museum (Natural History), London), J. Drexel (South Australian Department of Minerals and Energy, Parkside), Belinda Gillies, Barry Wilson, Don Anderson and Frank Coffa (Museum of Victoria, Melbourne), Frank Ippolito (American Museum of Natural History, New York, for the reconstructions of *Meiolania*), Noel Kemp (Tasmanian Museum, Hobart), B.E. Leake (University of Glasgow), Donald Russell and Mr. Laissus (Museum d'Histoire Naturalle, Paris), Charles Schaff (Museum of Comparative Zoology, Harvard University, Cambridge), Sam and Catherine Wintercooke (Murndel Station) and Rhys Walkley (Black Rock).

In discussions between Frank Knight and several artists in American museums and universities some were especially helpful, in particular Jay Matternas, Rudolph and Richard Zallinger and Chester Tarka. Frank's visit was made possible by a C.S.I.R.O. overseas study award. The Department of Earth Sciences of Monash University, Clayton, and the Commonwealth Scientific and Industrial Research Organization, Division of Wildlife and Rangelands Research, Canberra provided bases where much of the research leading to this publication was carried out. The respective institutions supporting each of the authors of the various chapters are also to be thanked. Steve Morton, Monash University, prepared many of the photographic illustrations from negatives, improving their quality greatly. Monica Leichester and Pam Hermansen prepared much of the final manuscript and Reid Macdonald critically read it. Our thanks to Tom Rich for his help in preparation of the index and with editorial details.

During the writing of this book the funding provided by the Australian Research Council, Monash University, the Museum of Victoria, the Australian Museum, the Commonwealth Scientific and Industrial Research Organization, the National Geographic Society, the Australian Army, Earthwatch, the Ingram Trust, Western Mining, Utah Mining, the Institute for Aboriginal Studies, the Australian National Parks and Wildlife Service, Imperial Chemical Industries, Atlas Copco, Ingersoll-Rand, Safeway, the Sunshine Foundation, as well as the respective institutions of the authors and editors have been crucial in making possible the research and field work relevant to this book. We are all deeply indebted to these groups and to hundreds of volunteers, both local and international, who helped with field work and preparation of the fossil material reported herein.

And finally, we owe a deep gratitude to our Australian publisher, Derrick Stone, and Emily Wilkinson and Ed Tenner at Princeton University Press. They have all shown keen interest and patience and given us their support throughout the production of the two editions of *Kadimakara*.

Preface

This book is about Australian discoveries; about people and fossils. It is about the people who discovered the fossils of animals and plants that have inhabited our continent for over 600 million years. It tells how the fossils were found and, more importantly, gives insights into how palaeontologists view their own processes of interpreting what the animals and plants were like when alive, and the conditions under which they lived so long ago.

Until little more than a generation ago, by world standards Australia was held to have a very poor fossil record. Certainly, in the early days of exploration numbers of fossil sites were found yielding the remains of giant marsupials and birds. But, as fossils, these were all relatively young — only from the last million years or so.

A mere tip of an iceberg. The great extent of our fossil history back into time was unrepresented, except by tantalizing scraps from here and there in our museums and a few exciting complete fossils that indicated that there was a rich history — if only extensive deposits could be found by someone.

As this book shows, in the last 25 years all that has changed, and today the information about Australian fossil history is pouring out in numerous articles in books and technical journals, in films and newspapers. We now know that abundant records of even the earliest forms of life occur in Australia equal to those found anywhere. Palaeontologists from all over the world welcome opportunities to work in Australia, with Australians, because the remains of our earliest past also tell of the past history of the world.

This book is different from most other books about fossils because it consists of accounts written by many of the people who have made this change possible. Each segment was written by a person who through her or his enthusiasm and persistence — sometimes under conditions of great physical difficulty — have made discoveries. Their enthusiasm is infectious. It makes reading the book rather like the experience of eavesdropping on a congenial group of collectors of rare and beautiful things in a hotel lounge after an antique exhibition, or in a college common room during a congress, or around a camp-fire after a day's work in the field. On these occasions, sooner or later, someone says 'did I ever tell you. . .' — and then if you are lucky (as we are in this book) yarn will follow fascinating yarn throughout the evening.

To a listener, even if ignorant of the subject matter, such stories are fascinating because most of them reveal that mysterious interplay of happy chance and alert knowledge that we call 'serendipity' after those three Princes of Serendip who, in an old story, went on a journey making marvellous discoveries by a combination of luck and logic.

As Pat Rich, to whom we are indebted for the initial idea of the book, shows in her interpretation of the history of vertebrate palaeontology in Australia, serendipity played its part from the very beginning — from that magic moment in 1831 when Mr Ranken attached a rope to a rock to let himself down into a cave and, the rock breaking, discovered the first Australian fossil bone.

The account that unfolds from that point is common in its general features to all the Australian natural sciences. It shows how, in a small pioneering society, adventurous, inquisitive, and acquisitive people found things that were interesting and important — and sent them on for study to experts (some of whom had even asked them to keep their eyes open) who could use them to reveal secrets of their new, but ancient, land.

At first, there were few among the colonists trained in natural history and most experts were outside the Australian colonies. However, among the medical men, theologians, and public servants, local expertise began to appear. Little by little, as educated communities gathered and put down roots, and increased in affluence, groups of like-minded people formed societies devoted to science. They swapped information. They established museums, and in due course, governments formed instrumentalities (such as geological surveys) with palaeontological functions, and ultimately, universities were established as centres from which indigenous work, as sophisticated as any in the world, began to flourish.

Vertebrate palaeontology in Australia has always been international. It was born as a science in a time of wars between Britain and France; and developed in Australia in a time in which, despite war, the bonds of science persisted. In its subsequent development, even with the increasing availability of experts within Australia, Australian scientists and institutions continued to involve the best skills that were offered either by enticing them to Australia, or by involving them in far away places.

Examples in the early days of Australian palaeontology are not hard to find. They occur in numerous scientific papers and the great monographs of eminent overseas scientists such as Professor Richard Owen (1877), Professor Baron von Ettinghausen (1888), Arthur Smith Woodward (1890, 1908), and in the work of settlers like the Rev. W.B. Clarke (who arrived from England in 1839), Gerard Krefft (who arrived from Germany in 1852), and Professor Frederick McCoy (who arrived from Ireland in 1854). In recent times the best examples are to be found in the authors of this book itself. They include, in addition to native-born and trained Australians, and Australians trained overseas, Americans, British, Canadians and Dutch some of whom have become Australians and others who, by virtue of modern transport, are able to share their expertise with local colleagues in the laboratory, and even in the field, without becoming resident. Nowadays this international cooperation takes place to a degree that would have been impossible a century before when a journey from the other side of the world could take as long as six months.

Today, it can truly be said that the history of Australian vertebrate palaeontology is an example of Baron von Mueller's hope for Australian science expressed in 1890, that 'while we endeavour to follow the cosmopolitan course, as adopted in the northern world, we would cherish some predilection for maintaining a command over the fields of indigenous work in these far southern regions, without any wish however of monopoly'.

Another factor that makes this book interesting is that its subject matter lies squarely within the period encompassed by two of the greatest revolutions of thought that have taken place in natural science, namely, that of the evolution of organisms by natural selection, and of the evolution of the earth's surface by continental drift and ocean floor spreading.

The theory that animals and plants have evolved over countless ages had been put forward and had gained some acceptance among scientists before the time of exploration that this book covers. But the rational exposition of the process occurring as the result of natural selection probably first found expression in the mind of Charles Darwin in the decade after which Ranken found his bone. It is even possible that the account of the Australian fossils that was published in Edinburgh immediately after that discovery provided some of the information that helped the idea to form in Darwin's mind. Because of this timing, the discoveries of the first 50 years of Australian vertebrate palaeontology were made and written up in the ferment of arguments about the mechanism of evolution. Such famous names in science, describers of Australian fossils and supporters of evolutionary theory as Richard Owen, Thomas Henry Huxley and Gerard Krefft, were involved in the furious arguments about its mechanism — although the fossils themselves were to provide no solutions. An explanation did not come until long afterwards when the science of genetics made its contribution — but that is a different story.

By contrast with the main arguments about natural selection that took place so long ago, the revolution in thought brought about by continental drift has involved most of the writers of this book. It has all taken place within their generation. Although the idea of explaining similarities of ancient forms in the faunas and floras of the southern continents was evoked during the last century, the idea that the continents had nuzzled up against each other in the Antarctic was generally held to be far-fetched and unsupported by any conceivable geological mechanism. However, a little more than twenty years ago, geophysical work on palaeomagnetism, much of which was carried out at the Australian National University, showed beyond doubt, that continental drift had occurred. Moreover, it is still going on today at a rate of about 8cm per year away from the mid-ocean ridges.

When my acceptance of the certainty that continental drift had taken place sufficiently recently to affect the Australian marsupials was first stated in 1964, it was met by scornful disbelief. Many of those who heard it, reacted almost as though the whole foundations of their science had been shaken. Which they had!

However, it did not take long for the evidence to be assessed and as a result of this new perspective, we, who had formerly interpreted fossil distributions solely on the basis of migration routes and land bridges found that a whole new range of explanations presented themselves. It was almost as though we had been believers in a flat earth and had discovered that it was round. To the ancients the earth was flat because they could see it to be; to us the continents had been fixed because we could detect no movement and could dream of no mechanism. In this book a lucid modern account of continental drift is given by Elizabeth Truswell and Gerald Wilford, two people who have been deeply involved in establishing the environmental consequences to Australia of that slow movement northwards through geological time.

My father once told me that he thought that he had lived in the most exciting period of human history. As a small boy driving a buggy he had seen the first motor car driven in that part of northeastern Victoria in which he lived. In his lifetime he flew in an aeroplane faster than sound. On reading this book, I am sure that you will get the feeling as I have, that we, too, are living in an exciting and accelerating world of scientific discovery in which new ideas and comprehensions are being generated — not only by the pace of technology — but by the energy, knowledge and enthusiasm of explorers who live among us.

David Ride.

Canberra,
21 May 1984.

Preface to the second edition

Australian vertebrate palaeontology has continued to grow in both number of workers and new discoveries since *Kadimakara, Extinct Vertebrates of Australia* was first published in 1985. It is the intention of this brief resumé to point to the major achievements in this field during the past five years and to provide a short bibliography that highlights an astounding amount of literature which has accumulated during that time.

Australia's vertebrate fossil record is still richest during the Devonian Period, the Late Mesozoic and the Late Cainozoic eras. There are still great gaps, ghastly blanks, that separate those periods of better records. Such gaps contain only rare fossils, or none at all, and provide a fascinating challenge to palaeontologists willing to persevere, despite frequent failure.

During the last five years some of those gaps have begun to yield to dogged determination, such as the early Cainozoic, where a new locality, Boat Mountain in southeastern Queensland, has produced primitive marsupials (Godthelp, Archer & Hand, 1990), the oldest for the continent, that are associated with other vertebrates. Another exciting discovery, Australia's oldest mammal and first Mesozoic mammal, *Steropodon*, was found amongst an opal collector's treasure trove of Early Cretaceous age from Lightning Ridge, N.S.W. (Archer, Flannery, Ritchie & Molnar, 1985). *Steropodon* had fully-enamelled teeth, unlike those of the living platypus, but the pattern of cusps on the teeth is similar to that of the dentine, juvenile teeth that are lost soon after an *Ornithorhyncus* baby hatches. Thus, the link can be made between the Cretaceous and modern platypi. But to what other group are these strange egg-laying mammals related? There has been a debate as to whether they are closest to the therian mammals (which include both marsupials and placentals) (Archer, Hand & Godthelp, 1989; Bonaparte, 1990; Kielan-Jaworowska, Crompton & Jenkins, 1987), or the 'prototheres' that encompass a variety of primitive, Mesozoic mammals (Archer *et al.*, 1985), which essentially gave rise to no modern mammals. The therian-supporters seem to have the upper hand at present.

From similar-aged Early Cretaceous rocks deposited in a shallow marine environment of today's western Queensland (Toolebuc Formation) the first bony remains of birds have been recovered on the Australian continent (Molnar, 1986). These primitive forms, *Nanantius eos*, appear to be enantiornithines, a group known from Mongolia, Mexico and Argentina, but the Australian occurrence is oldest.

The complex Riversleigh area in northwestern Queensland has continued to yield a variety of new forms that are filling out details on the marsupial family tree as well as hinting at where some of the branches come together (Archer *et al.*, 1989; Archer, Hand & Godthelp, 1988). Much of the Riversleigh area, as well as the Bullock Creek sites in the Northern Territory, are of Oligocene to Mid-Miocene in age and are part of a vast limestone plateau that has been incised by limestone-depositing streams and lakes and has been affected by karstic activity. In these carbonate-rich tombs, millions of vertebrates have come to rest, and the exquisite preservation of the material in an uncrushed state makes these fossils important for detailed anatomical study, despite the fact that not many are articulated.

Another new set of discoveries, along the southern Victorian coast, have yielded the first varied array of Late Mesozoic terrestrial vertebrates, including several genera of small, endemic herbivorous dinosaurs (the hypsilophodonts). These tiny dinosaurs appear to have adaptations for coping with the low light conditions (such as large optic lobes of the brain and large eyes) prevailing at this time when Australia lay far south, alongside Antarctica. In these polar faunas along with a diminutive *Allosaurus*, the big, crocodile-like labyrinthodont amphibians made their *last stand*; their occurrence here is tens of millions of years after the group became extinct elsewhere in the world (Rich *et al.*, 1988; Rich & Rich, 1989).

Labyrinthodonts not only put in their last appearance in Australia, but also probably their first. An Early Devonian tetrapod trackway in the Grampians of western Victoria, discovered only recently (Warren *et al.*, 1986), is the oldest record of tetrapods, the first land-dwellers, anywhere in the world, and most probably was made by a labyrinthodont. The occurrence of these tracks, as well as the Late Devonian tracks from Genoa River in eastern Victoria, has forced palaeontologists to rethink the classic idea that labyrinthodonts, and thus tetrapods, originated in the northern hemisphere (Panchen, 1977).

A new direction in Australian vertebrate palaeontology, emphasized in the past five years, has been championed by Dr. Sue Turner; that of microvertebrate biostratigraphy. Tiny scales, spines and other miscellaneous parts left behind by vertebrates can be most useful in dating rocks, particularly in the Palaeozoic. Turner has summarized and carried out research on micro-remains, especially of the jawless thelodonts, the armoured placoderms, the spiny acanthodians and has noted that much more effort needs to be placed on the higher bony fish throughout geologic time (Turner, 1986; Turner, Pickett & Myers, 1985; Turner, 1990; Young *et al.*, 1987).

Continuing study of the beautifully-preserved fish of Devonian age from Gogo Station in Western Australia has led to an enriched view of the marine fauna of this time on the Australia continent (Long, 1988a, b). New discoveries of Palaeozoic fish from central Australia (Young, 1985) as well as continuing detailed work with material from known sites in eastern Australia (particularly New South Wales and Victoria) have clearly shown that the appearance and disappearance of many placoderm fish through time is different in Australia than in Europe, North America, East Greenland and Russia. In addition, certain forms that occur in Australia simply are unknown on the Euramerican and Siberian continental fragments of the time and vice versa. A likely explanation of this seems that East Gondwana (which included Australia, Antarctica, India) and the closely apposed Shan-Thai and North and South China fragments did not come in close contact with the Euramerican-Siberian continents until very late in the Devonian (Long, 1990; Young, 1987), and it was only at this time that faunal exchange between the two areas became possible.

During the past five years several compendia of papers have been published that emphasized the Australian vertebrate fossil record in one way or another, mainly in a technical vein. The most recent was the *Proceedings of the de Vis Symposium* edited by Sue Turner, Tony Thulborn and Ralph Molnar (1990), which presents thirty-three papers dealing with Phanerozoic Australasian vertebrate fossils, emphasizing a re-analysis of original material first studied by Charles de Vis, a late nineteenth century–early twentieth century palaeontologist based at the Queensland Museum. Mike Archer edited two volumes (1987) that deal primarily with fossils, emphasizing new finds of Australian diprotodont marsupials, but encompassing a number papers on extra-antipodean recent and fossil forms (mainly didelphimorps) and even a short paper on a rhinoceros that had been incorrectly identified as a diprotodontian form New Caledonia. Two further books nearing completion are revisions of older volumes, both of which should be available by early 1991: Archer and Clayton's second edition of *Vertebrate Zoogeography & Evolution in Australasia (Animals in Space & Time)* to be published by Hesperian Press (Carlisle, Western Australia) and Rich *et al.*'s 2nd edition of *The Fossil Vertebrate Record of Australasia* (edited originally by Rich and Thompson, 1982) to be titled *Vertebrate*

Palaeontology of Australasia now in press with Chapman and Hall (London). More general books include *The Fossil Book* P. V. Rich, T. H. Rich and M. A. Fenton published by Doubleday in 1989, which contains major sections on Australian vertebrate fossils and *The Antipodean Ark* edited by Hand and Archer (1987) that illustrates and briefly comments upon many of Australia's extinct vertebrates.

All in all, the past five years have provided amazing new insights into Australia's past and the next five promise to be equally exciting, with the hope of filling in more and more of those ghastly blanks.

REFERENCES

Archer, M., ed., 1987. *Possums & Opossums. Studies in Evolution.* 2 vols., Surrey Beatty & Sons Pty Ltd, Chipping North.

——, and Clayton, G., eds., 1984. *Vertebrate Zoogeography and Evolution in Australasia (Animals in Space and Time).* Hesperian Press, Carlisle.

——, Flannery, R.F., Ritchie, A. and Molnar, R.E., 1985. First Mesozoic mammal from Australia – an early Cretaceous monotreme. *Nature* **318:** 363-366.

——, Godthelp, H., Hand, S. and Megirian, D., 1989. Fossil mammals of Riversleigh, northwestern Queensland: Preliminary overview of biostratigraphy, correlation and environmental change. *Aust. Zool.* **25:** 29-65.

——, Hand, S. and Godthelp, H., 1988. A new order of Tertiary zalambdodont marsupials. *Science* **239:** 1528-1531.

——, Hand, S. and Godthelp, H., 1989. Dentition of the Oligocene/Miocene ornithorhyncied genus *Obdurodon* and the phylogenetic relationships of monotremes. *Abs.* Conference on Australasian Vertebrate Evolution, Palaeontology and Systematics, Sydney.

Bonaparte, J.F., 1990. New Late Cretaceous mammals from the Los Alamitos Formation, northern Patagonia, and their significance. *Nat. Geog. Res.* **6:** 63-93.

Godthelp, H., Archer, M. and Hand, S.J., 1990 (submitted). Australia's oldest known (Palaeocene/Eocene) marsupials. *Nature.*

Hand, S. and Archer, M., eds., *The Antipodean Ark.* Angus & Robertson, North Ryde.

Long, J.A., 1988a. Late Devonian fishes from the Gogo Formation, Western Australia. *Nat. Geogr. Res.* **6:** 436-450.

——, 1988b. The extraordinary fishes of Gogo. *New Scientist* 1639: 40-45.

——, in press. The long history of Australian fossil fishes. *In:* P.V. Rich, R.F. Baird, J.M. Monaghan and T.H. Rich, eds., *Vertebrate Palaeontology of Australasia.* Chapman and Hall, London.

Kielan-Jaworowska, Z., Crompton, A.W. and Jenkins, F.A., 1987. The origin of egg-laying mammals. *Nature* **326:** 871-873.

Molnar, R.E., 1986. An enantiornithine bird from the Lower Cretaceous of Queensland, Australia. *Nature* **322** (6081): 736-738.

Panchen, A.L., 1977. Geographical and ecological distribution of the earliest tetrapods. *In:* M.K. Hecht, P.C. Goody and B.M. Hecht, eds., *Major Patterns in Vertebrate Evolution.* Plenum, New York.

Rich, P.V., Rich, T.H. and Fenton, M.A., 1989. *The Fossil Book.* Doubleday, New York.

——, ——, Wagstaff, B., McEwen Mason, J., Douthitt, C.G., Gregory, R.T. and Felton, A.E., 1988. Biotic and geochemical evidence for low temperatures and biologic diversity in Cretaceous high latitudes of Australia. *Science* **242:** 1403-1406.

Rich, T.H. and Rich, P.V., 1989. Polar dinosaurs and biotas of the Early Cretaceous of southeastern Australia. *Nat. Geog. Res.* **5**(1): 15-53.

Rich, P.V. and Thompson, E.M., 1982. *The Fossil Vertebrate Record of Australasia.* Monash Univ. Offset Print. Unit, Clayton.

Turner, S., 1986. Vertebrate fauna of the Silverband Formation, Grampians, western Victoria. *Proc. R. Soc. Vic.* **98:** 53-62.

——, in press. Palaeozoic vertebrate microfossils in Australia. *In:* P.V. Rich, R.F. Baird, J.M. Monaghan and T.H. Rich, eds., *Vertebrate Palaeontology of Australasia.* Chapman and Hall, London.

——, Pickett, J. and Myers, B., 1988. The age of marine sediments from Tumblong, south west of Gundegai. *Quart. Notes Geol. Surv. N.S.W.* **1:** 12-15.

——, Thulborn, R.A. and Molnar, R.E., 1990. *Proceedings of the de Vis Symposium. Mem. Qd. Mus.* **28**(1): 1-376.

Warren, A.A., Jupp, R. and Bolton, B., 1986. Early Devonian tetrapod trackway. *Alcheringa* **10:** 183-186.

Young, G.C., 1985. New discoveries of Devonian vertebrates from the Amadeus Basin, central Australia. *Bur. Min. Res. J. Geol. Geophys. Aust.* **9:** 239-254.

——, 1987. Relationships between the northern and southern vertebrate faunas during the Middle Palaeozoic. *Int. Symp. Shallow Tethys* **2,** Balkema, Rotterdam: 79-85.

——, Turner, S., Owen, M., Micoll, R.S., Laurie, J.R. and Gorter, J.D., 1987. A new Devonian fish fauna, and revision of post-Ordovician stratigraphy in the Ross River Syncline, Amadeus Basin, central Australia. *Bur. Min. Res. J. Geol. Geophys. Aust.* **10:** 233-242.

P.V. Rich, 15 June 1990

Addenda and Corrigenda

p. 11, l. 17	'it is still going on today (for Australia) . . .'
p. 15, l. 29	'during his short history . . . approximating a mere 2 million years for the genus in which he belongs, *Homo.*'
p. 39, l. 21	'(roughly 2 to 5 million years ago)'
p. 39, l. 22	'67 million', change to '66 million years ago'
p. 62, l. 28	'in younger time periods', change to 'in older time periods'
p. 63, l. 11	'remnant', change to 'remanent'
p. 70, l. 27	'In the northern hemisphere', change to 'Outside of Australia'
p. 72, l. 1	'In Australia, the record . . .' should be first sentence in the next paragraph.
p. 75, l. 29	'deterioration', change to 'change'
p. 80, last line	delete 'broadleaf'
p. 84, l. 31	*'Rhaetosaurus'*, change to *'Rhoetosaurus'*
p. 86, 4th para.	replace from line 4 'yet the Australian biota . . . prevailed' with 'and the Australian biota reflects these cool conditions in both flora (of conifers, pteridosperms, ferns, ginkgos and the rare angiosperms) and fauna, dominated by the large-eyed, big-brained (constructed with enormous optic lobes) hypsilophodontid dinosaurs and cold-water-loving insects (Rich, T. H. and Rich, P. V., 1989. Polar dinosaurs and biotas of the Early Cretaceous of southeastern Australia, *Nat. Geog. Res.* **5**(1): 15-53). Tree rings in woody conifers suggest a seasonality and oxygen isotope studies further point to a mean annual temperature near freezing.
p. 101, last para.	'The surprising variety . . .', change to 'The variety of the Stairway . . . in existence and diversifying before Middle Ordovician times.'
p. 111, left fig.	Cap. added at end 'in dorsal view'.
p. 124, l. 19	'club-mosses', change to 'tree ferns'
p. 138, l. 2	'it is possible to judge that he had material from more than one animal, but one cannot determine if material came from one or two different skulls.'
p. 154, fig.	add at end 'Komodo Dragon and *Megalania* drawn to same scale.'
p. 164, l. 11	'sebecosuchian (a kind of mesosuchian) . . .'
p. 165, l. 6	'its kin and convergent forms . . .'
p. 165, l. 16	'(and the larger giant goanna, . . .)'
p. 166, l. 2	'in the early part', change to 'in the middle part'
p. 166, l. 6	delete 'more'
p. 177, l. 6	delete 'very'
p. 188, fig.	'in preparation for fossilization', change to 'preceding fossilization'
p. 190, l. 3	'which Stirling & Zietz (1896) . . .'
p. 195	'Gondwanic', change to 'Gondwanan'
p. 202, l. 2	replace reference – '(Rich, *et al.*, 1987. The Pliocene and Pleistocene flamingoes of Australia. *Mem. Qd. Mus.* **25**(1): 207-225)
p. 202, last line	replace reference – (Baird, R. F. and Rich, P. V., in press. *Palaelodus* (Aves: Palaelodidae) from the late Cenozoic of Australia. *Contrib. Sci. Nat. Hist. Mus. Los Angeles Co.*)
p. 204, l. 1	'three years ago.', change to 'a decade ago.'
p. 205, l. 23	replace reference – (Baird, R. F., 1985. Avian fossils from Quaternary deposits in 'Green Waterhole Cave,' South-eastern South Australia. *Rec. Aust. Mus.* **37:** 353-370.)'
p. 206, l. 30	change reference – (Rich, P. V., McEvey, A. R. and Baird, R. F., 1985. Osteological comparison of the scrub-birds, *Atrichornis*, and the Lyrebirds, *Menura* (Passeriformes: Atrichornithidae and Menuridae). *Rec. Aust. Mus.* **37** (3,4): 165-191.)
p. 209, l. 10	delete '(from 2-5 million years old)'
p. 210, l. 3	'our', change to 'the'
p. 210, l. 24	'Mid- to Early Miocene', change to 'Early to Middle Miocene'
p. 216, 4 lines from bottom	change reference – (Pledge, N. S., 1986. A new species of *Ektopodon* (Marsupialia: Phalangeroidea) from the Miocene of South Australia. *In:* M. O. Woodburne and W. A. Clemens, eds., Revision of the Ektopodontidae (Mammalia: Marsupialia; Phalangeriodea) of the Australian Neogene. *Univ. Calif. Publs. Bull. Dept. Geol.* **131:** 43-67.
p. 217, l. 11	change reference – (Woodburne, M. O. and Clemens, W. A., 1986. Revision of the Ektopodontidae (Mammalia: Marsupialia; Phalangeriodea) of the Australian Neogene. *Univ. Calif. Publs. Bull. Dept. Geol.* **131.**
p. 218, l. 19	delete 'Mid-'
p. 220, l. 39-40	delete 'More recently this transition . . . 1974).'
p. 222, l. 8	'precludes', change to 'preclude'
p. 225, l. 5	'challenged, and Owen (1871) wrote in . . .
p. 225, l. 12	'131 and more than 40 . . .'
p. 236 fig.	'the *Palorchestes* seen in the . . .' – omit comma
p. 244, fig.	add to end 'Front of track to upper right.'
p. 246, l. 14	'herbivores (e.g. *Palorchestes* sp.) are almost as rare as the giant rat kangaroos, all carnivores . . .'
p. 248, l. 23	'permit possums to turn into lions'
p. 249, l. 20	'.2 million', change to '1.6 million'
p. 256, l. 18	'hipposiderid may only recently have arrived'
p. 264, 3 from bott.	'fly-wire screen or even smaller mesh'

Introduction

The bones and teeth of ancient vertebrates can tell many tales. They can give the palaeontologist, both amateur and professional, insights into the relationships of animals to one another, both living and fossil. They can also provide a view into past climates, geographic and ecologic barriers, and even ancient geographic arrangements of continents, which have not always been where they are today. If vertebrate fossils were unknown or ignored, a rich array of life that has existed for more than 500 million years would be undocumented and the course of evolution of backboned animals would be much less understood.

So, to have any depth of understanding of Australia's history, its fossil vertebrates, its *Kadimakara* or animals of the Dreamtime, must be scrutinized and as much information as they can provide extracted. In Australia this record is long. It begins with a small, 'four-eyed', jawless fish, *Arandaspis,* found in Ordovician rocks in the Northern Territory, and it continues for more than 450 million years up to the present. Some time periods boast better records than others, especially when compared to that of the rest of the world. The Devonian Period, for example, has a superb record, even by world standards, of armoured fossil fish, especially from locales in Victoria, New South Wales, and Western Australia. The Quaternary record of marsupials is, likewise, relatively good.

But it is clear, that on the whole, the Australian fossil vertebrate record is not as good as that of many other continents such as North America and Europe. Why is this the case? The reasons are numerous and extremely frustrating, yet challenging, to the 'bone hunters' on the Australian continent.

Firstly, the record is somewhat limited because there are very few professional palaeontologists that are out in the field looking for fossils. By the mere fact that the population of the Australian continent is small, so is the number of professionals in the field of vertebrate palaeontology. There are only a few paid positions for people in this field, mainly in museums and universities. A few other vertebrate palaeontologists are hired by such bodies as State Geological Surveys, the Bureau of Mineral Resources, or institutions such as the Institute for Aboriginal Studies and the Commonwealth Scientific and Industrial Research Organization.

Other problems that limit the number of vertebrate fossils in Australia are related to topography, climate and geochemistry. Australia is a very flat continent, and there are few areas where mountain ranges expose long sequences of rocks that might bear vertebrate fossils. Unfortunately, in the few areas where mountains exist, such as the Eastern Dividing Ranges, thick vegetation may cover the rock outcrops making the task of searching for vertebrate fossils difficult, if not impossible. Eroded outcrops with freshly exposed surfaces are rare, but even so, some fossils have been recovered from such areas as the Late Devonian fish and amphibian trackways known from the southeast of Australia.

Arid climate leads to little vegetation and generally encourages erosion, but in those parts of Australia with a low rainfall regime, topography is generally flat and featureless, and often remote and difficult to get to. The Lake Eyre Basin in particular, and interior Australia in general, are such areas, and despite being limited to low outcrops along the edges of the numerous salt pans, these places have produced the best record of Tertiary marsupials in Australia. But, as Richard Tedford relates in a later chapter of this book, the remoteness of this terrain hampered the discovery of vertebrates for some time, with pre-Pleistocene forms only coming to light in the mid-twentieth century.

To even further complicate the story, many of the rocks that may have originally contained fossil vertebrates have been chemically altered since they were deposited. Often Mesozoic and Cainozoic sediments have a deep red colour that grades downwards into chalky, white rocks. About all that remains in these sequences are iron oxides (the red colour), silica, and white clay (kaolinite). Everything else, including the calcium-phosphatic bone, has been dissolved and moved elsewhere in solution. So, the vertebrate record has been 'erased'. This sort of chemical weathering has affected much of Australia's sediments several times, probably representing periods of much higher rainfall, and perhaps more tropical conditions.

Once fossils are found, they come with biases that the palaeontologist must take into account before interpreting their meaning. Many of the fossil accumulations have been brought rather unnaturally together by the action of streams, scavengers, wind and many other environmental factors (Behrensmeyer and Hill, 1980). Streams, for instance, sort and concentrate medium-sized to large animals that have resistant bones or teeth. So, birds with hollow bones, small and young animals are rarely represented in stream-deposited sediments. Animals that are rare in living faunas, such as carnivores, and those that do not frequent sites of deposition, such as lakes and streams, are less likely to get preserved than those that are abundant, such as herding herbivores and rodents and those that do frequent wetlands, such as ducks and rails.

Palaeontologists must also be careful when trying to determine if a fossil animal is new to science. They must have a firm grasp on what the closest modern relative to the fossil looks like, thus a grasp of the variability of the skeletons and teeth of the living relatives. So, it is imperative to have good collections of modern skeletons to compare the fossils with. The modern bones give clues to the natural variability within a species due to sexual dimorphism, and juvenile *vs* adult. They also will mirror differences brought about by disease which can radically change the shape of bones and teeth.

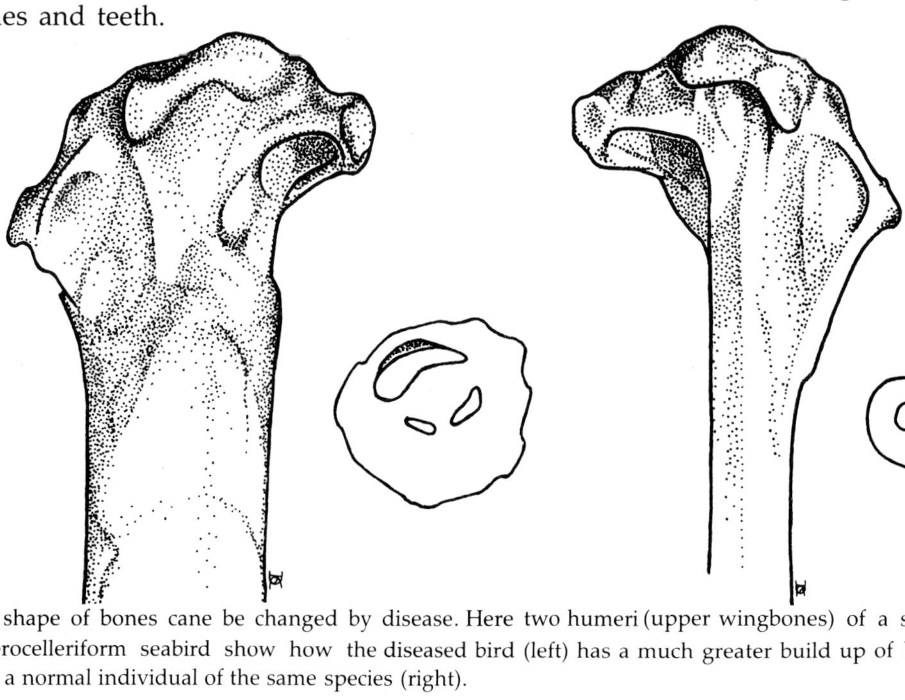

The shape of bones cane be changed by disease. Here two humeri (upper wingbones) of a single species of procelleriform seabird show how the diseased bird (left) has a much greater build up of bony material than a normal individual of the same species (right).

Once fossils are located, other problems raise their ugly heads, but are not insoluble. First, the fossils must be properly collected, and an entire chapter of this book is devoted to this subject. Perhaps the worst problem of all, however, is the lack of monetary support for the proper curation and storage of often rare, unique, and scientifically quite valuable specimens. Support is often lacking for public display of such fascinating fossils. Museums have quite often found it necessary to seek funds from the private sector, because state and federal aid for natural history displays in the major institutions within this country is pitifully poor. Such groups as Esso Australia, The Myer Foundation, Cathay Pacific and Trans Australia Airlines, Dow Corning, Kelloggs, Kodak, and McDonald's Family Restaurants amongst others have provided substantial help in exhibit support. It is hoped that such support can continue for display of not only spectacular overseas exhibitions, such as the recent *Dinosaurs from China*, in Melbourne (the Museum of Victoria) and Sydney (the Australian Museum), but also of Australia's own truly magnificent ancient and extinct fossil vertebrates. Perhaps more public funds can be directed towards such display efforts that can only better the Australians' understanding of their rich past.

But, why bother about vertebrate fossils in Australia, anyhow? Perhaps the best person to ask about the importance of vertebrate fossils is a 6 year old. The answer one is likely to receive is a long list of the names of all the dinosaurs that the child knows, starting with *Tyrannosaurus*. Certainly the hundreds of thousands of people who queued to view the *Dinosaurs from China* held in 1982-1983, are also clear evidence of an innate interest that people, young and old alike, have in things of the past. Such public interest should be nurtured, not discouraged, and should include a large Australian, not just an overseas, component.

Vertebrate fossils in Australia do indeed have many tales to tell. They, together with those of the rest of the world, can make man contemplate his significance, or insignificance, amongst the vast array of beasts, both living and dead. What has happened to these vertebrates, extinction or persistence, can provide a backdrop upon which he can evaluate the changes that have occurred during our short history on earth, something approximating a mere 2 million years. To evaluate this in an Australian context is an exercise that is quite new in the popular literature (Quirk and Archer, 1983). In fact, most Australians, 6 year-olds and adults alike, are far more likely to be acquainted with British or North American fossils, than they are to know much, if anything, about Australian fossils. This book has made an attempt to remedy such a terrible lapse and introduce to the public one of its unique and intriguing natural wonders, its fossil vertebrates.

Priest-Geologists and Knighted Explorers:
A short history of the discovery of vertebrate fossils in Australia

Bunyips have existed in Aboriginal legends for probably thousands of years. Part of their existence is most certainly a product of fertile imagination, the essence of myths and legends worldwide. But some aspects of stories about these mythical monsters that inhabited deep waterholes and roamed the billabongs at night, may stem from a 'truth'; the fossilized bones of some of the enormous marsupials, birds and reptiles that lived in Australia in times past. These animals were unlike anything, especially in size, that lives in Australia today. When shown the bones of these extinct animals, such as *Diprotodon*, Aborigines at times in the past have frequently identified them as bones of the bunyip (Barrett, 1946; Dugan, 1980). Other legends, such as those of some western Victorian Aborigines, allude to *Mihirung paringmal*, birds which were 'taller than the mountains' and lived when the volcanic peaks in western Victoria were erupting — perhaps only a few thousand years ago. In rock paintings in the Cape York Peninsula and rock carvings at Mootwingee in New South Wales and at Ku-ring-gai Chase in Sydney's backyard, these mythical animals occur again and again, these *Kadimakara* or animals of the Dreamtime (Gregory, 1906); or *Kadimurka* of H.Y.L. Brown (1894).

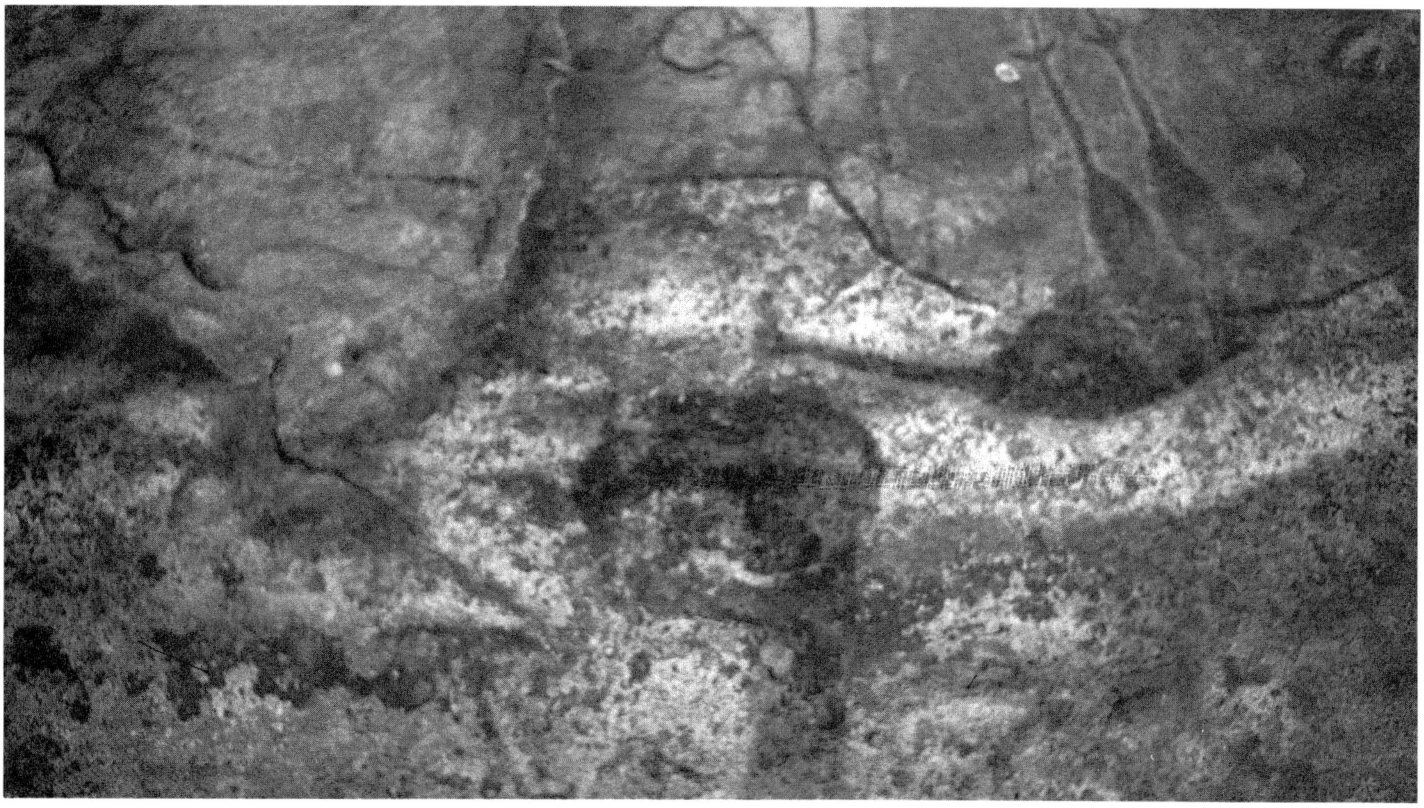

A giant bird, perhaps *Genyornis*, from Quinkan country on the Cape York Peninsula.
(FOUND BY PERCY TRESIZE AND PHOTOGRAPHED BY XENIA DENNETT)

Needless to say, Aboriginal Australians were the first to see some of these extinct backboned animals as well as their bones, for they certainly overlapped in time (Gillespie *et al.*, 1978). It was much later, however, that British and European naturalists and explorers began to understand just what these fossils represented (Rich *et al.*, 1982). They, and the palaeontologists from all parts of the world who followed them, began to unravel the complex tapestry of the antipodean continent's past.

DISCOVERIES IN THE ANTIPODES. GRIST FOR THE EUROPEAN MILL.

The Nicolas Baudin Expedition, and the less fortunate Matthew Flinders Expedition, in the early nineteenth century were the first serious attempts to scientifically document Australia's natural resources in detail. The Baudin expedition, splendidly outfitted with both scientific personnel and equipment, returned a vast amount of natural history material to France, amongst which were living specimens of the King Island Emu and a clam, *Trigonia*, long thought to be extinct. Lamark, renowned for his evolutionary theories, but also the leading invertebrate zoologist in Europe, was, along with many others, impressed by the similarity of many forms living in Australia to taxa that were known only as fossils in Europe. Australia became known then, as it is still today, the land of the *living fossil*, a refugia for forms that have become extinct on other continents, principally those of the northern hemisphere.

Nicholas Baudin, leader of a French natural history expedition at the beginning of the nineteenth century, so important in introducing this unknown land to European eyes.
(COURTESY OF THE MUSEUM D'HISTOIRE NATURELLE, PARIS)

Artist's impression of the ships used by Nicholas Baudin in his 'voyage de découvertes' around the south and east coasts of Australia.
(COURTESY OF THE MUSEUM D'HISTOIRE NATURELLE, PARIS)

The Flinders Expedition of 1801-1803 ended in shipwreck and the loss of almost all specimens except a few that Robert Brown, a shipboard botanist, had taken ashore with him in Sydney when he left the expedition.

Neither expedition collected any vertebrate fossils, but they certainly set the stage for later discoveries. Their pioneering efforts at exploring the unknown coasts of Australia began a scientific search of the continent that is still underway. It was not until the interior of the continent was penetrated sometime later that the first intriguing vertebrate fossils appeared.

Major Sir Thomas Mitchell led one of the first inland exploration trips in 1831, which stopped for a while in the Wellington Valley of New South Wales (Lane and Richards, 1963). There, he and George Ranken, a local colonist, spent some time exploring the caves that abound in the area. Before Mitchell visited the Wellington Valley, Ranken (Ranken, 1916) had already found bones in these caves and had taken them to Sydney intending to send them to Robert Jameson in Edinburgh (Anderson, 1933). Somehow, before leaving for Scotland, they were briefly discussed in the *Sydney Gazette* by the Rev. J.D. Lang. Eventually, however, the bones were despatched, reached Edinburgh, and were examined by Jameson in the same year that Mitchell visited the Wellington Valley. From Jameson the collection was forwarded to William Clift at the Royal College of Surgeons in London, where bones and jaws of wombat, kangaroo and native cat were recognized.

It was Mitchell's 'discoveries', however, that roused the interest of overseas palaeontologists, and other scientists alike, and established that fossil vertebrates did exist in Australia. Mitchell's bone collecting began on a humorous note as his diary reveals —

'The pit (Breccia Cave) had been first entered only a short time before I examined it, by Mr. Ranken, to whose assistance in the researches, I am much indebted. He went down by means of a rope, to one landing place, and then fixing the rope to what seemed a projecting portion of rock, he let himself down

another stage, where he discovered, on the fragment giving way, that the rope had been fastened to a very large bone, [a gigantic bird femur, probably of the family Dromornithidae] and thus these fossils were discovered'
(Mitchell, 1838, p. 362)

Ranken evidently survived the fall, and he and Mitchell shipped the bones and teeth overseas for study by foreign experts, a recurring pattern that characterized Australian vertebrate palaeontology for most of the nineteenth century. Although Mitchell's and Ranken's material from the Wellington Caves was examined and reported on by Georges Cuvier at the Paris Museum, Pentland in Britain, and Darwin was aware of it when he visited Australia in 1836, it was Sir Richard Owen, the best known British comparative anatomist of the time, who described it in detail. He, like the Frenchman Lamark before him, proposed that 'it was necessary to search Britain's secondary (oolitic) [Mesozoic] formations to find specimens analogous to Australia's recent marsupial fossil forms' (Moyal, 1976).

The young Sir Richard Owen holding part of the leg of a New Zealand moa. The outstanding comparative anatomist of the day, he became the world authority on Australian and New Zealand fossil vertebrates.
(COURTESY OF THE BRITISH MUSEUM (NATURAL HISTORY), LONDON)

Sir Richard Owen near the end of his career with a reconstructed New Zealand moa.
(COURTESY OF THE BRITISH MUSEUM (NATURAL HISTORY), LONDON)

Over the next few decades, Sir Richard made Australian and New Zealand vertebrate palaeontology his own domain. He wrote prolifically and introduced the world to a wide array of giant marsupials, birds and reptiles now absent from these lands, and never before imagined until found as fossils. He was aided in his efforts by many resident Australians: Leichhardt provided bones from southern Queensland in the 1840s; Clarke and Stutchbury recovered bones from their northern surveys of the Darling Downs and other spots they dealt with; local graziers, as a result of digging wells or surveying property, sent material to the Australian Museum or state geological surveys for identification, and much of this eventually appeared on Owen's desk.

Fossil material went to other foreign experts such as William Clift (1831), Conservator of the Hunterian Museum, which still holds a sizeable Australian vertebrate fossil collection. T.H. Huxley (Darwin's 'bulldog'), a staunch advocate of evolutionary theory, even published on Australian fossils (1862), as did a variety of other Britons and Europeans — Pentland (1832), Gervais (1848-1852), Hochstetter (1859), and Lydekker (1887, 1896) to mention only a few.

These early collections of fossil vertebrates from *Terra Australis* had their effects on European science. Despite the fact that the majority of the Wellington Cave fossils were described by Owen, a firm anti-evolutionist, they gave some support to Darwin's ideas instead. Kathleen Dugan (1980), a science historian interested in this period of 'foreign expert' science in Australia, has pointed out that identification of the Wellington bones as animals closely related to living Australian species led to the proposal of the 'Law of Succession'. This law suggested that fossil animals in any particular area are 'succeeded' by other animals that are closely related to them (no matter what the environmental conditions). This flew in the face of the special creationist view that certain animals were 'created' to be perfectly suited to their environment, and that if such environmental conditions were present, then certain predictable kinds of animals should be there too. Thus, one would expect the same kinds of animals in the tropical belts of Australia and Africa. Although many alternative explanations can certainly be thought of today, in the mid-nineteenth century, the fact that the Wellington Valley fossil kangaroos resembled living ones in Australia, was a great bother to anti-Darwinians, as was enjoyably noted by Mitchell in his writings:

> 'I understand Buckland's [a British geologist and thorough anti-evolutionist] nose is put completely out of joint by the bones from Australia, their not being those of lions and hyaenas is, I find, a fact which is considered in England to entirely upset his theory. And I have now heard from the best authority that the fact of their fossil bones belonging to animals similar to those now existing has worked a great change in all their learned speculating on such subjects at home'
> (Ranken, 1916 in Dugan, 1980).

GOLD RUSHES, SYDNEY GENTRY AND SCIENTIFIC SOCIETIES: THE BEGINNINGS OF A HOME-BASED SCIENCE

The first half of the nineteenth century was definitely a time when Australia, the colony, provided a steady stream of data, including vertebrate fossils, that was being processed in Europe. Bones were collected in Australia, but the speculation about the collections took place in the learned societies and private clubs of Paris and London.

This began to change, however, as members of this European scientific community took up temporary or even permanent residence in Australia. Certainly by the end of the century near-independence in certain fields of palaeontology had been attained through the efforts of such men as Ralph Tate in Adelaide and Robert Etheridge in Sydney, both determined advocates of Australian science standing on its own two feet. 'Foreign experts' were looked upon more and more as equals, rather than superiors, and were often debated with rather than unquestionably believed.

Several events in Australian history led to this semi-independence. One, of course, was the discovery of gold, mainly in southeastern Australia. This provided an economic incentive for setting up one of Australia's early geological surveys; the Victorian Colonial Survey, in 1852. A.R.C. Selwyn, as director of the survey, was given the task of documenting the geology of the gold-bearing regions, probably in the hope of predicting where more gold was to be found. He had been trained by some of the most prominent geologists of Britain, thus well prepared for the task of recording the stratigraphic succession, nearly totally unknown in Australia at this time. He produced well over two score of detailed geological maps of Victoria before political turmoils managed to dissolve the Survey for a while. But in the meantime his Survey was able to *locally* carry out and publish its own scientific work and to *locally* train geologists, who later served to initiate other surveys and institutions in Australia. Other States followed suit by setting up similar surveys (Johns, 1976), and thus reports and collections of fossils, including many backboned animals, began to grow.

Even before the state surveys were underway, repositories of fossil objects had been established in Australia, which augmented the surveys and eventually served as the 'factual' base for an independence in natural sciences, including palaeontology. Since the early voyages to Australia by the French and British, and before them the Dutch, considerable European interest in Australian natural history specimens had led to a lively trade in Australian oddities, both for commercial and scientific gain. Sizeable collections of Australian materials accumulated in London initially through the efforts of Sir Joseph Banks, as well as in the Hunterian Museum, Glasgow. Naturally these Museums came to hold many of the types for the newly discovered natural curiosities. But the tide began to change with the arrival in Sydney in 1826 of Alexander Macleay, sent as Colonial Secretary to the Government of New South Wales, a position he held for a decade. He brought with him, from Europe, a fine library and an extremely fine collection of insects, thought to be amongst the best insect collections in all of Europe.

When Macleay arrived in Sydney a museum of sorts was already in existence; a small room in the Colonial Secretary's Office, initiated by the colony's first scientific society, the Philosophical Society of Australasia (Branagan, 1979). Each of the seven members paid £5 to organize the collection and purchase books, and Major Goulburn provided the room. Not only were Australian specimens on display, but contacts with foreign museums had provided overseas objects as well. To this collection Macleay added his own and his enthusiasm, which nurtured the beginnings of the Australian Museum, which then became an obvious storehouse for antipodean specimens, gradually encouraging residents to retain material in Australia and to be more aware of their native fauna. When this first museum came into being Sydney was still quite small, with a population approaching 20,000 — and yet by 1837, this small museum had public hours every Tuesday and Friday from 11 to 4; exhibits of native fauna and flora, including 317 species of birds (Fletcher, 1893) as well as ethnological and geological specimens. By 1832 the government provided £200 annually. It was a beginning.

Soon after the Australian Museum was established, others followed suit, namely the National Museum of Victoria in Melbourne in 1854 (Pescott, 1954), the Queensland Museum in Brisbane in 1855 (Mack, 1956), the South Australian Museum in Adelaide in 1856 (Hale, 1956), and somewhat later the Tasmanian Museum (Hobart), the Queen Victoria Museum (Launceston), the Western Australian Museum (Perth) and the Northern Territory Museum (Darwin and Alice Springs). Together with these were a number of private museums, such as the Kyancutta Museum (South Australia). Most of these collections were incorporated into the state museum collections at a later date. Collections continued to grow, but the museums were not just places for storage of Australia's natural wonders and antiquities; they also provided public education facilities, places for scientific research and exchange of ideas, all in an Australian setting.

Originating generally slightly later than the surveys and museums, the universities (Hill, 1981), also contributed to the independence of science and with it the study of vertebrate fossils in Australia. Frederick McCoy, an important figure in the founding of the National Museum of Victoria (now the Museum of Victoria), also held the chair of Professor of Natural Science at the University of Melbourne from 1854-1899. He was especially interested in palaeontology and worked occasionally on vertebrate fossils. J.W. Gregory, the intrepid Scottish geologist, who led an early expedition composed of university students riding camels into the Lake Eyre Basin, also held a post at Melbourne University as Professor of Geology in the early twentieth century, continuing the commitment of Melbourne University to palaeontology. Gregory also served as Director of the Geological Survey of Victoria, demonstrating the often recurring pattern in the early history of geology in Australia where one person, such as McCoy or Gregory, would serve in several institutions (such as surveys, museums, universities), often simultaneously. This reflected how small the scientific community was, and how talented people often had to wear many hats.

A young Frederick McCoy, instrumental in establishing palaeontology at Melbourne University and important in the early history of the Museum of Victoria.
(COURTESY OF THE MUSEUM OF VICTORIA, MELBOURNE).

An older Sir Frederick McCoy. The picture was presented to Kershaw in October 1898.
(COURTESY OF THE MUSEUM OF VICTORIA, MELBOURNE).

J.W. Gregory, the enthusiastic Scotsman who had connections with both Melbourne University and the Geological Survey of Victoria. In the summer of 1901-1902 he led a group of Melbourne University students on camels into the Lake Eyre Basin, the 'Dead Heart of Australia', where they made an early collection of Pleistocene fossil vertebrates. An avid field man, he died in his late 60s while retrieving his field notes from a stream on the upper reaches of the Amazon.
(COURTESY OF THE MUSEUM OF VICTORIA, MELBOURNE).

T.S. Hall, McCoy's student at Melbourne University, and the first locally-trained palaeontologist in Australia, worked mainly on fossil animals without backbones. He, like Chapman, however, found time to work with fossil whales in the late nineteenth and early twentieth centuries.
(COURTESY OF THE MUSEUM OF VICTORIA, MELBOURNE).

A.M. Thompson, who occupied the Chair of Geology at the University of Sydney from 1866, had enough interest in fossil vertebrates that he spent considerable time in the Wellington Caves, a factor which probably hastened his early death in the 1870s. Ralph Tate, who taught palaeontology with gusto and enthusiasm at the University of Adelaide beginning in the early 1870s, certainly sparked an infectious interest in his students. He also adamantly supported Australian independence from 'colonial palaeontology', exhorting his students to rely on their own opinions as worthwhile, and encouraging them to accompany him in the field by always assuring that a large keg of beer was part of the 'collecting' gear.

With the rise of training institutions, which provided permanent indigenous positions, the first Australian-educated students of palaeontology were produced. T.S. Hall, who was trained by F. McCoy and worked with the Victorian Geological Survey under Selwyn, was the first of many. Although locally-trained vertebrate palaeontologists did not appear until the twentieth century, palaeontology by then had its independence. For the sub-discipline of vertebrate palaeontology, Australian maturity came a bit later, primarily due to the lack of workers in this country and the absence of financial support for their work.

A final factor which led towards independence of scientific thought in Australia (Branagan and Townley, 1976) was the development of local scientific societies. Patterned after similar groups, such as the Royal Society of London, such societies gave the educated gentry a chance to exchange information at meetings and perhaps more importantly, to give local scientists a ready and rapid source of publication. One of the earliest of the societial publications was the *Tasmanian Journal of Natural Science* published by the Royal Society in Hobart, set up by 1841. There followed a series of such societies, and although some were short-lived, there seems to have been at least one, or even a few, current at any one time after that.

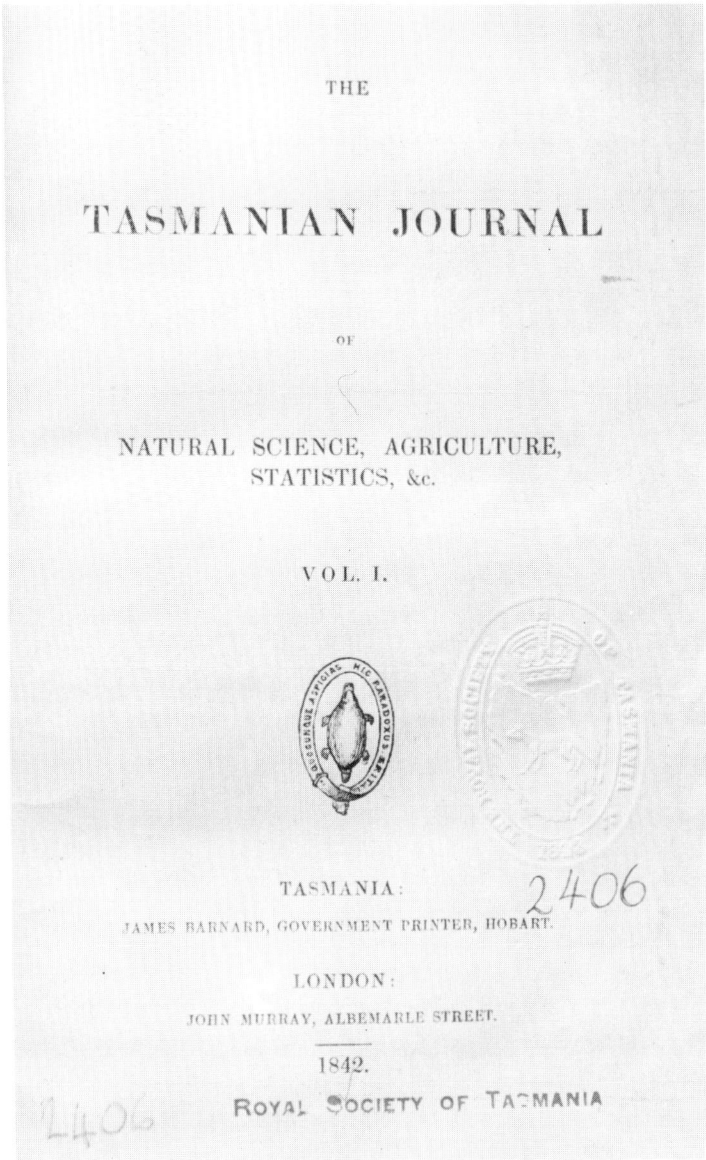
THE

TASMANIAN JOURNAL

OF

NATURAL SCIENCE, AGRICULTURE,
STATISTICS, &c.

VOL. I.

TASMANIA:
JAMES BARNARD, GOVERNMENT PRINTER, HOBART.

LONDON:
JOHN MURRAY, ALBEMARLE STREET.

1842.

Reproduction of the front page of the initial issue of the *Tasmanian Journal of Natural Science,* an Australian-produced scientific journal allowing local publication of scientific topics.
(PHOTOGRAPH: N. KEMP)

During this period of transition from dependence to independence in the latter half of the nineteenth century, there were a number of scientists and collectors, many of whom were interested in backboned animals that significantly influenced a move towards intellectual equality with European colleagues. These included the Rev. William B. Clarke, Alexander Macleay, Edmund C. Hobson, the Rev. Julian Tennison Woods, Frederick McCoy, Gerard Krefft, Robert Etheridge Jr., Charles W. De Vis, E.C. Stirling, A.H.C. Zietz and J.W. Gregory, to mention only a few.

The Reverend W.B. Clarke (1798-1878) arrived in Sydney in 1839 and took the position as Rector of Willoughby, North Sydney. He was an avid fossil hunter and geologist and took every opportunity to escort visiting scientists on trips around his chosen residence. He had an ongoing correspondence with well-known palaeontologists and geologists such as Sedgwick and Murchison in Britain and Dana in the United States, so never worked in an isolated atmosphere. Besides collecting, he published much in local newspapers and both local and international geological journals, especially on the rocks around the Sydney area (Branagan and Townley, 1976). He summarized most of his geological and palaeontological ideas in his 'The Sedimentary Formations of New South Wales' that appeared in several editions from 1867-1878. Although he published much on stratigraphy and general geology, most of the vertebrate fossils that he collected or that were sent to him were sent overseas for study.

Also a part of the Sydney community and a contemporary of Clarke's, was Alexander Macleay (1767-1848). He arrived in Australia in 1826. His large house at Elizabeth Bay with its fine library (the best palaeontological library in Australia at the time) plus his natural history collections, served as an intellectual hub in this colonial town. Although he wrote articles for the local newspapers about fossils, he published no substantial scientific papers. After his arrival, his influence was paramount in the early success and growth of the Australian Museum and encouraging scientific investigation in:

> 'a convict settlement ... racked with dissention between free immigrants and emancipists, businessmen and farmers, army and government, colony and colonial office ... an environment conducive [only] to ... activities [such as] those directed to individual survival and aggrandisement.' (Strahan, R., 1979, p.6)

It was none too easy a task to foster or pursue science in a society that was very small, nearly half unlettered convicts, and cut off from rapid communication with the centres of scientific advancement. At this time such endeavours were restricted to those educated men who had both the training and the means to support themselves and their assistants.

During these years the first 'native born' vertebrate palaeontologist appeared; Edmund C. Hobson (1814-1848) born in Parramatta and later trained in Britain as a physician. He published a series of papers on vertebrate fossils from the Mt. Macedon area in Victoria. He was known too as the first Victorian naturalist and kept notes on birds and other aspects of natural history. His last paper, written on the subject of the jaw of *Diprotodon*, was published posthumously (Vallance, 1978).

Julian Tennison Woods (1832-1889), the Roman Catholic priest-geologist, led a varied and somewhat controversial life that took him from England to Australia, Europe and Asia (including China) and back again to Australia where he died in 1889 (Press, 1979). Collecting in the mid to late Cainozoic deposits in southeastern Australia, he wrote many scientific papers based on the material that he had collected or had collected for him such as the large bird bones collected near Penola, South Australia, some of the first dromornithids, probably *Genyornis*, ever found in Australia. He had an active correspondence with other palaeontologists in Australia and with overseas professionals, such as Sir Charles Lyell, the most prominent geologist of the time. He used what few chances there were for intellectual conversation within his parishes when he stopped for a while on properties such as that of Samuel Pratt-Winter at Murndel near Hamilton, Victoria. Murndel was perfect for Woods, as it combined an abundance of Miocene (marine) fossils, a superb library and a well-educated land-owner to offer an intellectual spar. Although he described and named some of the material he collected, Woods quite often sent specimens to Melbourne or to London to the 'experts' for a final decision, in part using them as trade items.

Sir Frederick McCoy (1823-1899) came to Melbourne in 1854 from the Geological Survey of Ireland to take up the post of Professor of Natural Science at the University of Melbourne. Simultaneously, he was crucial in setting up the National Museum of Victoria (now the Museum of Victoria) where he was director for 45 years (Pescott, 1954). Even though set in his ways, sometimes dogmatic, his contributions towards the building of both institutions cannot be denied. His prime interest was in palaeontology, thus he wrote on vertebrates as well as invertebrates. Despite his writing and his long residence in Melbourne, however, he remained a 'British' geologist and quite often relied on foreign expert opinions on Australian fossils.

Henry Yorke Lyell Brown (1844-1924), a Nova Scotian, was trained initially at the Royal School of Mines London, and after working in Canada and New Zealand trained under Selwyn in Victoria. He worked briefly in New South Wales before taking a job in 1882 as government geologist of South Australia, where he remained for 30 years. At that time South Australia included the Northern Territory. For transport he used camels, horses and boats. His reports and maps include not only observations relevant to geology, mining and water resources, but also on the environment, past and present. He noted such things as the habits of the Aborigines, including their cave paintings; the nesting habits of the Lowan or Malleefowl, *Leipoa ocellata*; and the ecological separation of Red Kangaroos and Euros (Brown, 1890).

H.Y.L. Brown in the early twentieth century. Brown was responsible for the discovery of many vertebrate fossils from South Australia and other parts of the arid centre of Australia.
(COURTESY OF THE SOUTH AUSTRALIAN ARCHIVES, ADELAIDE).

H.Y.L. Brown en route to Charlotte Waters in the 1880s with E. Warman and an Aboriginal guide, Billy.
(COURTESY OF THE SOUTH AUSTRALIAN ARCHIVES, ADELAIDE).

A map prepared by H.Y.L. Brown, Government Geologist, of the Lake Eyre Sub-basin, one of the areas rich in fossil mammals and birds of the Cainozoic Era.
(COURTESY OF THE NATIONAL LIBRARY OF AUSTRALIA).

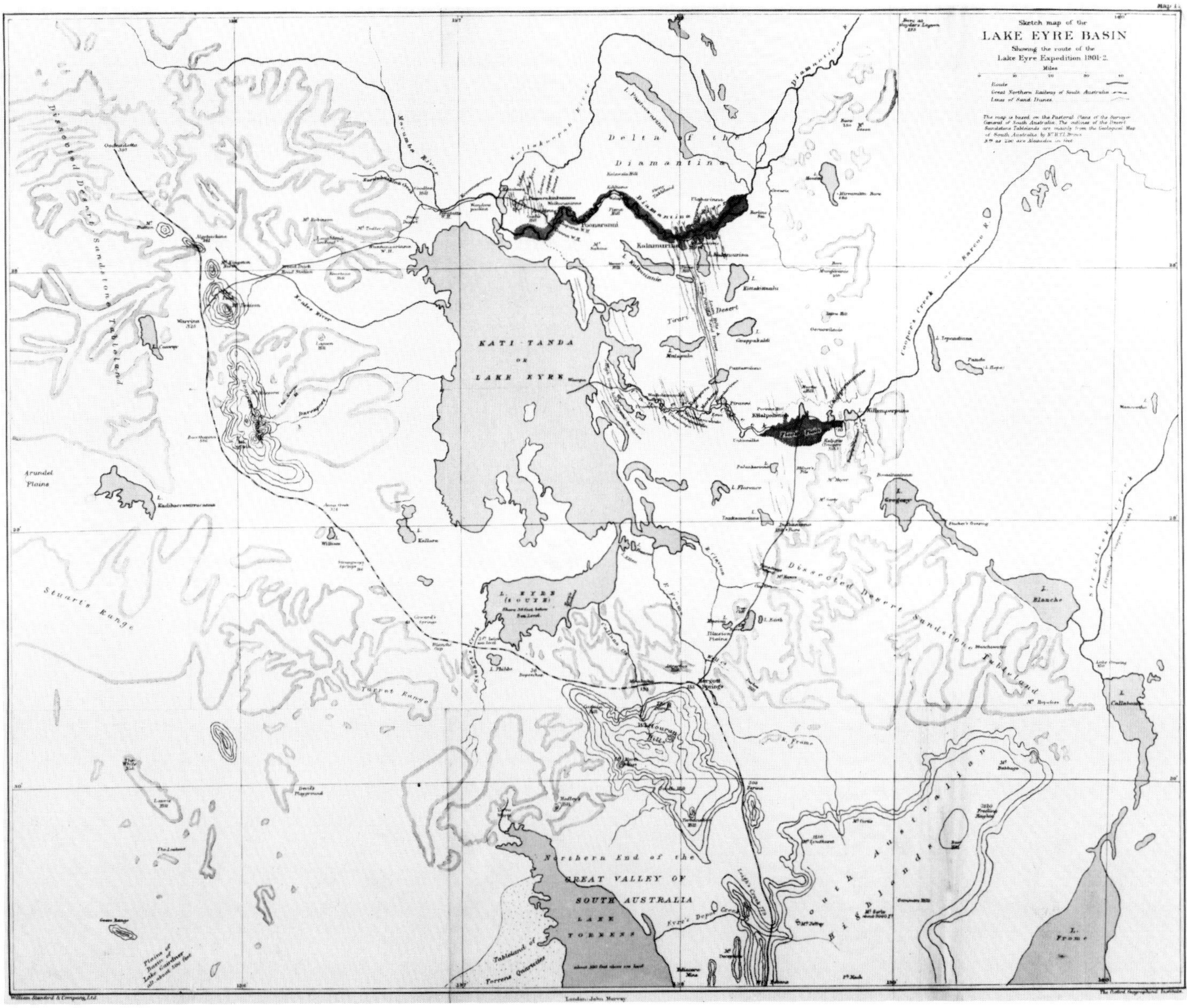

J. W. Gregory's map of the Lake Eyre Area from his expedition in the early twentieth century (from *The Dead Heart of Australia*).

Northeast of Lake Eyre, Brown (1892, 1894) collected fossil remains of *Diprotodon*, *Megalania*, and an extinct crocodile, ?*Pallimnarchus*, also known from the Darling Downs, as well as a few bones of extant species of pelican and cormorant. The following quote is from Brown (1892) and was later noted by Gregory (1906): 'The natives account for the presence of these bones by saying that they belong to the cadimurka, a large fish which lives in the bottom of the waterholes, and which has consequently never been seen by them.'

Brown (1894) reported on the collecting at Lake Callabonna by Hurst supported by the then Adelaide Museum with a permit from the South Australian Government. Hurst found about 80 specimens of *Diprotodon* including complete skeletons with foot bones and casts of feet and fur as well as leg bones of a giant bird, and bones and teeth of a large wombat and a kangaroo. He concluded his report (dated 27 June 1893) as follows: 'In view of the importance of preserving these relics of a bygone age for the future scientific exploration I would recommend that the whole area of the lake be reserved for that purpose, and to prevent the indiscriminate digging up and removal of portions of the specimens.' This recommendation was implemented 30 November 1901 by the South Australian Government.

Two workers who did not rely unquestioningly on foreign expert opinions were associated with the Australian Museum in the latter half of the nineteenth century and early part of the twentieth; Johann Ludwig Gerhard Krefft (1830-1881) and Robert Etheridge Jr. (1847-1920). Both strongly believed in the ability of Australian scientists to make reliable decisions without the constant necessity of overseas advice. Etheridge, an imaginative and careful scientist (together with Ralph Tate at the University of Adelaide), especially supported the idea that the Australian stratigraphic sequences might not be a direct mirror of those in the northern hemisphere.

Krefft is renowned for his ongoing public disagreement with Sir Richard Owen over the nature of the fossil marsupial *Thylacoleo* and was thus an obvious believer in the value of a local opinion. Krefft originally had sent Owen material for study. He ceased, however, when Owen interpreted the jaws and teeth of *Thylacoleo* to be those of a carnivorous animal. Krefft disagreed, claiming the beast to be a plant-eater something akin to a giant rat kangaroo or a 'lion in phalanger hide' (Archer, 1982a). Krefft further disagreed with Owen and many European colleagues, plus most Australian colleagues, in supporting Darwin's theory of organic evolution. Recent work on near-complete specimens of *Thylacoleo* has suggested that Owen's interpretation is closer to the truth. But Krefft's conviction to form his own opinions despite opposition from the top ranks of overseas professionals is one of the factors that led to an independent Australian palaeontology in the twentieth century.

During the last decade of the nineteenth and early twentieth century, Charles De Vis (1829-1915) (Queensland Museum), E.C. Stirling (1848-1919) and A.H.C. Zietz (1874-1922) (South Australian Museum) and J.W. Gregory (1864-1932) were all active collectors and describers of large quantities of bones coming into their institutions from private collectors, miners in the opal fields, graziers, survey geologists — a wide variety of sources both amateur and professional.

Amongst these Victorian gentlemen, Gregory stands out as a great organizer and collector, which he remained until he died from drowning, as he tried to recover his field notebooks lost as his canoe overturned in the upper reaches of the Amazon — when he was 68! His collections of Quaternary material from the Lake Eyre Basin provided the first good glimpse of the vertebrate history of Australia's arid centre. With Melbourne University students and camels he collected along Cooper Creek in the heat of the summer of 1901-1902, a trip which he recounted in various newspaper reports and in the well-known *The Dead Heart of Australia* (1906). It was not until

Professor J.W. Gregory (left) examining a section of Carboniferous rocks near Glasgow in the late 1920s after he returned to Scotland from Australia.
(PHOTOGRAPH PROVIDED BY PROF. B.E. LEAKE, COURTESY OF THE UNIVERSITY OF GLASGOW).

Caricature of J.W. Gregory in Osmar White's '*The Super Roo of Mungalongaloo*' indicates how Gregory's expedition into central Australia captured public interest.
(COURTESY OF JEFF HOOK).

half a century later that the full potential of this area was realized when R.A. Stirton (see next chapter by R.H. Tedford) followed somewhat in Gregory's footsteps. But, Gregory had made the initial exploration under difficult conditions.

E.C. Stirling and A.H.C. Zietz based at the South Australian Museum are remembered most for their excavations of giant marsupials and birds at Lake Callabonna in South Australia. One of the first excavations of its kind, where whole animals were being recovered in numbers, it gave a glimpse of what animals like *Diprotodon*, a ponderous quadrupedal marsupial, and *Genyornis*, a giant ground bird, looked like *in toto*. In most other localities bones were disarticulated, not associated, because of the jumbling that occurred in cave, stream channel, and even swamp deposits. Stirling and Zietz were not simply field collectors, but also set about to describe what they had found, producing a series of fine, large format monographs on a variety of the Callabonna vertebrates. Sir Richard Owen, who had worked so long and hard on *Diprotodon* would have been envious of such work, or would perhaps have done the work himself had he the chance. Ironically, the great anatomist had died in 1892 the very year that bones were discovered at Lake Callabonna and would never know what the feet of his treasured *Diprotodon* resembled. Callabonna provided the answer that Owen could not supply.

E.C. Stirling, Director, or Honorary Director, of the South Australian Museum from 1889-1914. He was also involved in the excavation and later monographic descriptions of many of the fossil vertebrates from Lake Callabonna.
(PHOTOGRAPH COURTESY OF NEVILLE PLEDGE, SOUTH AUSTRALIAN MUSEUM, ADELAIDE).

A.H.C. Zietz, Assistant Director of the South Australian Museum from 1888-1909, and one of the excavators at Lake Callabonna when vertebrate fossils were first found there.
(PHOTOGRAPH COURTESY OF NEVILLE PLEDGE, SOUTH AUSTRALIAN MUSEUM, ADELAIDE).

Charles De Vis was yet another late nineteenth century vertebrate palaeontologist — but unlike Gregory, Stirling and Zietz did little field collecting himself. He worked instead, on specimens forwarded to him by other collectors, such as Brown and Gregory. He was happy to work on all kinds of vertebrates, as was often the case because of the too few workers in Australia. He came to Australia from Manchester, living in a number of places including Rockhampton, where he worked as a librarian for some time. He published articles on geology in the newspaper under the name of 'Thickthorn', then later took up a curatorship at the Queensland Museum. There he published profusely and named many a new form mainly from Pliocene and Quaternary-aged deposits. His comparative collections were small, and perhaps his communication with the remaining scientific world was hampered by distance. Because of this, his idea of the degree of variability within a species was often not realistic, thus leading him to set up many new, supposedly extinct species that upon review have been shown to be bones of living forms. Many of the original De Vis names are invalid. When the final evaluation of De Vis is made, however, the collections he worked with and the isolation he encountered in nineteenth century Queensland are factors worth remembering.

Certainly by the beginning of the last decade of the nineteenth century, Australian-based vertebrate palaeontologists were collecting, describing and contemplating Australian fossils, and not automatically shipping them overseas. Still there was little funding for this science, either for collection and study or for the hiring of vertebrate palaeontologists in permanent positions. As a result, students were not trained in this discipline in Australia, and little local expeditionary work was carried out.

Frederick Chapman, primarily an invertebrate palaeontologist, also described fossil whale material from Australia. He concentrated on fossils from his home state, Victoria, during the early part of the twentieth century.
(COURTESY OF THE MUSEUM OF VICTORIA, MELBOURNE).

The original building that housed the National Museum of Victoria (now the Museum of Victoria) on the campus of Melbourne University. (COURTESY OF THE MUSEUM OF VICTORIA, MELBOURNE).

Moving the treasures from the old National Museum building at Melbourne University to new quarters in Russell Street, where the Museum of Victoria stands today. (COURTESY OF THE MUSEUM OF VICTORIA, MELBOURNE).

Inside the original National Museum of Victoria. The great ape display which McCoy purchased to counter Darwin's evolutionary ideas can be seen at the end of the hall. (COURTESY OF THE MUSEUM OF VICTORIA, MELBOURNE).

AUSTRALIAN INDEPENDENCE AND INTERDEPENDENCE, ON AN INTERNATIONAL SCALE

During the twentieth century work in vertebrate palaeontology in Australia has greatly expanded, especially since the 1950s. The greatest activity occurred in palaeoichthyology and palaeomammalogy, with both of these disciplines providing biostratigraphic information useful in unravelling sequences of rocks ranging from highly deformed Devonian sediments to flat lying, monotonous Late Tertiary stream channels that mimic the underlying Cambrian limestones.

During this period advancement has been decidedly influenced and enhanced by a few enthusiastic workers and the discovery of important new fossil fields. For the first time, too, as information accumulated the summaries of the data appeared (in part in Chapman, 1914a, 1914b, 1934; Hills, 1958; Ride, 1964; Rich and Thompson, 1982; Archer and Clayton, 1984).

Names that stand out amongst the many vertebrate palaeontologists who worked or are still working during the twentieth century are E.S. Hills (Aust.), R.A. Stirton, (U.S.A.), R.H. Tedford (U.S.A.), D. Ride (Aust.) and E.D. Gill (N.Z.). Hill's (1958) review of the entire field of Australian vertebrate palaeontology and Ride's (1964) summary of Australian palaeomammalogy were the first real attempts to draw together the rapidly accummulating data. All contributed significantly in the field and/or in description of the newly found material. Hills' specialty was Devonian fish while Stirton, Tedford and their associates finally made the breakthrough in locating the first concentration of Tertiary mammals in Australia, in the Lake Eyre Basin where H.Y.L. Brown and Gregory had trekked with camels. Stirton and Ride also promoted vertebrate palaeontology in Australia and had direct and indirect influence (by way of training students and cooperating with collaborators) on many of the research workers involved in Australian vertebrate palaeontology today. Gill, through his enthusiasm as a collector, and his rapid-fire pen also promoted vertebrate palaeontology in the Antipodes, especially in Victoria.

Just as the Wellington Caves, Darling Downs, and Lake Callabonna discoveries stand out as significant in the nineteenth century, several significant finds mark the twentieth century as well: Gogo, a rich Devonian fish locality in Western Australia; the Devonian armoured fish locales in western New South Wales and southeastern Victoria; the Cretaceous terrestrial and marine sequences containing reptiles in southwestern Queensland and Victoria; the Cainozoic vertebrate-bearing clays and sands of the Great Artesian Basin, together with the Quaternary cave deposits of southern and eastern Australia are the prime jewels.

The discovery of the Lake Eyre Basin vertebrates is momentous because it not only produced a variety of new mammals and other backboned kin but led to a real acceleration of activity in vertebrate palaeontology in Australia. Ruben Stirton came to Australia from the University of California at Berkeley in 1952 to find pre-Pleistocene mammals in quantity. He did. From those expeditions, generated by his enthusiasm came students, funding, and a momentum that continues to the present. Unlike much of the earlier work, little has been written about these mid-century expeditions, and because of their importance and the chance to mirror the workings of exploratory work on this continent, they are worthy of a close look. Such a look is given in the next chapter by Dick Tedford, one member of these expeditions.

Certainly by the 1970s Australian vertebrate palaeontology had 'come of age' with palaeontologists looking to their overseas colleagues as equals and co-workers. This period from 1970 to the present has been characterized by a number of joint expeditions involving such groups as the South Australian Museum (Adelaide), the Univer-

Gogo, a remarkable Devonian fossil fish locality in Western Australia worked on by staff of the Western Australian Museum and the British Museum (Natural History).

Packing specimens away according to a grid reference that recorded where they were found in the Devonian reef.
(COURTESY OF THE BRITISH MUSEUM (NATURAL HISTORY) LONDON).

Camp

Loading up the barrels containing the fish for transport to Perth.
(COURTESY OF THE BRITISH MUSEUM (NATURAL HISTORY) LONDON).

sity of California (Berkeley and Riverside), the Bureau of Mineral Resources (Canberra), the American Museum of Natural History (New York), the Smithsonian Institution (Washington, D.C.), the Queensland Museum (Brisbane), the Museum of Victoria (Melbourne), the British Museum (London), the Field Museum of Natural History (Chicago), the University of Texas (Austin), the University of New South Wales (Sydney), Monash University (Melbourne), and the Australian Army. This *modus operandi* so differs from the colonial days where material was collected by local collectors and shipped overseas for study by people such as Owen. Now field work and theoretical work are carried out by both groups, locals and 'foreign experts' and there is little attempt to fit the *unruly* Australian record into the European pattern.

There is an ever-increasing desire to thoroughly document the record of vertebrates in Australia, the beginnings of enough 'flesh and blood' (in the form of both home-grown and foreign-trained) and funding (both private and governmental) to carry this out. No better an example of the results of both independence and commensal co-operation in all these areas than the palaeontological chapters in Mike Archer's recent (1982b) *Carnivorous Marsupials* published in Australia by the Royal Zoological Society of New South Wales.

Four vertebrate palaeontologists who worked in northern South Australia beginning in the 1950s: Alden Miller, Richard Tedford, Paul Lawson and R. A. Stirton.
(COURTESY OF THE PAUL LAWSON).

The Stirton Years 1953-1966
A search for Tertiary mammals in Australia

On 27 July 1953 at Lake Palankarinna in South Australia, R.A. Stirton ('Stirt') and I of the University of California had completed our search for fossils in the northwestern-most outcrops of Tertiary rocks and drove down the lake edge to pick up G.D. Woodard of the University of Adelaide. My field notes record the following:

'We were rather disappointed that these deposits had not yielded mammal remains and were ready to give the show away. Pulled up across a green clay bench where Woodard was prospecting to be greeted by a "top of the morning" doff of hat, big smile and outstretched hand bearing fragments of kangaroo bones and teeth. Geoff had done it and we had rolled over his hunting ground, a bench strewn with mammal bones, turtle shell and crocodile teeth. This was it, the first Tertiary mammals from Australia!'

Actually, these bones were the first diverse array of Tertiary mammals, because such fossils as *Wynyardia*, a primitive marsupial from Tasmanian marine rocks of Miocene age, were already known. This youthful elation was tinged with relief, for our nine-month stay as Fulbright scholars in Australia in search of Tertiary mammals had come within 10 days of being a complete failure. In fact, we were saved from failure three days before by a chance happening known as 'bone diggers luck'. Lake Palankarinna is one of many saltpans isolated in the sea of sand ridges east of Lake Eyre and hidden from all except from aerial view. We had seen it only by chance from a towering dune crest where we had gone to eat our lunch in the wind to escape the bush flies!

Woodard's locality, later the Woodard Quarry, proved to be just barely of Tertiary age, representing the Pliocene or last epoch (2 to 5 million years ago) of the Tertiary Period, which began 67 million years ago. This together with the Quaternary Period (about the last 2 million years) constitutes the Age of Mammals or the Cainozoic Era. The quarry yielded the remains of a rabbit bandicoot, a wallaby and a small diprotodontid, all new genera, plus crocodile teeth, turtle shell fragments, fish bones and the remains of yabbies. The day of discovery was climaxed by finding a tiny koala maxillary at another site (the 'Tedford Locality') in older deposits, lower in the rock sequence at Lake Palankarinna.

1953 was the beginning of a concerted search for Tertiary mammals in Australia, and the initial discovery at Lake Palankarinna in the eastern Lake Eyre Basin had justified the opinions of Sir Douglas Mawson, F.W. Whitehouse and C.W. Bonython, who suggested to us that the region was promising. We had also chanced upon J.W. Gregory's *Dead Heart of Australia* (1906), which described a journey with a similar purpose to our own. Gregory, following up on discoveries made by the pioneer South Australian geologist H.Y.L. Brown, had made a large collection of fossil vertebrate remains from the lower reaches of the Cooper and Warburton rivers near Lake Eyre at the turn of the century. These proved to be Quaternary remains, but they focused attention on the fossil potential of this area, and we retraced some of Gregory's steps that year before the chance sighting of Lake Palankarinna.

Of the pioneer naturalists who described Australian fossil vertebrates in the first half of the century, those remaining active in 1953 were E.S. Hills, R.T. Wade and L. Glauert, but others were beginning to make contributions to the field. David Ride, while at Oxford, began a re-evaluation of Robert Broom's Wombeyan Caves collection,

The initial excavation of the Woodard Quarry, Lake Palankarinna, South Australia, July 1953. The holotype lower jaw of the diprotodont *Meniscolophus mawsoni* Stirton, 1955 is in the centre foreground between the brushes. (R. Tedford).

Looking southeast across the bluffs at the Tedford Locality, western shore of Lake Palankarinna, South Australia. White deposits are dolomitic mudstones within the Etadunna Formation (medial Miocene) overlain by darker sandstones of the Tirari Formation (Plio-Pleistocene).

then thought to be Tertiary. Jack Woods had begun a review of the large and important collections from the Darling Downs that Charles De Vis had brought together at the Queensland Museum. Woods later proved that the materials De Vis had obtained from the western Downs near Chinchilla represented an older fauna and one probably of Pliocene age. Edmund Gill at the National Museum, Victoria, using the fluorine content of fossil bones, was able to show that the famous skeleton of *Wynyardia* from Tasmania and certain fragmentary fossil marsupials from Victoria were really contemporary with the marine deposits in which they had been found. He thus established a Tertiary (Oligocene-Miocene) age for fossils that had lain in collections unrecognized, most since the last century, as the remains of truly ancient marsupials. Stirton and Gill collaborated in 1953 to study the remains that Gill's investigations had indicated were of Tertiary age. This was a most important project, despite the fragmentary nature of the material, because it provided an independent estimation of the geological age of the fossil marsupials by reference to a marine geochronology that could be correlated with the international geological time scale.

Other aspects of our work in 1953 emphasized the Quaternary or Pleistocene faunas. We knew that despite the wealth of information on Pleistocene faunas known through the nineteenth century work of Sir Richard Owen, Gerard Krefft, Charles De Vis and Frederick McCoy, there were few sites that had been carefully excavated so that the relative ages of the remains could be established. Little was known about the ages of these faunas or even whether the species found at a given site represented contemporary forms. Such data is vital to modern studies. When Norman Tindale of the South Australian Museum showed us the fossil material that he and Joseph Birdsell (University of California) had collected in 1939 from blowouts in the shore dune (lunette) of Lake Menindee on the Darling River, we recognized the potential of this area for providing a rich Late Pleistocene fauna, whose association and possible coexistence with human remains might be established. We spent a month there, guided by Tindale, and parts of the subsequent field seasons of 1957, 1958 and 1961, but we did not materially advance the case for contemporaneity of man and extinct marsupials at that site. We did prove however, that a large suite of fossil marsupials of extinct and extant species had lived together about 20,000 years ago in western New South Wales. I based my Ph.D. dissertation on the geology of the site and its diverse kangaroo fauna. Lake Menindee thus became an important reference locality in studies of Quaternary mammals. Jeannette Hope and associates at the Australian National University are reinvestigating the Menindee Lakes area at the present time.

The famous Lake Callabonna site was also on our itinerary in 1953, and we were the second party, after 60 years, to make a large collection there. In this work, as well as an earlier unsuccessful attempt that year to find Tertiary mammals west of Lake Eyre, the South Australian Museum joined us materially and lent us the assistance of Paul F. Lawson. Thus began Paul's long association with the search for fossil mammals in South Australia. His involvement in all aspects of the field work from conception through execution usually included the essential logistic magic that made the work in remote areas successful and enjoyable. He also had a knack for finding the best or most interesting fossils.

The South Australian Museum was most anxious that we find the Lake Callabonna locality, since the area had been set aside as a Fossil Reserve under state administration. No one at that time had confirmed the location of the site, and it was not known whether it was actually within the Reserve! Old hands from the north claimed that it had been buried and could no longer be found. It took us four days to find the site, but there was no doubt about it when we did, as related by Stirton in his 1954 narrative of our work. We only had time to sample the locality and try some

Paul F. Lawson at Lake Pitikanta in 1958 excavating a small marsupial skeleton from the Etadunna Formation. (R. Tedford).

modern collecting methods in 1953, but this did provide fresh material, all of which was *Diprotodon*. The *Diprotodon*-producing lake beds crop out over several square kilometres of the present saltpan floor, in some places skeletons were so abundant as to allow one to literally step from the back of one animal to the next without putting a foot on intervening ground!

The original collectors in 1893 had found giant kangaroos, giant wombats and a giant mihirung bird as well as *Diprcdoton* at Lake Callabonna. I was fascinated by the thought of getting skeletons of these and others unknown, as well as unravelling the geological story of the lake's history, so that when the opportunity presented itself to join forces in 1970 with the Smithsonian Institution and the South Australian Museum, we were again at work at Lake Callabonna. A narrative of that work is included in an article in *Australian Natural History* (Tedford, 1973), and it suffices to say here that the site was even richer than expected, yielding skeletons of the giant marsupials and birds that were occasionally associated with impressions of hide, hair and feathers. We also found trackways of *Diprotodon,* and abundant plant remains including wood, cones and pollen that allow a reconstruction of the ancient vegetation surrounding the site. Fossil fish, snails and insects were also found. Our exploration revealed that the fossiliferous ground extended throughout the central part of the present lake floor. Literally thousands of individual, large, extinct mammals were bogged at this site and preserved in the laminated lake clays. The work so far has barely scratched the surface of the information to be gathered here.

In 1954 'Stirt' expected to spend most of a field season excavating the Woodard Quarry and carrying out other work at Lake Palankarinna. A large party including Paul Lawson and William Riedel of the South Australian Museum; William Cassidy, a Fulbright scholar; Woodard; and Leslie Marcus from the University of California was established at the site. When Stirton arrived 11 days later, they were able to tell him that the Woodard Quarry had been worked out and no extension could be found. It had been a deep pocket in a Pliocene stream channel, preserved from erosion before deposition of the overlying red sands and clays. The party had had little luck at the Tedford Locality, so after exploratory work on the lower Warburton River, they split-up; Stirton, Woodard and Marcus went on into Queensland and eventually to the Darling Downs for a first-hand examination of that classic ground. Paul Lawson had a hunch that the party had overlooked something at Lake Palankarinna and he spent a few days there before returning to Adelaide. A lucky blow from his geology hammer rolled out a tiny koala jaw from a site south of the Tedford Locality, but at the same geological level. This became the type specimen of the oldest known koala (mid-Miocene in age).

In 1954, another palaeontological Fulbright scholar, Ernest Lundelius from the University of Texas, came to Australia with his wife Judy. They explored cave deposits of coastal western Australia and the Nullarbor Plain, finding a wealth of fossil remains that clearly reflected the response of the mammals to the environmental changes related to the shift from the glacial (ice-age) Pleistocene to today's climates. The Lundelius's returned to Australia in 1963 accompanied by William Turnbull of the Field Museum of Natural History and his wife Priscilla. They extended the work on the Nullarbor and explored Victoria, parts of New South Wales and Tasmania for Tertiary fossils. They were most successful at a site that had been discovered by Edmund Gill in western Victoria near Hamilton. Gill and Stirton had described the site and its few mammal remains in 1957, but Lundelius and Turnbull made a much larger collection that included 20 species of marsupials, some belonging to living genera and others to wholly extinct forms. The Hamilton locality is directly overlain by basalt and a sample of this rock was submitted to Ian McDougall of the Australian National University. His isotopic age determination of about 4.5 million years provided the first direct date for a fossil mammal locality in Australia. In 1967 Turnbull and Lundelius worked again at this site, and in recent years Tom Rich and parties from the Museum of Victoria have greatly augmented these collections. Today the Hamilton fauna stands as an important reference in comparative studies.

The failure in 1954 to find other Tertiary sites teased Stirton. When I returned to graduate studies in 1956, he was easily persuaded to send me back for another look. Plans were made for a joint venture with the South Australian Museum that would combine fossil hunting and a biological reconnaissance of the northeastern part of the state. Both the 1953 and 1954 parties had collected small vertebrate fossils from Lake Palankarinna, and we thought we might try the technique of soaking the fossil-bearing matrix in water and washing the fine sediments through screens so that the fossils were retained on the mesh. We had the koala locality particularly in mind. We had also stumbled on the report of a scholarly well-digger, G.L. Debney, who in 1881 reported to the Royal Society of South Australia on his geological observations while drilling water bores north of Cooper Creek from Killalpaninna westward nearly to Lake Eyre. He found fossil bones and wood at a saltpan in the middle of the sand-ridge desert 40 kilometres northwest of Killalpaninna. This seemed a lot like the situation at Lake Palankarinna. Brian Daily had joined the staff of the South Australian Museum as Curator of Fossils and Minerals, and he was able to confirm that such saltpans did exist. Brian persuaded the RAAF to let him examine the only comprehensive airphoto coverage then available for that remote area. He carefully examined the photos for the tell-tale serrated pattern of rock outcrops along the edge of the numerous saltpans buried in the sand ridge country east of Lake Eyre. These outcrops were plotted on a tracing of the air photo mosiac and the map he produced served as our guide to potential discoveries for many years.

Brian Daily, Paul Lawson and Harry Bowshall of the South Australian Museum and I had a very full schedule for 1957, one that would determine whether more detailed work at Lake Palankarinna would yield useful results, and whether there were other localities in the eastern Lake Eyre Basin that contained Tertiary fossils. In several ways the 1957 trip was one of the most significant undertaken east of Lake Eyre, not in the quantity of fossils obtained, but in the discoveries made that set the stage for all future work there. We found that we could collect fossils from the Lake Palankarinna deposits by the washing technique, but no significant fossils were obtained because we had neither the time nor the equipment to process the tons of matrix required for success. Field parties in the 1970s came equipped and had success with the method. Work of this kind continues there.

Brian Daily, plotting the expedition's course, lower Strezlecki Creek, 1957. (R. Tedford)

Lower jaws and bones of the wallaby *Prionotemnus palankarinnicus* Stirton, 1955 *in situ* in the Mampuwordu Sand, Lawson Quarry, Lake Palankarinna, South Australia. (R. Tedford, 1961).

We found another locality not far from the Woodard Quarry and at the same geological level that yielded abundant marsupial bones of the same types found at the previous site. We actually had two sites, the Daily and Lawson quarries which subsequent work proved to be parts of one Pliocene stream channel. We did find rich fossil deposits approximately where Debney had indicated and Daily's map showed outcrops. Two small saltpans, later called Lakes Kanunka and Pitikanta, bordered by high dunes yielded two fossil faunas, one above the other, and thus in direct geological superposition. The older, from green and dolomitic clays, contained strange little diprotodont marsupials of a previously unknown and primitive type. A younger stream channel cut through the red clays and into the green contained a variety of animals more modern in appearance than those from the Woodard Quarry, yet more archaic than the widely known Late Pleistocene forms.

Brian Daily's careful geological work, carried on while the bone diggers were exhalting over each new fossil, began to make sense out of the geological history of the region. At Lake Palankarinna the direct superposition of the rock enabled him to

Looking south along the bluffs on the north-western side of Lake Palankarinna, South Australia. Vehicle on right is below the Lawson Quarry; tractor just to left of mid-view at base of bluff is at Keane Quarry; the Woodard Quarry is on the flats at the foot of bluffs just to right of farthest bluff, 1962. (R. Tedford).

Looking west across Lake Kanunka, South Australia. Bluffs expose fossiliferous Tirari and Etadunna formations; Stirton Quarry lies behind the hill at the right hand end of the outcrop. (R. Tedford, 1957).

recognize four major episodes of deposition separated by periods of non-deposition, erosion and soil formation. The oldest rocks were sandstones that contained no fossils. They were enriched with iron-oxide and silica and were then thought to be Cretaceous in age, thus older than 67 million years. Overlying these sandstones were green claystones, fine sandstones and dolomitic clays and limestone, a series of lake deposits, containing mostly aquatic vertebrate remains (turtle, crocodile, lungfish, and water birds, including flamingoes) and rare marsupials, notably the tiny koala. These deposits had been folded into a broad downwarp (syncline) before the stream channels containing the Woodard Quarry were cut and filled with debris mostly derived from the underlying deposits. Thus Brian had proved that a major break in the geological record intervened between the Woodard and Tedford localities, a point not recognized in our previous work. The youngest unit in the stratigraphic sequence exposed at Lake Palankarinna is unfossiliferous red clays and sand bearing a thick gypsum-cemented cap rock upon which the present day dune topography was developed. At the tiny lakes 64 kilometres northwest of Lake Palankarinna we found the same sequence of strata, beginning with the green and dolomitic clays. These clays yielded the same aquatic vertebrates as at Lake Palankarinna, but marsupial remains were much more abundant. The green clays were overlain by the red clays and sands, without the intervening Woodard Quarry channels. The red clays were in turn deeply trenched and filled by stream-borne sands containing the geologically younger fauna mentioned above. These deposits were cemented into the cap rock gypsum. Thus there was a clear correlation of depositional events between these two distant sites indicating a common geological history and sequence in time for the fossil faunas we had discovered.

Later we explored outcrops along the lower Warburton River around the abandoned Kalamurina Station and found again a similar sequence: the green clays and limestones with aquatic vertebrate remains above silicified sandstones; the red clays which contained stream channels and marsupial bones, all strongly gypsum cemented. Now, however, a younger unit was encountered, well exposed in the walls of the present-day river valley, composed of stream-deposited sands, locally rich in large marsupial bones, particularly the giant *Diprotodon*. These deposits were not strongly gypsum cemented, but they were overlain by the present dune-field. They were the source of the fossils collected from the bed of the Warburton River by J.W. Gregory in his turn-of-the century work with his students from Melbourne University.

From our camp at Kalamurina I wrote to Stirton to tell him of our success and to outline the stratigraphic evidence that placed four distinct fossil assemblages in relative geological time. When we published our results in 1961, we drew on the experience of an additional field season, but the outline was unchanged from the 1957 discoveries. In keeping with codified practice in geology, the major rock units were named after geographic locales: Woodard had already used the term Etadunna Formation for the green and dolomitic clays; the name coming from a nearby station on the Birdsville Track. This station was long our base of operations thanks to the kindness and hospitality of the Oldfield family. The channel sands containing the Woodard, Daily and Lawson quarries were called the Mampuwordu Sand from a site nearby at Lake Palankarinna. The unfossiliferous red clays and sands were named the Tirari Formation after the Aboriginal tribe that inhabited the area east of Lake Eyre where these deposits crop out widely. Finally, the overlying fossil-rich channels were named the Katipiri Sands for an outcrop at a waterhole of that name in the lower reaches of Cooper Creek.

Thus, we had four fossil-bearing rock layers stacked one on top of the other, but

Green Bluff, Warburton River, South Australia looking downstream. White limestone band is in the green Etadunna Formation claystones and is overlain by fossiliferous Pleistocene deposits. (R. Tedford, 1957).

we did not know their ages in terms of the geological time scale. At first we had only the probable Late Pleistocene age of the *Diprotodon*-bearing youngest assemblage based on correlation with radiocarbon dated faunas of similar type from elsewhere in Australia. The silicified sandstones were thought to be Cretaceous, and our other faunas were scattered between these rather wide limits. Remember that all the species, and in some cases the genera, we had found were new to science, and that none had been found with other fossil animals that could be referred to the standard geological column. At first we resorted to an educated guess which followed the assumption that the rates of evolution of these now extinct herbivorous marsupials might approximate those established for comparable placental mammals elsewhere in the world. Thus, we suggested a Late Oligocene or Early Miocene age for the fauna of the Etadunna Formation, a Pliocene age for the Woodard Quarry and a Pleistocene age for the two faunas from the Katipiri Sands. In this scheme the geological succession that yielded mammals east of Lake Eyre covered only a small part of the last 25 million years.

Such was the status of our work when in 1958 'Stirt', Paul and I returned to capitalize on the discoveries of 1957. We explored Lake Ngapakaldi a few miles north of Pitikanta and found that concentrations of marsupials in the Etadunna Formation there gave us our first real look at our oldest fauna accordingly named the Ngapakaldi Fauna. We worked the Daily and Lawson quarries and connected them, finding more complete material than in the Woodard Quarry, but no greater diversity of mammals. We called this the Palankarinna Fauna. I did a lot of detailed geological work everywhere we went, drawing section after section of rocks in the hope of testing and extending the ideas Brian Daily had proposed. We discovered a *Diprotodon*-bearing channel cut deeply into the Tirari Formation at Lake Palankarinna, and like its equivalent at Lake Kanunka it bore a gypsum cap. At the latter site we opened an excavation, eventually called the Stirton Quarry, and obtained a diverse suite of marsupials we named the Kanunka Fauna. Our explorations carried us again down into the lower reaches of Cooper Creek, where J.W. Gregory had collected, and we recognized there, as we had on the lower Warburton, that the typical Katipiri Sands were deposited in an ancient riverine tract cut deeply into the Tirari Formation and covered by the sand ridges of the modern desert. Like Gregory and H.Y.L. Brown, we made our best collections from Malkuni and Katipiri waterholes, where the river in flood had winnowed the fossils from outcrops of the Katipiri Sands along the river banks into gravel bars at the ends of these permanent waterholes. We called this *Diprotodon*-rich assemblage the Malkuni Fauna.

These conclusions were summarized in the first comprehensive review of the evidence published by Stirton, Alden Miller and myself in 1961. Since the beginning of our collecting in this region, we had been impressed with the very abundant representation of fossil birds in all these vertebrate faunas, usually more abundant

R.A. Stirton ('Stirt') in 1958 at Stirton Quarry, Lake Kanunka, South Australia. The holotype jaw of the extinct kangaroo *Troposodon kentii* Campbell 1973 is *in situ* in the Tirari Formation at his right hand. (R. Tedford).

than mammals, in contrast to the reverse situation in most fossil deposits around the world. Accordingly, Alden Miller, a University of California ornithologist with a family tradition in the study of fossil birds, was a welcome addition to our field party.

In 1961 'Stirt', Alden, Virginia Miller and I, with Paul Lawson and Harry Bowshall returned to the Lake Eyre Basin. Our work alternated between exploration for new sites and exploitation of those previously discovered. We explored outcrops indicated by Daily's map on the eastern limb of Kallakoopah Creek and nearby saltpans. The results were of more geological than palaeontological interest. We returned to prospect the Warburton River near Kalamurina, so that I could show Stirton the evidence behind our 1957 conclusions and we could improve our sample of the Malkuni-like fauna obtained there. Most of our work was divided between the little lakes and nearby Ngapakaldi. We used a winch-powered, man-hauled, 1 metre scoop to take a part of the overburden off the Stirton Quarry at Lake Kanunka. A plough was used to crack the gypsum cap rock off the top of the Tirari Formation at the Daily-Lawson Quarry, so that the scoop could be brought into action to cut this unit down to the fossiliferous level in the Mampuwordu Sand. The Ngapakaldi Quarry was opened up at that lake and the whole party got a chance to collect the skeletons of the strange little diprotodontoid, later called *Ngapakaldia*, from the green clays that entrapped them 15 million years ago (as discussed later in this book by Mike Plane). The intense physical labour was always rewarded by the excitement of collecting remains of very ancient animals whose bones and teeth had been previously unknown to science. It is a unique experience today, given the extent of palaeontological exploration in other lands, to be collecting fossils in which every day's work brings to light totally new kinds of animals!

Morning tea at Lake Ngapakaldi, South Australia in early August 1961 during early work at Ngapakaldi Quarry. From left, R.A. Stirton, H.J. Bowshall, Alden and Virginia Miller. (R. Tedford).

These were the thoughts that urged us back into the field the following year. By that time I had joined the staff of the Department of Geology at the Riverside campus of the University of California, and our party was augmented by Michael Woodburne from the Berkeley campus, in addition to 'Stirt' and Paul. By the time we reached Adelaide to outfit our trip, Brian Daily had received a small collection of fossil marsupials from Alcoota Station in the Northern Territory. This collection was made by Keith Rochow, of the Alice Springs office of the Bureau of Mineral Resources, and Alan Newsome then of the Northern Territory Wildlife Management. One look

Removing overburden at the Stirton Quarry, Lake Kanunka, South Australia in 1961. Alden Miller at scoop, Tedford and Paul Lawson at winch. (R.A. Stirton).

Removing overburden at the Lawson Quarry, Lake Palankarinna, South Australia, 1961. Paul Lawson guides the scoop, Stirton driving the vehicle. (R. Tedford).

Stirton Quarry, Lake Kanunka, later stages of excavation in 1961 with Stirton in the quarry. Blocks and ledges behind vehicle and in right background represent the gypsum cemented base of the fossiliferous Tirari Formation stream channel overlying the Etadunna Formation. (R. Tedford).

Two partial skeletons (both headless) of the small diprotodont *Ngapakaldia tedfordi* Stirton 1967 in Ngapakaldi Quarry, Lake Ngapakaldi, South Australia, 1962. Skeleton in upper right of view lies on its back, front feet to the right, skeleton at lower left lies on its right side, facing left. (R. Tedford).

Lawson Quarry, Lake Palankarinna, South Australia, near end of excavations in 1961. Kangaroo long-bones on either side of geology hammer in base of Mampuwordu Sand. Top of the Mampuwordu Sand is about level with the bottle tops, vertical at face is basal Tirari Formation.

convinced us that the collection of small diprotodontid teeth was of Tertiary age and we resolved to visit that site rather than do further work at Lake Callabonna as planned for that year.

Waiting in Marree in 1962 was a tractor with backhoe, presented by an interested Adelaide contractor, Vincent P. Keane and a large four-wheeled trailer to be drawn by the tractor. This equipment saved us hundreds of man-hours of backbreaking labour in cleaning the overburden from the Lawson and Ngapakaldi quarries. We also dug trenches to reveal the geology beneath sand-covered slopes. In this way we discovered another bone-bearing Mampuwordu channel at Lake Palankarinna between the Woodard and Daily quarries. This was appropriately named Keane Quarry and proved to contain remains of a large diprotodontid, *Zygomaturus*, that was poorly represented at the other sites.

While we were working the Ngapakaldi Quarry, Paul continued exploration along the eastern shore of the saltpan and found a site north of the quarry where several fossil turtle shells and pebbles of silicified sandstone were exposed. He took the backhoe to the site and dug a trench well inland of the shells, exposing the edge of the filled ancient stream channel that contained the turtles. A few days later under Paul's urging, the whole party began to follow the outline of this channel and soon encountered a jaw of a tiny fossil possum. When we reluctantly left that site in 1962 we had collected a variety of marsupials, birds, turtles, fish and impressions of fossil leaves from various units in the channel fill. The fauna was a completely new one, unlike any from the underlying Etadunna Formation and included the first teeth of the weird possum *Ektopodon* (discussed later in the book by Neville Pledge) that we originally thought was a monotreme because the construction of its teeth seemed not to follow the pattern of more advanced mammals. The rest of the mammal fauna included a small diprotodontid, small kangaroos, a koala, a wombat and small possums, all of a more modern stamp than comparable forms from the Ngapakaldi Fauna in the Etadunna Formation. The leaf impressions included forms closely resembling *Eucalyptus* species; we called this the Leaf Locality. The backhoe was invaluable in working out the geological relationship of the Leaf Locality channel. The channel was cut into the Etadunna Formation and its basal gravel included large fragments derived from that older formation as well as debris (silicified sandstone) from the distant edges of the Lake Eyre Basin. Clearly the channel was part of a well integrated drainage system of a type unknown in the older rocks. We eventually called the channel fill the Wipajiri Formation, its fauna the Kutjamarpu. Thus the backhoe had helped us find a fifth fossil vertebrate fauna, interposed between the Ngapakaldi and Palankarinna faunas, presumably therefore of Late Miocene age. The Wipajiri Fauna was an assemblage of more modern type that included forms similar to those which now occur in Australia's eucalypt forests.

More surprises awaited us at Alcoota, situated about 120 kilometres northeast of Alice Springs in the upper part of the Sandover River drainage that eventually joins the Georgina and loops around the eastern end of the MacDonnell Ranges in its long course toward Lake Eyre. Keith Rochow and Alan Newsome showed 'Stirt', Mike and me the fossil site. Bones were exposed in a gilgai where Keith had found them. The Aboriginals had long known of the locality and had told Alan about it. The best preserved material was revealed by the rabbits, whose warren at the spot was littered with bone and tooth fragments brought up from below. Carefully following the rabbit burrows we encountered a stratum packed with bones about two feet from the surface. We had only time enough to excavate a test pit of about two cubic metres, but what riches there were! I took out one block that contained five jaws stacked on top of one another! The bones, jaws, teeth and skull fragments were so numerous that in

Backhoe in operation at Lawson Quarry, Lake Palankarinna, South Australia in 1962. Lawson operates hoe, M.O. Woodburne collects material left exposed the previous year. (R. Tedford).

R.A. Stirton (left) and R.H. Tedford at Leaf Locality excavation, Lake Ngapakaldi, South Australia in 1962. Fossil leaf-bearing shales have been stripped off the top of the ancient stream channel by the backhoe and the collectors are working on the sands, conglomerates and clays that fill the deeper channel where fossil vertebrates occur. (M.O. Woodburne).

M.O. Woodburne picking out small bones and teeth from Leaf Locality pebbly sands that had been washed through screen wire in ground-water pools dug at Lake Ngapakaldi, South Australia. (R. Tedford).

some cases they rested on one another without intervening sediment. It took great patience to collect this material doing the least damage to the tightly interleaved bones. The leg bones of giant mihirung birds formed the framework of the bone heap that contained the limbs, jaws and skull parts of three diprotodontids of medium size, two kangaroos, a marsupial lion and crocodiles. This assemblage is now known as the Alcoota Fauna. We had barely scratched the surface of this unique locality, whose fauna seemed more closely allied to the Palankarinna than the Kutjamarpu.

M.O. Woodburne excavating at Paine Quarry, Alcoota Station, Northern Territory, in the initial work in 1962. Bones can be seen drying after shellacing in left foreground and in the excavation to the right of whiskbroom. (R. Tedford).

Woodburne and I had to return to the university, but Stirton went on to the Wau and Bulolo goldfield district in Papua, where placer gold mining early in the century had turned up fossil marsupial bones. Some of these had been described in 1937 by C.W. Anderson of the Australian Museum. Geoff Woodard had been sent by the University of California in 1955 to survey the situation and his report, together with a small collection, had prompted another look. Mike Plane of the Bureau of Mineral Resources office at Wau was to be Stirton's host. He had been investigating the geological occurrence of these fossils and had located some of the old placer claims where material could still be obtained. The abundance of new fossils and the geological importance of the finds made a review of these faunas desirable. Plane brought his collection to Berkeley, where he eventually obtained his masters degree using the fossils as the subject for his thesis. The Awe Fauna, as the fossil assemblage was called, shows that New Guinea was biologically a part of Australia in the Pliocene with diprotodontids and kangaroos similar to those from Alcoota. Volcanic ash interbedded with the fossil-bearing sediments was eventually dated by isotopic means as being between 3 and 5 million years old.

1962 was Stirton's last field season in Australia. Mike Woodburne and I returned in 1963 with the help of the Bureau of Mineral Resources. Woodburne, with John Mawby of Berkeley, made a large collection at Alcoota, and I joined Alan Lloyd of the Bureau in a wide-ranging reconnaissance of potential sites in southern Northern Territory and northwestern Queensland. The Alcoota collection provided enough material for Mike Woodburne's Ph.D. dissertation on the marsupials and an important part of Pat Rich's dissertation on the extinct dromornithid birds, the mihirungs. Alan Lloyd and I collected a lot of geological information, but found few bones. We did investigate a site on Riversleigh Station, in Queensland where the Gregory River leaves the old limestone plateau on its course to the Gulf of Carpentaria. Bones had been reported since early in the century from younger limestones deposited on the plateau edge, but no collections had been made. We were able to confirm the occurrence, and make the first collection of identifiable material, including small diprotodontids, a primitive kangaroo, mihirung birds, crocodiles and molluscs. The marsupials were similar to those of the Ngapakaldi Fauna, so we regarded the Riversleigh assemblage as being approximately of that age. From our camp at the edge of the Gregory, we could easily envisage the Miocene scene: towering tea trees and pandanus lined the deep river pools where freshwater crocodiles lived and in the evenings wallabies and emus came to drink. The bones of all these would eventually be deposited in the limestone river gravel and sealed with travertine deposited from the lime-rich water in a tropical environment; the resulting rock would thus resemble the Miocene bone-bearing limestone at Riversleigh.

'Stirt's' death in 1966 ended what might be called the 'Stirton years', the decade of intensive field search for a record of Tertiary mammals in Australia that centred on the Lake Eyre Basin, carried out by himself, his graduate students, and colleagues from the South Australian Museum. By the 1960s a number of resident vertebrate palaeontologists were at work; notably David Ride, then Director of the Western Australian Museum; Jim Warren, Chairman, Department of Zoology, of the newly established Monash University; Alan Bartholomai who succeeded Jack Woods at the Queensland Museum and became its Director; and Ken Campbell of the Department of Geology, Australian National University. All of these men promoted research in vertebrate palaeontology and through their university affiliations brought the first wave of Australian students into that science.

In 1967 the first symposium on Vertebrate Palaeontology to be held in Australia was sponsored by ANZAAS (Australia and New Zealand Association for the

Advancement of Science) in Melbourne. Half of the contributed papers were by Americans: John Cosgriff reviewed the Triassic amphibians from the Sydney Basin; Tedford, on behalf of Stirton and Woodburne, summarized the results to date on the Tertiary faunas of Australia; Lundelius and Turnbull reported on their successful work at Hamilton, Victoria and attempts to find more material at Smeaton, Victoria. Australian contributions included Jim Warren's notice of the discovery of rich bone beds of fish and other vertebrates in the Devonian of Victoria; his graduate student, Mike Waldman's study of the incredibly well-preserved aquatic vertebrate fauna of Cretaceous age from Koonwarra, Victoria; and Mike Plane's announcement of a new Miocene mammal locality at Bullock Creek in northwestern Northern Territory.

When the second symposium on Vertebrate Palaeontology was held by ANZAAS in Brisbane four years later, A.S. Romer, John Cosgriff and I were the only foreign contributors to a very full program of 14 presentations running the gamut of subjects from Palaeozoic fish to Quaternary mammals. Particularly interesting was the development of the Australian record of very ancient vertebrates from the Palaeozoic and Mesozoic.

The 1970s saw an enormous increase in field work and publication. Work was being conducted in all the states and across the entire geological record of vertebrates. This recent history has been covered briefly by Pat Rich in the preceding chapter. It has been my intent to recall the events in the search for a Tertiary record of Australian mammals during the mid-century where my personal involvement brings a contemporary account of an important phase in the history of vertebrate palaeontology in Australia.

TIMOR SEA
Darwin
Bonaparte Gulf
Katherine
Gulf of Carpentaria
CAPE YORK PENINSULA
Laura
Chillagoe
Bullock Creek
Fitzroy Crossing
Gogo
Riversleigh
Gregory River
NORTHERN
TERRITORY
Great
Barrier
Reef
Canning Basin
Army Downs
Telemon
Julia Creek
Richmond
Lucerne
Hughenden
Bowen Basin
Sandover River
Alcoota
Georgina Cr
Boulia
Winton
Great
QUEENSLAND
Alice Springs
Mac Donnell Ranges
Diamantina River
Artesian
Basin
Rockhampton
Mount Charlotte
Mount Watt
Birdsville
Burnett River
WESTERN AUSTRALIA
Carnarvon
Kallakoopah Cr
Carnarvon
National Park
Kolane
Maryborough
Durham Downs
SOUTH
Warburton
R.
Creek
Roma
Mary River
L. Eyre
Cooper
Darling
Chinchilla
Lowood
Etadunna
Strzelecki Creek
Downs
Brisbane R
Brisbane
GREAT VICTORIA DESERT
Coober Pedy
L. Eyre South
Texas Cave
Gore
Marree
AUSTRALIA
L. Callabonna
Lightning Ridge
Andamooka
White Cliffs
Moree
L. Frome
L. Torrens
Flinders Ranges
Broken Hill
Darling River
Leigh Creek
NULLARBOR PLAIN
NEW
Bugaldie
Coonabarabran
INDIAN
OCEAN
Yanchep
Koala Cave
Perth
Swan River
L. Menindee
L. Tandou
Willandra Lakes
SOUTH
Jemalong Gap
Wellington Valley
Lord Howe Island
WALES
Canowindra
Sydney
Adelaide
Murrumbidgee River
Yass
Wombeyan Caves
Mammoth Cave
Kangaroo Is
Murray River
Canberra
Wee Jasper
Taemas
L. Namba
L. Pinpa
L. Tarkarooloo
L. Yanda
Blanche Point
Port Noarlunga
Witton Bluff
VICTORIA
Hamilton
Bacchus Marsh
Mt Howitt
Pambula
Eden
L. Ngapakaldi
L. Pitikanta
L. Kanunka
L. Palankarinna
Mammalon Hill
Naracoorte
Henschke's Quarry Cave
Victoria Cave
Spring Creek
Melbourne
Genoa River
Buchan
Buchan Caves
Briagolong
Koonwarra
Mount Gambier
Green Waterhole
Otway Peninsula
King Island
Freestone Creek
SOUTHERN OCEAN
Pulbeena Swamp
Wynyard
TASMANIA
Hobart
TASMAN SEA

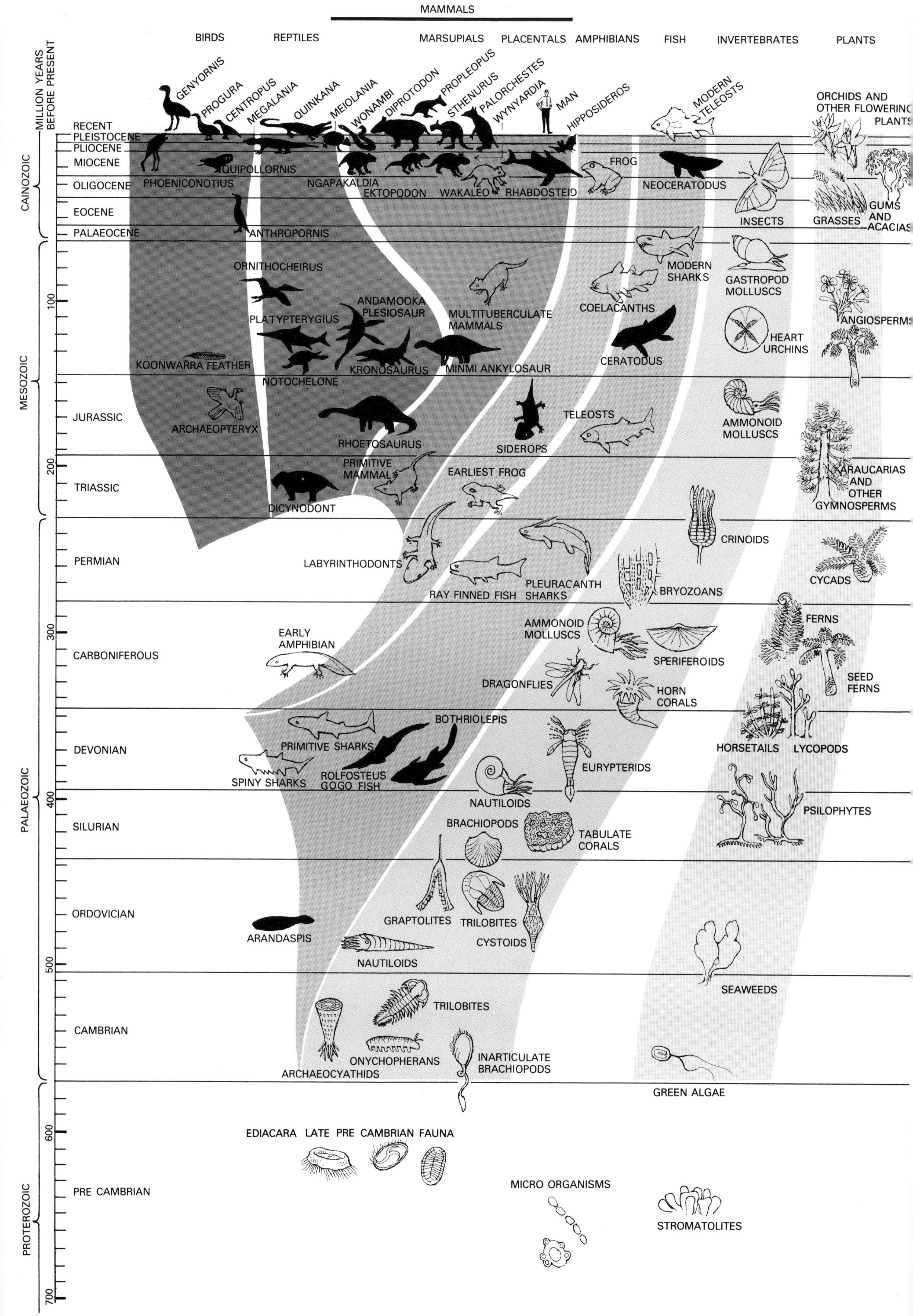
MAMMALS
BIRDS
REPTILES
MARSUPIALS
PLACENTALS
AMPHIBIANS
FISH
INVERTEBRATES
PLANTS
MILLION YEARS BEFORE PRESENT
CAINOZOIC
MESOZOIC
PALAEOZOIC
PROTEROZOIC
RECENT
PLEISTOCENE
PLIOCENE
MIOCENE
OLIGOCENE
EOCENE
PALAEOCENE
JURASSIC
TRIASSIC
PERMIAN
CARBONIFEROUS
DEVONIAN
SILURIAN
ORDOVICIAN
CAMBRIAN
PRE CAMBRIAN
100
200
300
400
500
600
700
GENYORNIS
PROGURA
CENTROPUS
MEGALANIA
QUINKANA
MEIOLANIA
WONAMBI
DIPROTODON
PROPLEOPUS
STHENURUS
PALORCHESTES
WYNYARDIA
MAN
HIPPOSIDEROS
MODERN TELEOSTS
ORCHIDS AND OTHER FLOWERING PLANTS
QUIPOLLORNIS
PHOENICONOTIUS
NGAPAKALDIA
EKTOPODON
WAKALEO
RHABDOSTEID
FROG
NEOCERATODUS
INSECTS
GRASSES
GUMS AND ACACIAS
ANTHROPORNIS
ORNITHOCHEIRUS
MODERN SHARKS
GASTROPOD MOLLUSCS
ANDAMOOKA PLESIOSAUR
MULTITUBERCULATE MAMMALS
COELACANTHS
ANGIOSPERMS
PLATYPTERYGIUS
HEART URCHINS
KOONWARRA FEATHER
KRONOSAURUS
MINMI ANKYLOSAUR
CERATODUS
NOTOCHELONE
ARCHAEOPTERYX
TELEOSTS
AMMONOID MOLLUSCS
RHOETOSAURUS
SIDEROPS
PRIMITIVE MAMMAL
EARLIEST FROG
ARAUCARIAS AND OTHER GYMNOSPERMS
DICYNODONT
CRINOIDS
LABYRINTHODONTS
CYCADS
RAY FINNED FISH
PLEURACANTH SHARKS
BRYOZOANS
FERNS
EARLY AMPHIBIAN
AMMONOID MOLLUSCS
SPERIFEROIDS
DRAGONFLIES
HORN CORALS
SEED FERNS
BOTHRIOLEPIS
PRIMITIVE SHARKS
HORSETAILS
LYCOPODS
EURYPTERIDS
SPINY SHARKS
ROLFOSTEUS GOGO FISH
NAUTILOIDS
PSILOPHYTES
BRACHIOPODS
TABULATE CORALS
GRAPTOLITES
TRILOBITES
CYSTOIDS
ARANDASPIS
NAUTILOIDS
SEAWEEDS
TRILOBITES
ONYCHOPHERANS
INARTICULATE BRACHIOPODS
ARCHAEOCYATHIDS
GREEN ALGAE
EDIACARA LATE PRE CAMBRIAN FAUNA
MICRO ORGANISMS
STROMATOLITES

The Setting
Changing Australian environments throughout geological time

Australia's fauna and flora have evolved against a backdrop of changing environments. The position of the continent has changed in relation to other continental masses and in relation to the earth's rotational poles. Its shorelines have been ephemeral features; at times in the past the continent has been so dissected by shallow seaways that it was reduced to several smaller landmasses; at other times it has ridden high above the surrounding oceans. Its topography has varied; at times low plains inundated by lakes have been widespread; in other periods there have been extensive areas of young and rugged mountains, some associated with the outpourings of large quantities of volcanic rock. These changing aspects of geography have been intimately linked with changing climates. Features of climates are both local, reflecting such factors as topography, and global, reflecting the thermal state of the earth as a whole, whether it was in a cold, glaciated phase or in a warm phase with temperate climates extending right to the poles.

As background to the descriptions of extinct vertebrates, this chapter outlines the environmental history of Australia through the Phanerozoic Era. That is the interval which commenced some 570 million years ago at the beginning of the Cambrian Period and which contains the major record of the evolution of life. There is, however, an increasing body of information which indicates that life began to evolve long before this, perhaps as long as 3.5 billion years ago. Prior to the Cambrian, however, the small size of organisms and the absence of hard parts, made preservation of the record of evolving life forms a rather improbable circumstance.

For each period of the geological time scale, a thumbnail sketch of Australia's position in the continually changing arrangement of the continents is shown, with a map that shows the major features of Australian geography. Each period represented by a map spans many millions of years, so the maps are a composite representation of the major features during that time span.

The techniques of unravelling past geography and past climates have developed rapidly in recent years, so that this chapter is prefaced by a brief description of the principles and methods used in creating these pictures of the past.

DECIPHERING THE GEOGRAPHY OF THE PAST

The last 20 years have seen a conceptual revolution in our understanding of the mobile earth. The idea that there have been large lateral shifts in the position of continents originated hundreds of years ago, but has only recently gained scientific respectability. Francis Bacon in 1620 was impressed by the apparent fit of the eastern and western margins of the Atlantic and suggested that the continental masses bordering that ocean may have drifted apart. Others followed with similar suggestions. The proposals of twentieth century proponents of continental drift, of Eduard Suess, of Wegener (1924) and Du Toit (1937), were received with less than total acclaim because geophysicists found difficulty in visualizing a mechanism for the movement of continents through the sea floor. The floor of the oceans was obviously rigid and known forces seemed insufficient for continents to force a path through it. The relatively recent revolution in understanding is the result of the accumulation of much information about the nature of the floor of deep oceans, information which has become available through post World War II developments in technology.

The new concepts have crystallized into what is termed the *global theory of plate tectonics* (Dewey, 1972; Kennett, 1982). This theory has as its basis the concept that the surface of the earth is divided into a number of rigid, relatively undeformable, plates. A single plate may be made up of both oceanic and continental regions. The plates are constantly in motion, both in relation to each other and to the axis of the earth's rotation. The plates themselves are in general devoid of earthquake and volcanic activity, but they are bounded by trenches, great fractures, or elevated ridges, all regions where earthquakes and volcanoes are concentrated. The motions of the plates are responsible for the formation of the earth's mountain ranges, its oceanic deeps and its major earthquakes.

At the boundaries between two moving plates, three types of interactions are possible, each with different geological consequences. First, where they move apart, molten basalt pours into the gap from sources beneath the earth's crust. The up-welling lava forms a ridge of new crustal material which is added to the plates; the present mid-ocean ridges have their origins in this process. Continents riding on the plates may be separated by the generation of new sea floor between them — by a process of sea floor spreading. Second, where plates converge or collide, one is consumed as it is thrust under the other and the crustal material sinks, thus forming the subduction zones associated with the deep ocean trenches. Uplift of mountain chains is a consequence of such under-thrusting, as light continental material rides

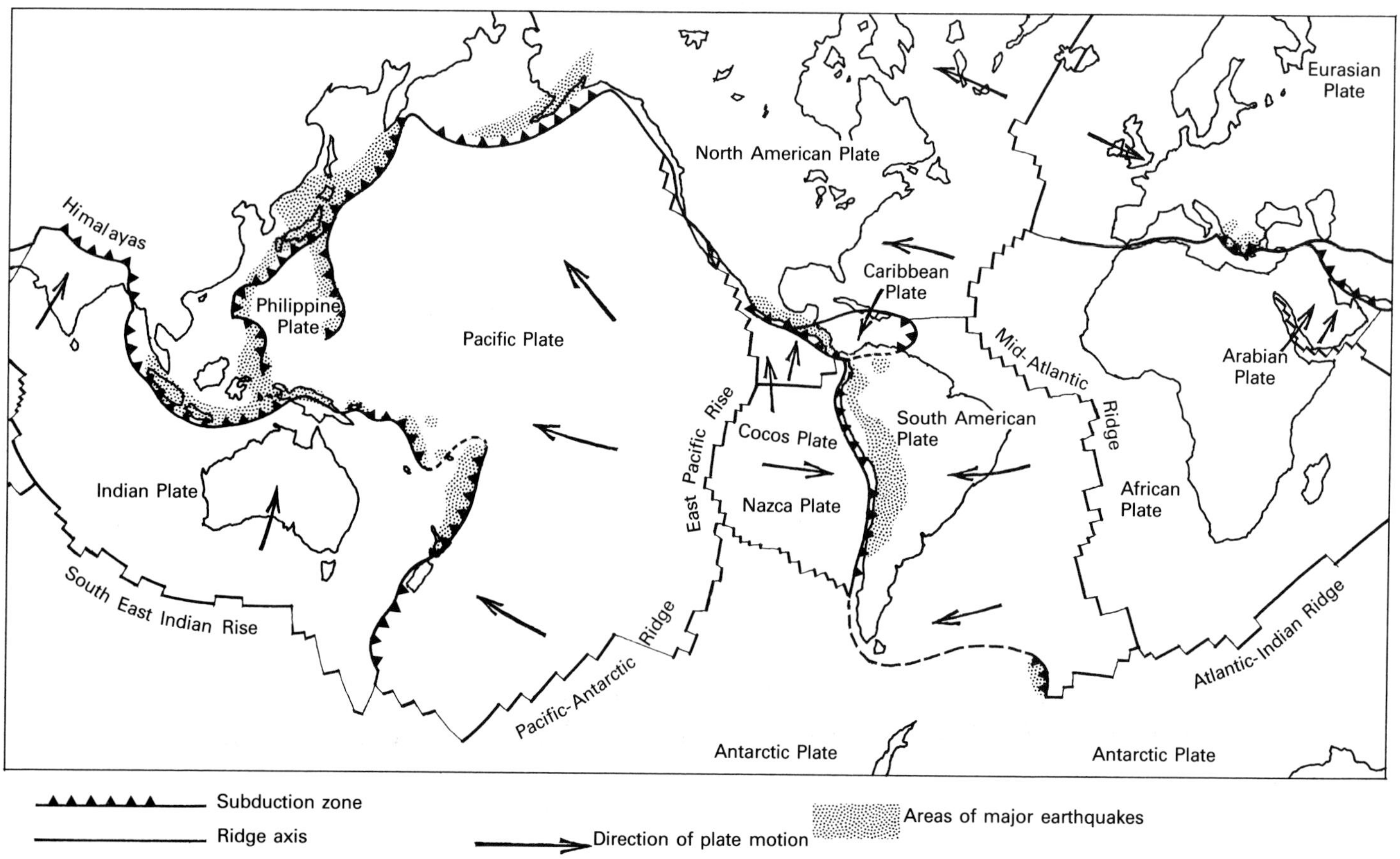

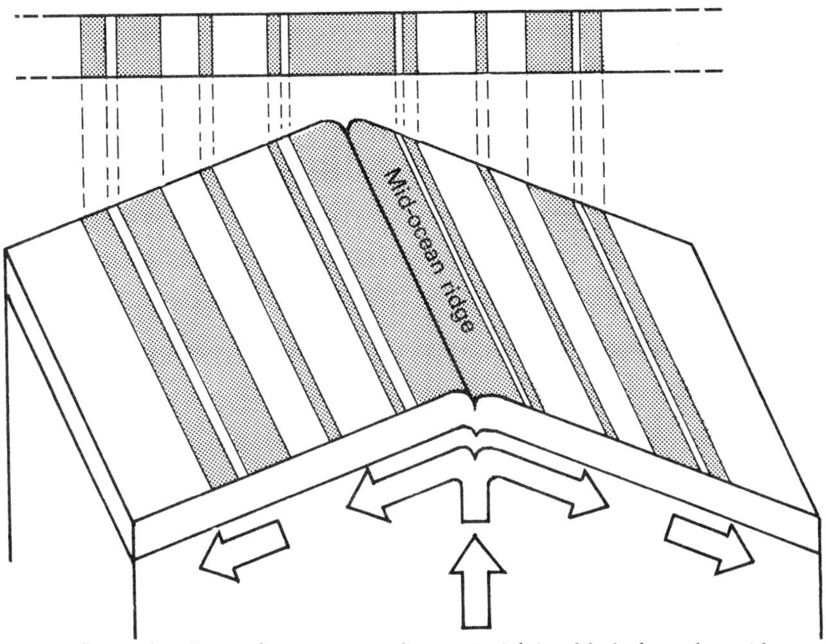

Model of the ocean floor showing where new molten material is added along the mid-ocean ridge. As this cools to a solid rock the iron minerals are 'frozen' in an orientation reflecting the prevailing magnetic field of the earth. Periodically this magnetic field changes, and each band on the ocean floor reflects this change. Because new material is being added on both sides of the ridge, the pattern on either side is a mirror image of the other.

high during the process and may suffer intense contortion. Third, where the plates slide past each other material is neither added to the plates nor destroyed and there are patterns of rectilinear faults, or transforms.

Now that the structural basis of the movement of continents — though not its driving force — is understood, it is possible to reconstruct the past positions of continents in relation to each other. The motion of plates has left a characteristic trace on the sea floor. Magnetometer surveys at sea reveal there a distinctive pattern of positive and negative magnetic anomalies or 'magnetic stripes'. The explanation for the stripes lies in the fact that as new sea floor forms at mid-ocean ridges through the extrusion of basalt, it becomes magnetized in the direction of the earth's magnetic field prevailing at the time the lava cools and solidifies. Yet more new sea floor material subsequently pushes the earlier-formed basalt symmetrically away from the ridge on each side. The fresh lava may carry the record of a different direction of magnetization. Zebra stripes of alternating normal and reversed magnetization are thus characteristic of the deep ocean floor, reflecting polarity changes that have occurred in the earth's magnetic field. The sea floor becomes systematically older on each side of and away from the ridge crest.

The pattern of magnetic polarity reversals through the last 160 million years is now reasonably well understood. The ages of individual reversals have been established, either through the isotopic dating of lavas, or by using minute fossils in sediments overlying them. Thus there has been established an efficient tool for dating the events reflected on the sea floor. Knowing the age of the sea floor that has been generated between once-joined continental masses enables determination of the time at which the split occurred between the two masses. Thus, in the Southern Ocean, the wide and stormy strip which now separates Australia from Antarctica, magnetic anomalies which are symmetrical about the Southeast Indian Ocean Ridge, the oceanic rise midway between the two continents, suggest that the oldest sea floor is of Eocene age (55 million years old). A more recent interpretation of the magnetic anomalies puts it even older— perhaps as old as the Late Cretaceous (80 million years ago). The first true oceanic crust was thus extruded between these continental masses as they began to drift apart, sometime before 50 million years ago.

One of the most striking aspects of the oceans to emerge during the recent phase of research is that all the present ocean basins are relatively young. The oldest deep ocean crust is only of Jurassic age, about 170 million years old. This old crust occurs along the margins of the Atlantic, and in the Indian Ocean to the northwest of Australia, indicating that these were amongst the earliest of the modern oceans to form. Much older rocks occur only on the continents.

This discovery of the relative youth of the modern oceans raises the question of whether the same processes of continental separation and convergence were operating before the Jurassic, when the present phase of sea floor spreading and tectonics began. The answer, at least in part, lies in mountain belts such as those of the Urals and the Appalachians. These tracts contain narrow zones where rock types are found that appear by their composition to be remnants of ancient ocean floors. These 'ophiolite zones' may well pinpoint the sites of vanished oceans. The mountain belts may thus mark the sites where ancient plates collided. By this reasoning, it seems likely that for the past billion years the process of sea floor spreading has been operative.

Another aspect of rock magnetism has been used to confirm the idea that the major continental masses have undergone substantial movement in the past. By measuring the palaeomagnetism of ancient continental rocks, it is possible to reconstruct the positions of the ancient poles with respect to those continents. An assumption is made that the magnetic pole has always been, as it is now, broadly coincident with the pole of the earth's rotation. Poles measured for successive geological periods are then joined by a line to give the path of the pole relative to that particular continent. (This is called the apparent polar wander path). When the polar wander paths from different, now widely separated continents are plotted, they show an incongruence or a mismatch that becomes more marked in younger time periods. When the polar wander paths are plotted for these continents in their reassembled positions, the paths for the different continents coincide, clearly demonstrating the former union of these landmasses.

The information that has come from the comparatively recent study of the sea floor has confirmed in detail what was suspected long before the advent of palaeomagnetism, that there have been major shifts in the positions of continents through time. Rearrangements of the continents that were visualized before the present phase of research were based on the matching of geological features on opposite sides of oceans, and on the matching of the configuration of coastlines. Alfred Wegener (1924) assembled information showing the similarities of geological structures, of fossils and of rock sequences on both sides of the Atlantic Ocean as evidence for the former union of continents on each side of that ocean. Geological information from many continents was used by Wegener to suggest that all the continents were previously united into a single supercontinent which he called Pangaea; in this landmass, Europe and Asia joined, and abutted Africa and the Americas. Wegener's world view was subsequently modified by the South African, Du Toit (1937), who suggested the former existence of a northern landmass — Laurasia — separated from a southern-Gondwanaland — by the east-west trending Tethys Sea. The essential unity of India, Africa, South America, Antarctica and Australia into the southern landmass was clearly demonstrated by the presence of Carboniferous to Permian rocks of glacial origin on all of those continents; the presence of the tongue-shaped leaves of the ancient seed-fern *Glossopteris* on the same continents was further proof of the former union between them.

UNRAVELLING PAST CLIMATES

At any place on the earth's surface, sedimentary rocks contain a record of past environments. They reflect two aspects of climate at the time of their deposition; they reflect an essentially local aspect, such as that determined by latitude, by local topography, by proximity to the sea or by the orientation of mountain ranges, and — they also reflect the global climate; whether the earth was in a glaciated phase or one in which ice caps were absent and the polar regions relatively warm. Clearly, these two aspects are inter-related and not always easy to separate in the rock record, but the increased knowledge of past geography makes the unravelling of these separate threads of climatic history easier. Latitude is probably the most fundamental of the factors determining climate; palaeolatitudes can be determined now with some accuracy by measuring the inclination of the remnant magnetism in ancient rocks, and the changing latitudes of the different continents can be traced through time.

When the broad framework of past geography is known, we can deduce past climates from distinctive rock types deposited both on the present continents and on the deep ocean floors. The evidence comes from the sediments themselves, from the plant and animal remains which may be incorporated in them, and from chemical information such as the ratio of particular isotopes. In continental deposits, the presence of evaporites (salt, gypsum and related minerals) implies an arid climate at the time of their formation. Coals, or peats, on the other hand, imply moist conditions at the site of their accumulation. A high water table, so that plant matter can be protected from decay, is essential in coal formation; coals can, however, form across a wide range of latitudes, provided local conditions are moist enough to produce abundant vegetation and to preserve it. Glaciers bear witness to their presence in the past, in accumulations of coarse boulder debris left behind as moraines, or in parallel grooves or striations etched in resistant rock by the passage of debris-laden ice across it.

Deposits in the oceans leave equally clear traces of the climates that prevailed at the time of their deposition. Reef limestones, which may preserve much of their original structure and the outlines of the original reefs, are excellent indicators of past climates. Modern reefs, with their diverse and abundant life forms, are restricted to the tropics and sub-tropics in waters warmer than about 21°C.

Reefs and reef limestones are deposits of shallow shelf areas of the oceans. In the deep oceans, the distribution of carbonate and siliceous oozes broadly reflects the former presence of warmer and cooler water masses respectively. Siliceous, diatom-rich oozes signal areas where cooler, nutrient-rich waters have welled up to the surface from great depths. The presence of coarse rock fragments in the usually fine-grained deep sea oozes may indicate the rafting of debris by icebergs.

Fossils provide a record of past climates in diverse ways. Organisms may show a clear anatomical relationship to particular climates — for instance, large leaves with smooth edges are usually associated with warm rainforests; the minute coiled shells of foraminifera show a pronounced tendency to coil to the left, rather than the right, in cool ocean waters. Also, it is possible to extrapolate backwards in time what is known of the environmental preferences of living plants and animals, that is, the temperature and rainfall regimes they now appear to favour. To do this does involve the assumption that these preferences have not changed significantly through time. This is an assumption that is open to challenge the further back in time one moves. Another biological tool that can be used to decipher past climates in rather broad ways is that of faunal or floral diversity gradients — in the modern world, biotic diversity is greatest in warm regions close to the equator and diminishes towards the poles.

In spite of the various sources of information, the picture that can be recreated of past climates must inevitably remain incomplete, and only the haziest of pictures can be drawn for any interval of geological time. It is possible, for instance, to gain information of a very broad kind on the amount of moisture that was experienced in a particular part of the world in the past, but it is very difficult to say at what times of the year the rain fell, and how much. It is possible, particularly by the use of oxygen isotopic data, to give figures for seawater temperature in the past, but it is not yet possible to show how these temperatures fluctuated throughout the year. This lack of clarity, this fuzziness, is a problem inherent in the nature of historical science — there are few ways of testing hypotheses about events that are long past.

AUSTRALIAN ENVIRONMENTS THROUGH THE PHANEROZOIC

PALAEOZOIC ERA

Cambrian (500-570 million years ago)

At the beginning of the Cambrian the Australian segment of Gondwanaland lay, according to palaeomagnetic evidence, north of the equator but largely within the tropical zone. As is evident from the world map, derived from Smith *et al.*, 1981, most of the continents occupied a broad band in the equatorial zone, hence they were surrounded by warm seas — temperature patterns on the earth were probably similar to those of the present.

Only the western sector of the present Australian continent was dry land; the sea occupied the eastern part, deepening to the east. Shallow shelf seas, with deeply lobed and irregular western shorelines, extended northwards from the vicinity of Adelaide to cover much of the Northern Territory and extend into the Bonaparte Gulf in the north of Western Australia (Cook, 1982). In some of the lobes of this seaway, access to open waters was at times restricted so the seawater evaporated, leaving wide salt pans behind. Near Mt Isa, seawaters from sources deep in the ocean welled up into the shallow shelf, carrying abundant nutrients and depositing sediments rich in phosphates. The range of environments in this area is typical of the Cambrian and includes intertidal environments; mud flats which were periodically flooded; shallow, more vigorous marine habitats; and quiet, deep waters where fine muds accumulated without disturbance.

In the northern part of the Northern Territory, and the Great Victoria Desert region of Western Australia, volcanoes were active and extruded enormous amounts of lava which consolidated into great horizontal sheets similar in size to the Deccan Traps of India. Smaller volcanoes were active in Tasmania, Victoria and New South Wales.

The seas of the Cambrian retreated to the southeast late in the period, and there was major uplift of the land in a belt running north through Adelaide.

The Cambrian saw the sudden and dramatic expansion of animal life, and the evolution of skeletons, which allowed records of this life to be preserved as fossils. Hard parts of great variety appeared in a wide spectrum of animals, providing support and protection against predators, and importantly, in the case of exoskeletons, a box inside which the body could function in a controlled environment. The evolution of skeletons allowed for a rapid expansion in the utilization of ecological niches on the sea floor.

The shallow seas around Australia were home to a multiplicity of organisms. Among these, the trilobites were the most spectacular group. These arthropods were to a large degree bottom-dwellers and mud-burrowers; they were present in some

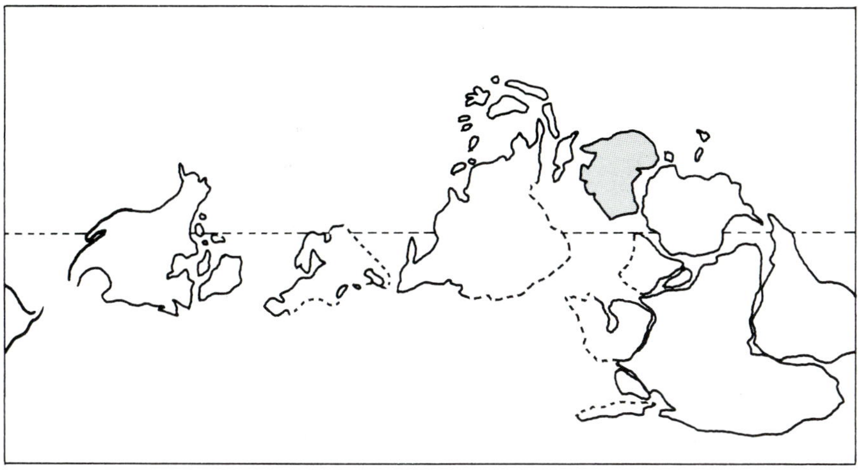

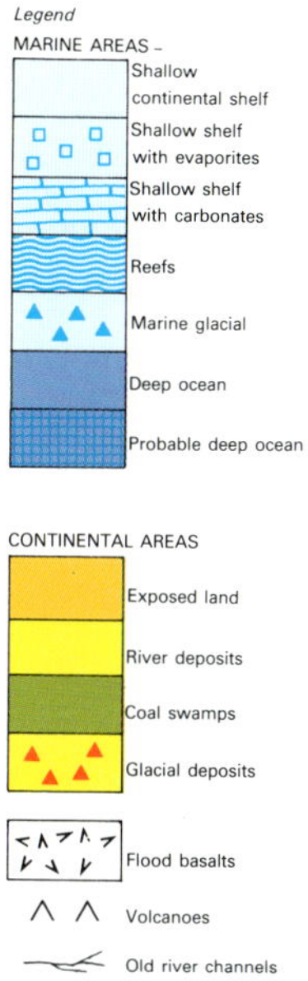

A	Adelaide	**D**	Darwin
AS	Alice Springs	**M**	Melbourne
B	Brisbane	**P**	Perth
C	Canberra	**S**	Sydney

CAMBRIAN

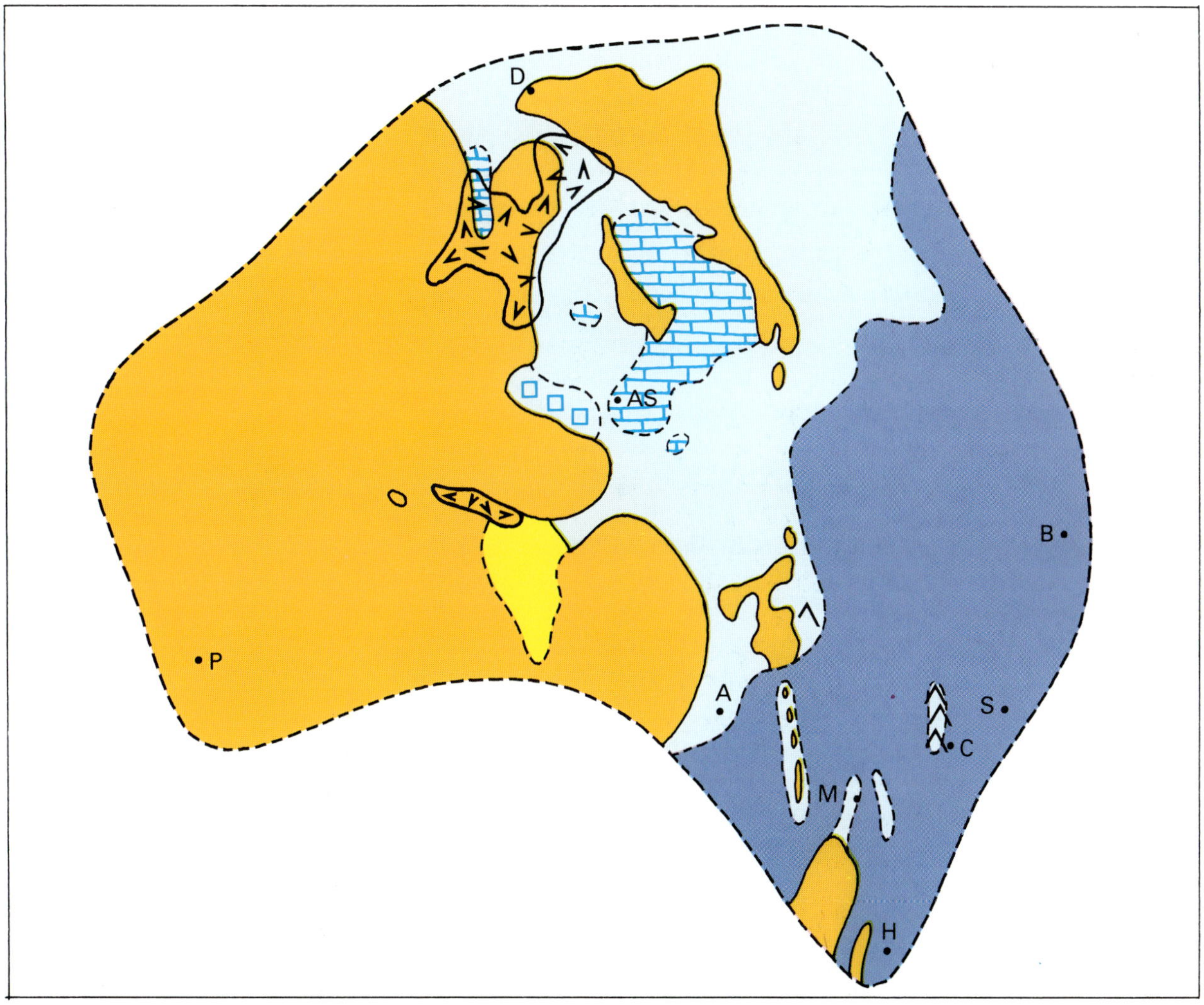

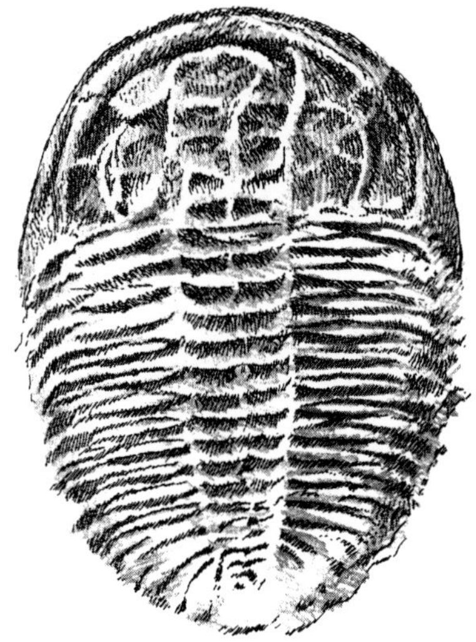

The external skeleton of *Xystridura saint smithi* from Cambrian marine rocks. This trilobite, as others, shed its exoskeleton as it grew. Trilobites, together with brachiopods, made up about 90% of Cambrian faunas, but became extinct by the end of the Palaeozoic. This specimen is about 60 mm in length.

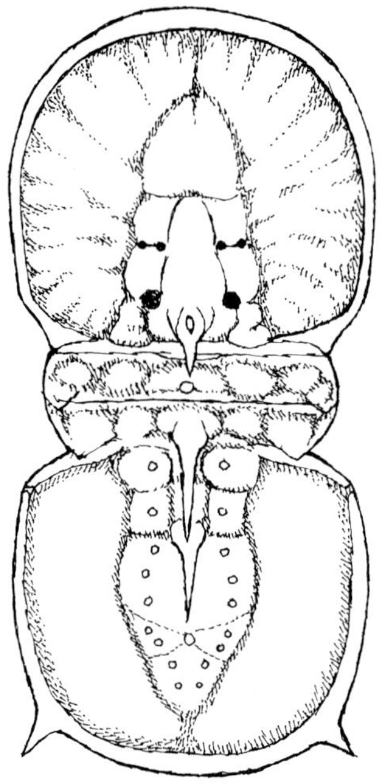

The Cambrian agnostid trilobite, *Goniagnostus*. These were small forms, usually less than 10 mm in total length.

diversity at the beginning of the Cambrian, suggesting a long period of evolution before their appearance in the fossil record. Late Cambrian species in Australia are diverse, and their closest affinities lie with faunas of northern China, Manchuria, Siberia and Kazakhstan: this probably mirrors the closeness of Australia to these areas in the Cambrian.

Other organisms in the shallow seas included the coral-like Archaeocyathina, which formed reef-like structures in seas to the north of Adelaide. The bottom-dwellers also included the brachiopods or lamp-shells — both articulate and inarticulate kinds — and the molluscs. The molluscs were present throughout the Cambrian, but diversified rapidly near the end of the period — this diversification included the appearance of the cephalopods — the squid-like organisms, which were to achieve great importance later in geological time. Earlier appearing molluscs included the rostroconchs, a primitive 'pseudo-bivalved' group, and the true bivalves or clams. The first vertebrate fossils are of Cambrian age but have not been found in Australia before the Ordovician.

The high diversity of animal life, the presence of lime-secreting algae and of carbonate sediments through wide areas of the Cambrian seas in Australia, are all features to be expected in shallow, warm water, bank environments. On land, only algae flourished in wet habitats. The vascular plants, which were later to colonize land surfaces, had not evolved from their marine ancestors, so the land areas of the Cambrian were probably barren.

Ordovician (435-500 million years ago)

In the succeeding Ordovician period, it seems that the Australian segment of Gondwanaland straddled the equator, and hence its east coast would have come under the influence of warm, westward flowing equatorial currents. Australia's warmth contrasted with frigid conditions experienced contemporaneously in polar African regions.

Within the area of the Australian continent, the seas of the Cambrian retreated at the end of that epoch. South Australia, western New South Wales and Tasmania were areas of intense mountain-building. At the same time the sea entered in the northwest, through the region of the Great Sandy Desert of Western Australia, the area today underlain by the Canning Basin. From the deeper oceanic areas of the eastern part of the continent, the sea also encroached westwards into central Australia. By the middle of the Ordovician the seas encroaching from both east and west joined to form a seaway extending right across central Australia. This, which has been dubbed the 'Larapintine Sea', separated a northern land area of low relief from a more rugged southern block of Gondwanaland.

In these shallow seas, sandy sediments were swept in from adjoining landmasses; they were probably deflected to the north and northwest by the west-flowing equatorial current. Cooler currents may have flowed in through the Canning Basin. Where the currents met, the upwelling of nutrients resulted in the deposition of phosphates: the deposits of the Stairway Sandstone of central Australia probably originated in this fashion. The seas of central and northwestern Australia retreated in the last stages of the Ordovician, and the rocks that had accumulated in them were uplifted and eroded. In the eastern part of Australia, however, deep seas persisted, with only occasional shallow banks developed over uplifted areas associated with volcanic eruptions.

The Ordovician seas were richly endowed with life. In contrast to the Cambrian, where bottom-dwelling organisms feeding on ingested mud were the common life forms, in the Ordovician there was a tremendous upsurge in the variety and number of suspension-feeding animals, taking food from the water. Brachiopods, bryozoans, bivalve molluscs and echinoderms all went through explosive evolution. The Ordovician saw the rise to prominence of the graptolites, colonial animals related to vertebrates, some of which floated in the water and filtered food from it. Their rapid evolution meant that they provided valuable fossils for subdividing rock sequences in terms of time; those which lived in the upper reaches of deep waters in Victoria provided a means of comparing Ordovician rocks there, with standard world time scales developed in Europe.

Molluscs evolved new ways of life in the Ordovician. The cephalopods became wandering opportunists, preying on a variety of smaller life forms; the chambered, straight skeletons of nautiloids are present in considerable diversity as fossils in most Ordovician sequences in Australia. Among gastropods, the bizarre horn-shaped opercula of the macluritacean *Teichiispira* are striking. In other groups, the first corals began to assume an important role; the first crinoids appeared. Trilobites persisted in abundance, and Ordovician forms include some of the largest known in Australia — one asaphid form from the Stairway Sandstone in central Australia reached a length of approximately half a metre. Some are present only as trace fossils; the curved 'nests' known as *Cruziana* probably reflect the burrowing activity of a trilobite.

The shallow Larapintine Sea of central Australia was home to the oldest vertebrates yet known from the southern hemisphere — the jawless fish, *Arandaspis*, lived in shallow, well-aerated waters above a sandy bottom.

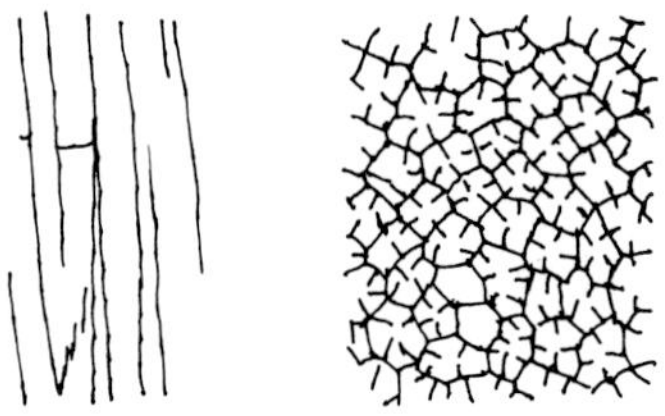

An Ordovician tabulate coral viewed from the side (left) and top (right). These corals had horizontal platforms, or tabulae, which divided up the colony of individual polyps and upon which the living animal rested. This colony is about 8 mm in width as viewed from above.

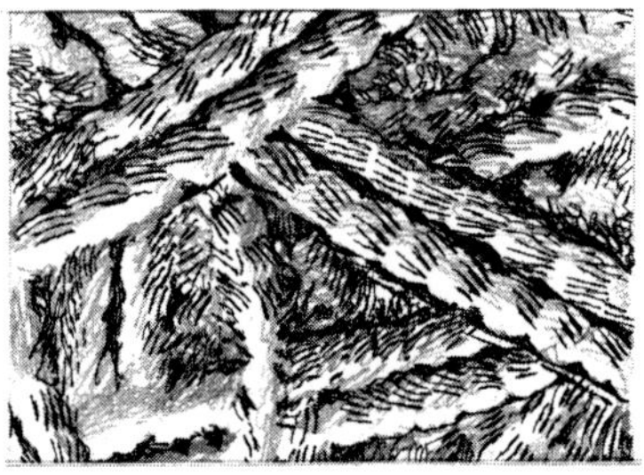

Cruziana, a trace fossil, probably the tracks (repichnia) of trilobites. Apparently the two parallel tracks were made by the passage of a single trilobite. Each parallel trackway is about 40 mm wide.

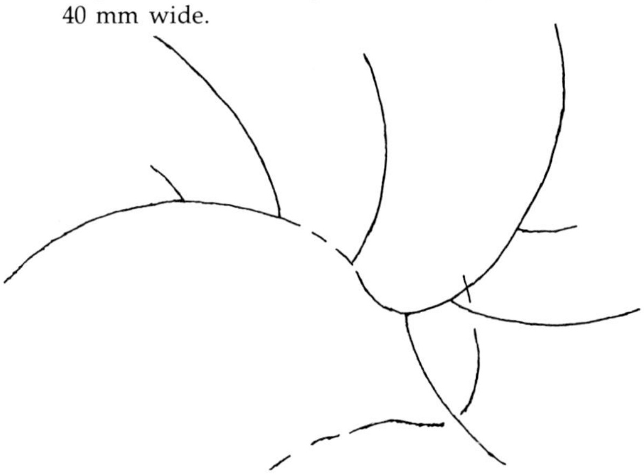

Leptograptus, a graptolite from the Ordovician of Australia. Ordovician times were the heyday of graptolites when they were extremely diversified and often had very short time-ranges. Thus they are extremely important in dating rocks of this time. They are often associated with black shales. A wide variety of graptolites became extinct at the end of the Ordovician. The width across this colony is about 85 mm.

Isograptus, a graptolite from Ordovician marine rocks of Australia. This specimen is about 20 mm across.

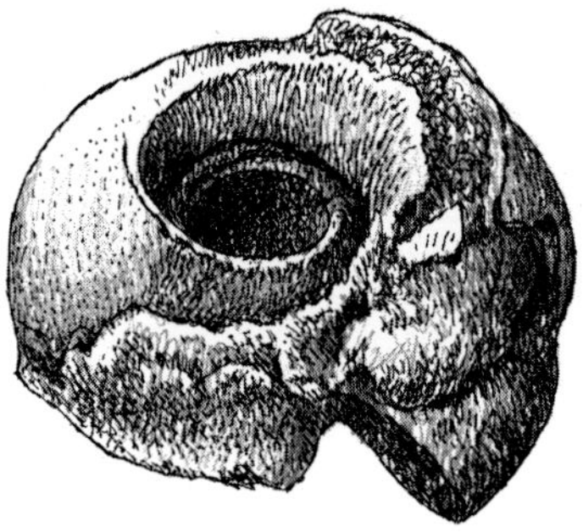

Teiichispira cornucopiae, a gastropod or snail from the Late Ordovician. This specimen is about 40 mm across.

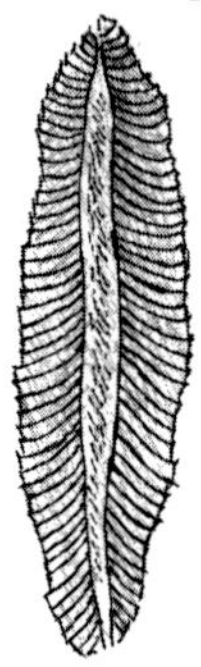

Phyllograptus, an Ordovician graptolite from marine rocks in Australia. Graptolites were both bottom-dwellers (benthonic) and floating animals (planktonic). The planktonic forms ranged widely and are important in determining similar ages of rocks on different continents of the world. Graptolites appear to be related to the ancestors of the vertebrates, as unlikely as their similarities may seem at first glance. This specimen is about 45 mm in length.

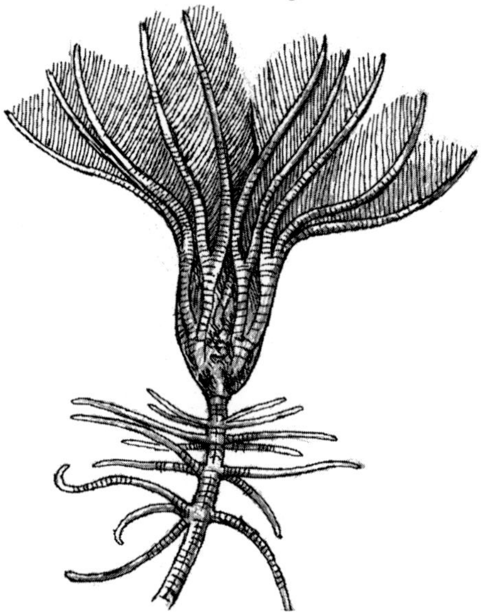

A crinoid or 'sea lily', one of the stemmed echinoderms (spiny-skinned animals including starfish, sea urchins and sea cucumbers). Crinoids and other stemmed echinoderms were very diverse during much of the Palaeozoic Era, but have left few living relatives. Crinoids came in a variety of sizes.

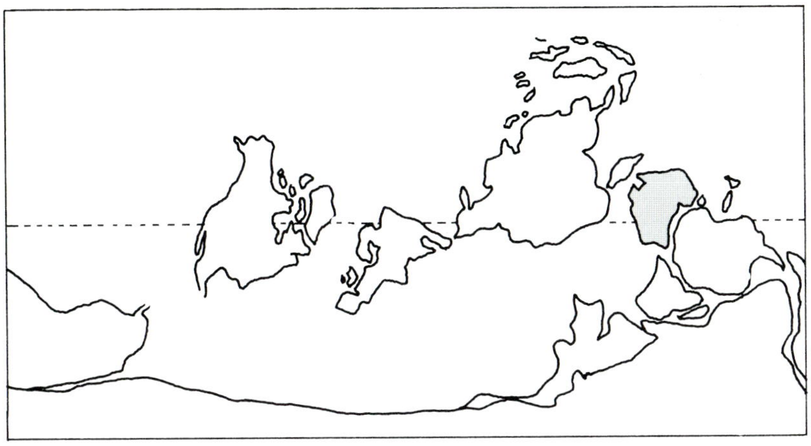

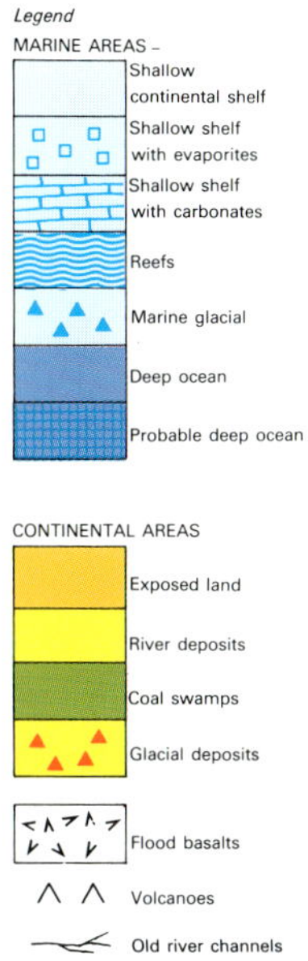

A	Adelaide	**D**	Darwin
AS	Alice Springs	**M**	Melbourne
B	Brisbane	**P**	Perth
C	Canberra	**S**	Sydney

ORDOVICIAN

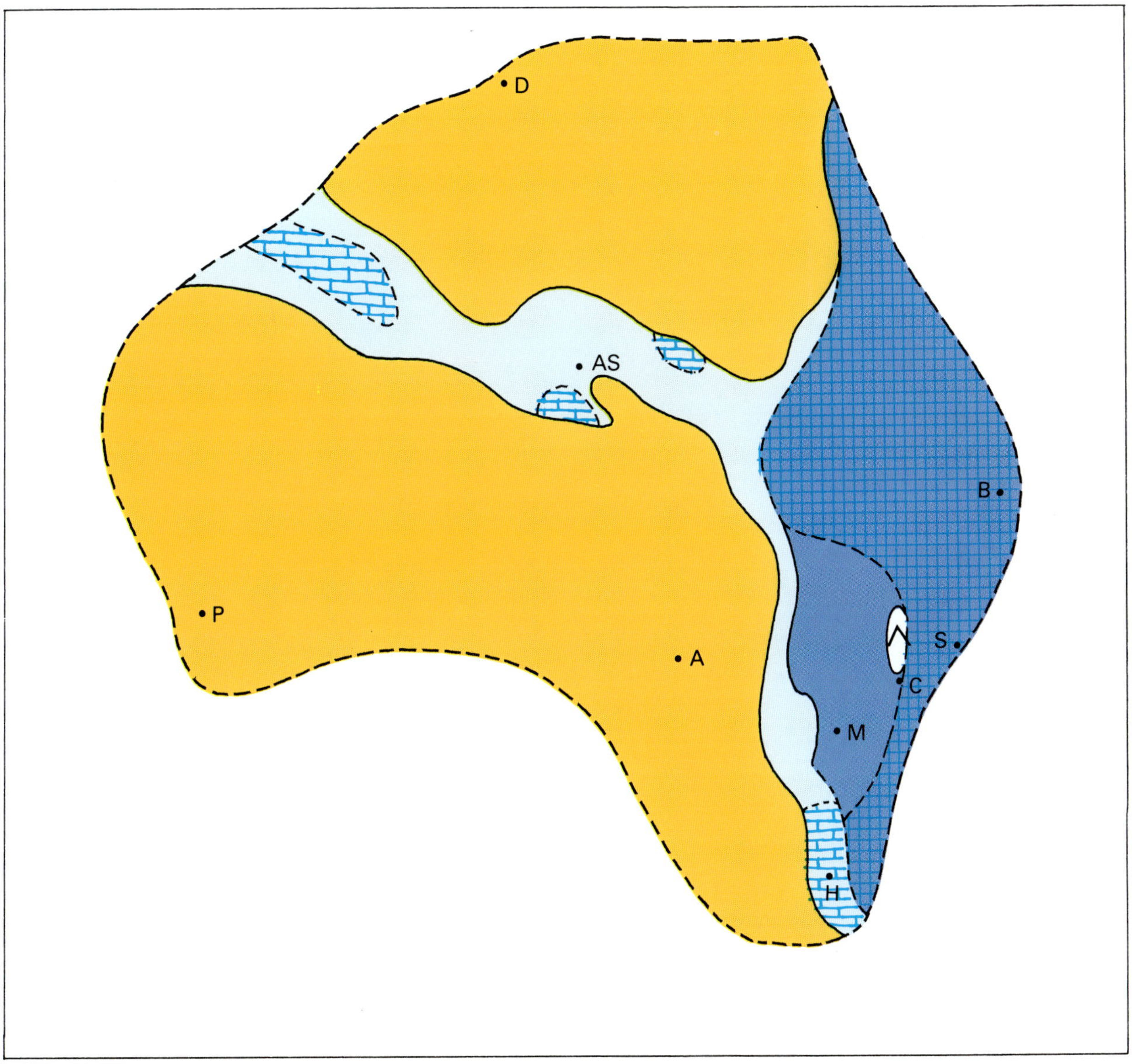

Silurian (395-435 million years ago)

For the duration of the Silurian Period, about two-thirds of the present Australian continent was emergent. At the beginning the land was uplifted and eroded in central Australia. Earth-movements off the then eastern seaboard resulted in a complex and changing pattern of deep-water troughs, narrow shelves and elongate islands. Volcanoes were particularly evident in the shallow seas, and were responsible for the growth of islands and, deep in the earth, for the intrusion of large masses of granite magma. Late in the Silurian the seas encroached on the western margin of the continent: in shallow seas there, sands and carbonate-rich sediments accumulated. Salt and other evaporites formed under dry climates in shallow lagoons.

Palaeomagnetic information shows Australia to have lain in near-equatorial latitudes. The warm-water aspect of fossil faunas found in shallow-water limestones is in clear accord with this geographic situation. In the seas off eastern Australia corals were widespread; the now extinct tabulate and rugose corals occurred in profusion, and formed low biostromes, structures approaching reefs in form — the Yass Basin of New South Wales contains some of the best known coral faunas. Brachiopods and trilobites were also abundant, the trilobites perhaps showing some decline in comparison with their Ordovician richness. Some were clearly sensitive to environmental factors, and fossil species are restricted to particular rock types. The knobbly-headed *Encrinurus* was apparently less environmentally sensitive, and was widespread on the Silurian sea floor. Near the surface of the deeper waters the graptolites continued to flourish, and are invaluable for subdividing Silurian rocks into recognizable time intervals. The graptolites are of cosmopolitan kinds, so that it is possible to recognize many of the graptolite zones of Europe in southeastern Australia.

On land, major environmental changes were occurring. The Silurian saw the emergence, on a world-wide basis, of the first land plants — true land plants in the sense that, for the first time, they were clearly adapted for growth and reproduction away from an aquatic environment. In the northern hemisphere the earliest vascular plants appeared in the second half of the Silurian; they were small, simply branched, leafless forms bearing sporangia at their tips, and are assigned to the Class Rhyniopsida. More complex forms, akin to the modern club-mosses, or lycopods, which bear

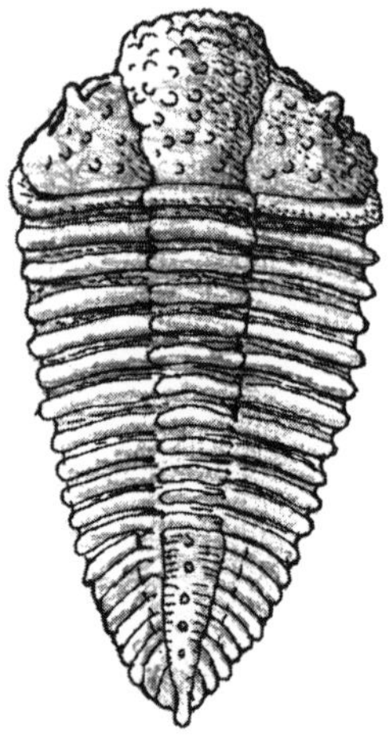

Encrinurus, a trilobite from Silurian marine rocks of Australia. Trilobites were mainly bottom-feeders but some were probably planktonic and wide ranging. This specimen was about 60 mm in length.

Baragwanathia, a lycopod land plant from Silurian and Devonian sediments. This specimen is about 90 mm long.

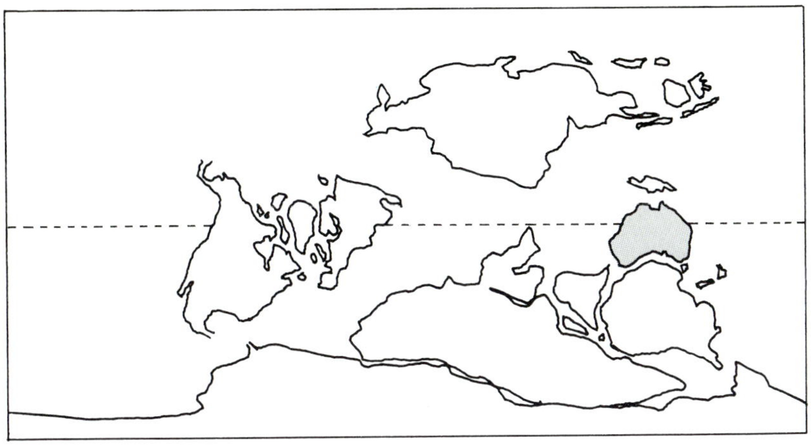

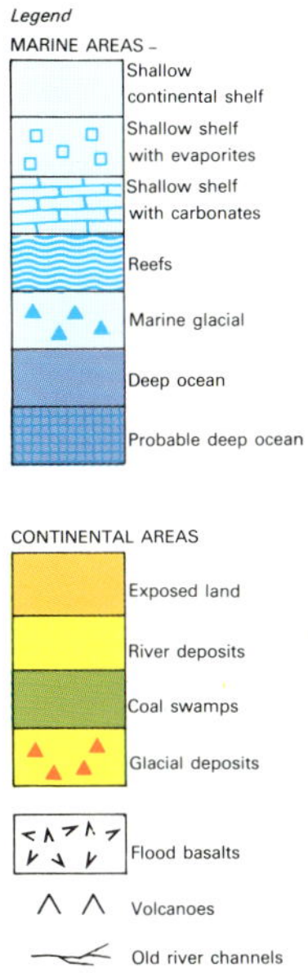

A	Adelaide	**D**	Darwin
AS	Alice Springs	**M**	Melbourne
B	Brisbane	**P**	Perth
C	Canberra	**S**	Sydney

SILURIAN

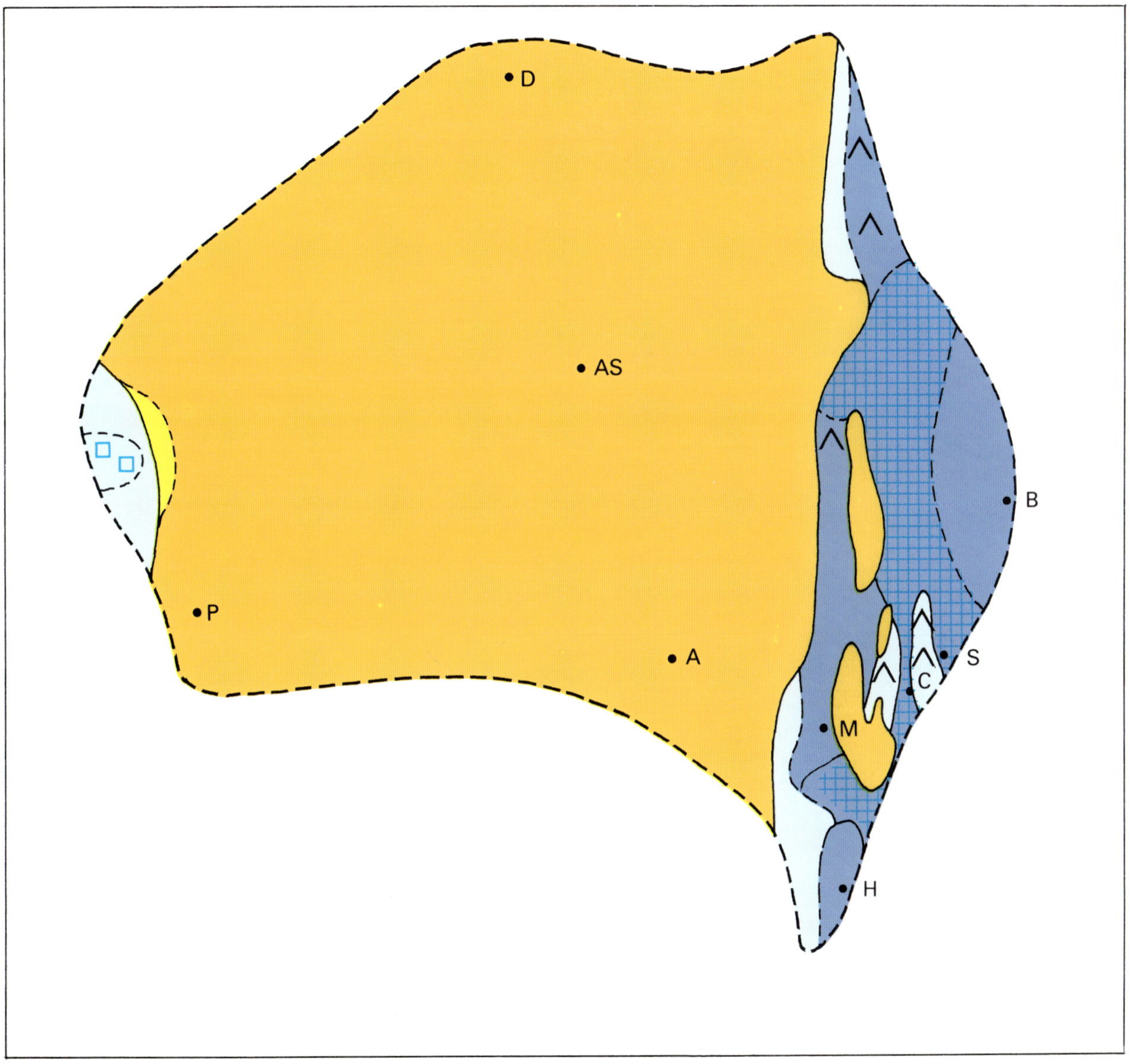

sporangia associated with small leaves, appear there later in the record. In Australia, the record of vascular plants differs in a way that challenges the current concepts of land plant evolution, concepts that have been developed from northern hemisphere evidence.

The oldest vascular plants in Australia are those of the *Baragwanathia* flora from marine beds deposited in the deep trough running north from Melbourne, and named for the most common plant, which has a fairly large (perhaps up to a metre long) axis bearing a dense cover of needle-like leaves with sporangia borne on some. This apparently rather complex plant grew in the Late Silurian, at the same time as much simpler forms flourished in the northern hemisphere. *Baragwanathia*, which may have grown semi-submerged in wet areas on the land surface, was accompanied by a number of other, simpler vascular plant forms. Land plants seem to have been fairly widespread in Australia towards the close of the Silurian, as suggested by the presence of spores in Silurian rocks in western New South Wales and Western Australia.

Devonian (345-395 million years ago)

The Australian sector of Gondwanaland remained the most tropical part of that supercontinent through the Devonian. The reconstruction of the earth shown in the figure is much simplified; information from rock types and fossils indicates that the landmasses of the northern hemisphere were disposed in a number of discrete blocks — Siberia, China and Kazakhstan were probably separate continental areas. Global climates were probably coolest in the early part of the Devonian. In comparison with other periods it was probably a fairly dry time, as indicated by the wide spread of evaporites. Overall, however, temperature gradients and humidity patterns were not greatly different from those of today.

In Australia, during the Early Devonian, earth movements and volcanic activity occurred along the eastern margin of the continent. While there remained deep ocean troughs near Melbourne and Bathurst, shallow seas in eastern Victoria and near Canberra supported reef growth. Shallow shelves with diverse faunas extended northwards through New England to north Queensland, flanked by deeper waters to the east. In the northwest of the continent, the sea invaded the flat-lying area of the Canning Basin — for a short time it again extended across central Australia — although whether these central Australian seas were connected to oceans in the east or west remains uncertain. In the northwest, there was poor circulation in the Early Devonian seas. The west coast situation was probably the cause of the excessive dryness in that region, where thick deposits of salt formed. Smaller deposits of the same kind formed near Adavale in central Queensland.

Later in the period earth movements caused the shallow seas of eastern Australia to retreat to the east. River-laid sands were deposited over much of southeast and central Australia: these incorporate many of the fish fossils found in Devonian rocks. Rivers draining the central Australian highlands flowed eastwards to reach the sea in the central Queensland area, or southeastwards through New South Wales. The presence of these rivers suggests a fairly high rainfall; as this is not in agreement with other aspects of Devonian geology in Australia, the suggestion has been made that these rivers originated in wetter latitudes far to the south, where Antarctica lay (Heckel and Witzke, 1977).

In the Late Devonian, massive reefs grew in the Canning Basin in Australia's northwest. The reefs, built of stromatoporoids (calcareous colonial organisms of uncertain affinity), of algae and of corals, have been little altered since their formation,

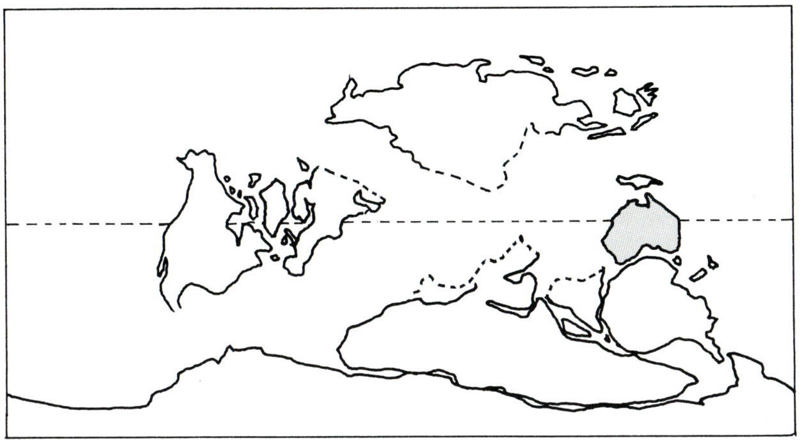

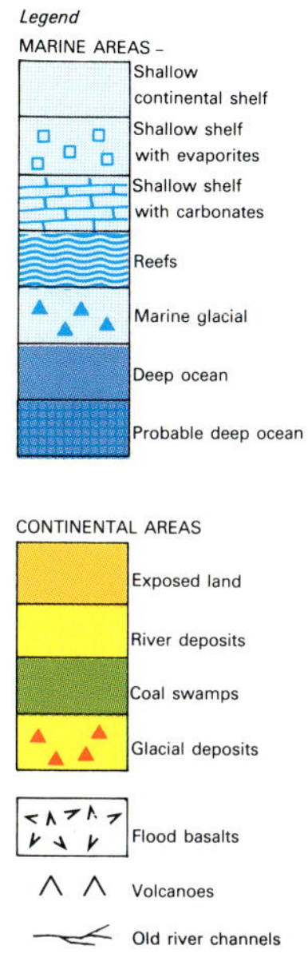

A	Adelaide	**D**	Darwin
AS	Alice Springs	**M**	Melbourne
B	Brisbane	**P**	Perth
C	Canberra	**S**	Sydney

DEVONIAN

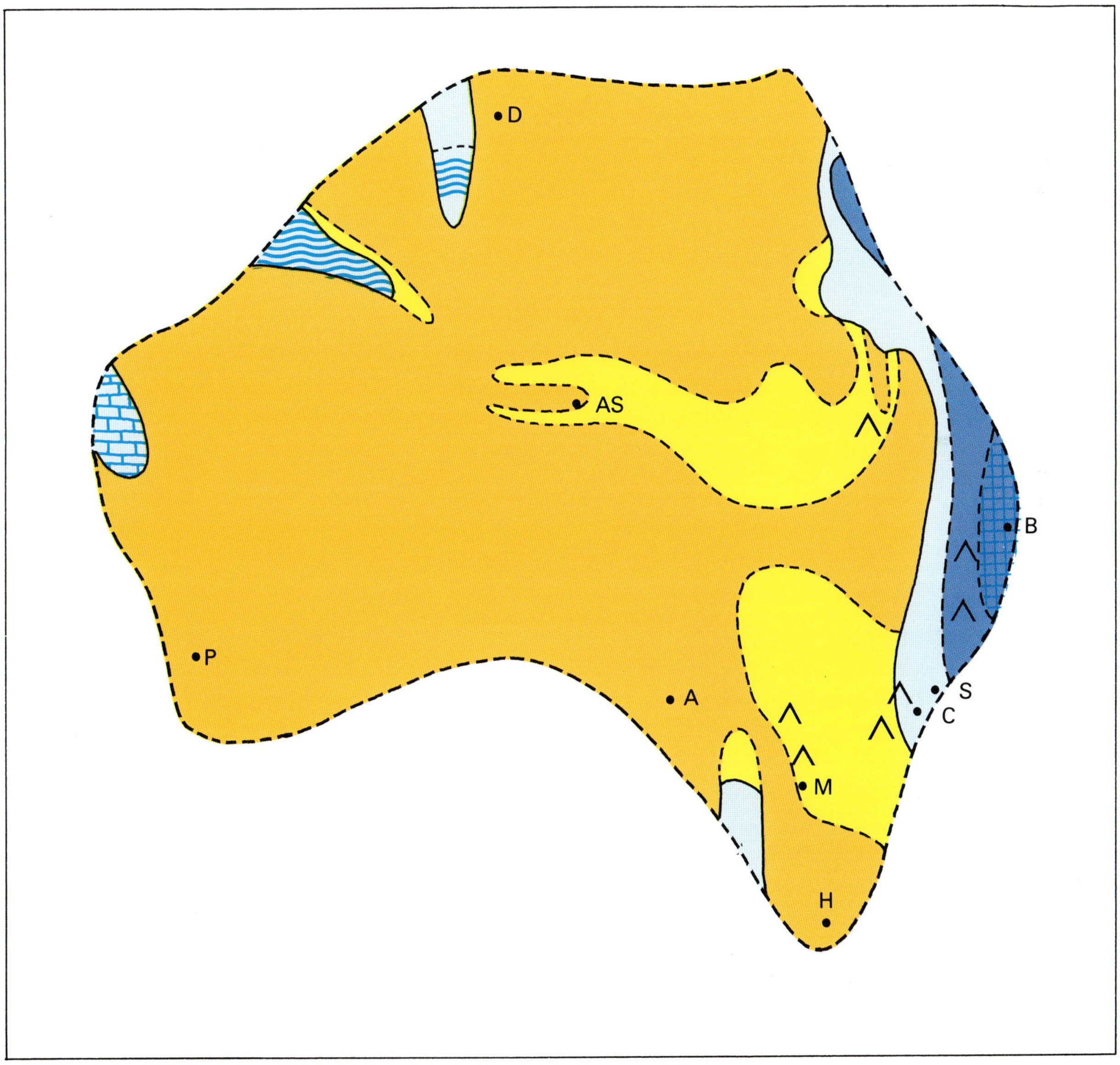

and stand today as a discontinuous line of ranges some 350 kilometres long. Rugose and tabulate corals abound in fore-reef and back-reef deposits. Perhaps curiously, the coral faunas are rich in individuals but poor in species, a condition which suggests that the area may have been close to the environmental limits for reef growth.

Faunas in the Australian region were a mixture of cosmopolitan and endemic elements. Most cosmopolitan in aspect were the ammonoid cephalopods, which originated probably in the Early Devonian. The coiled fossil shells found in Australia closely resemble those from Europe, North Africa and North America, a fact which makes the fossils ideal for dating purposes. The brachiopods and trilobites of south-eastern Australia show greatest affinity with those of the Old World. They differ, more particularly the trilobites, from those of the other southern continents, and have more in common with faunas of western Europe and the Mediterranean than with those of the 'Malvinokaffric' Province of Argentina, the Falklands and southern Africa. This distribution suggests that latitude then, as now, exerted a strong control on faunal distributions; Australia and western Europe lay within 20° of the Devonian equator, the Malvinokaffric Province lay much further south.

The freshwater fish, as might be expected, were more distinctively Gondwanan. Placoderm faunas from central and southeastern Australia have been grouped into a 'Wuttagoonaspid-phyllolepid Province' (Young, 1981), a province confined to the eastern part of Gondwanaland. The antiarchs show affinities between Gondwanaland

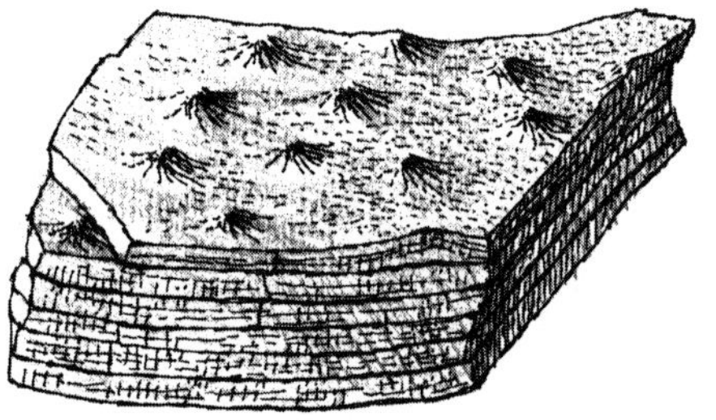

Stromatopora, a stromatoporoid from Devonian rocks. These organisms laid down layer after layer of carbonate material and at times during the Palaeozoic were important reef-formers. This specimen is about 45 mm wide and shows both the top and cross-section side views.

Phillipsastrea, a colonial rugose coral from Devonian marine sediments. Each dark circle represents the centre of the external skeleton of one animal that lived in close association with many others. Unlike the tabulates, the main skeletal walls in the exoskeleton of rugose corals were oriented vertically, not both vertically and horizontally. The dark circles in this specimen are about 15 mm in diameter.

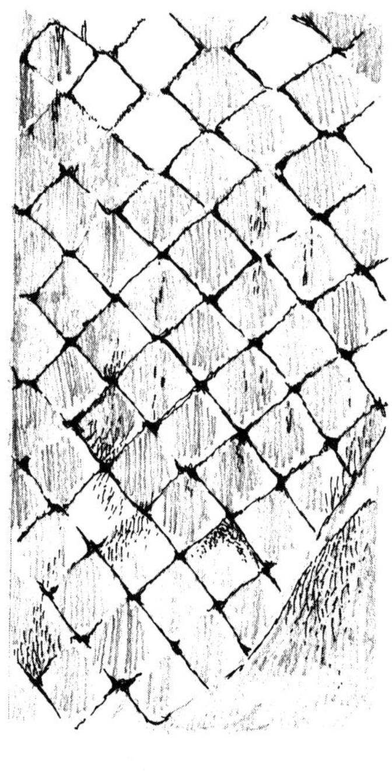

An impression of the stem of *Leptophloeum australe*, a lycopod or club-moss, from the Devonian. Today club-mosses are only small plants, but in the Palaeozoic they sometimes formed great forests. The width of this specimen is about 50 mm.

and southern China, suggesting that marine dispersal was possible sometime in the evolution of these predominantly freshwater forms.

On land, the *Baragwanathia* flora persisted in southeastern Australia into the Early Devonian. Later in the period the most characteristic plant was the large club-moss, *Leptophloeum*, whose distinctive trunk, with its diamond-pattern of leaf scars, is common in freshwater deposits. They may have been strandline plants, occupying a position similar to that of modern mangroves. Spores and pollen found dispersed in sedimentary rocks are quite varied by the end of the Devonian, suggesting a much more diverse flora than the record of larger plant remains indicates. It seems highly likely though, given the level of evolution of land plant reproduction, that vegetation was most common in wetter habitats, and the drier parts of the land surface were barren, or nearly so.

Carboniferous (280-345 million years ago)

In the early part of the Carboniferous, the Australian sector of Gondwanaland was located in tropical to subtropical regions. Shallow seas occupied areas of the eastern and western sectors of the present continent, in situations not greatly different from those of the Late Devonian. Shallow submarine shelves off northwestern Australia were warm, and limy sediments were deposited in the Canning Basin and Bonaparte Gulf regions and in the Carnarvon Basin, which is centred on the North West Cape. In eastern Australia volcanoes were active close to the Carboniferous shoreline and volcanic debris is common in a belt extending from the New England region of New South Wales, to north Queensland. In central Australia, uplift of the land, and its consequent erosion, caused riverine and lake deposits to form.

Early in the Late Carboniferous, Australia moved rapidly south towards polar regions and consequently experienced an overall lowering of temperature. The end of the Carboniferous was marked by a major period of glaciation which affected much of Gondwanaland. As a prelude to this major phase, alpine glaciers formed in the volcanic highlands bordering the sea in eastern Australia.

The onset of severe climatic deterioration had a pronounced effect on life in the Australian region. In the relatively warm seas of the Early Carboniferous, faunas of corals, brachiopods, bryozoans, bivalves, gastropods and ammonoid cephalopods flourished in shallow, marine shelf environments. In a general way these faunas have a cosmopolitan aspect. In southeastern Australia, the percentages of endemic forms among brachiopods are low in the early part of the Carboniferous. In the second part of the period, new, and very provincial faunas developed — these were distinctive high-latitude biotas. The brachiopods show a steep rise in endemic forms (Roberts, 1981). The faunas as a whole lacked compound corals, large foraminifera and ammonoid cephalopods, all of which occur in contemporaneous rocks in the northern hemisphere. In Australia the faunas became very impoverished — poor in species, although in some areas rich in the number of individuals.

On land, the floras were similarly affected by advancing cold. The succession of different fossil plant floras from the Sydney Basin best shows the changes occurring during this interval of dramatic climatic change (Retallack, 1981). The early part of the Carboniferous saw the dominance of the giant club-mosses — *Lepidodendropsis*. These floras were followed, in the middle part of the Carboniferous, by forests dominated by the now extinct seed-ferns or pteridosperms. Forms such as *Sphenopteridium* dominated coastal plain forests; other elements were horsetails or 'scouring rushes' and lycopods (club-mosses); in protected seashores were fringes of lepidodendroid 'mangroves'. Then came the chilling, and the advance of glaciers from high ground. The severe climates

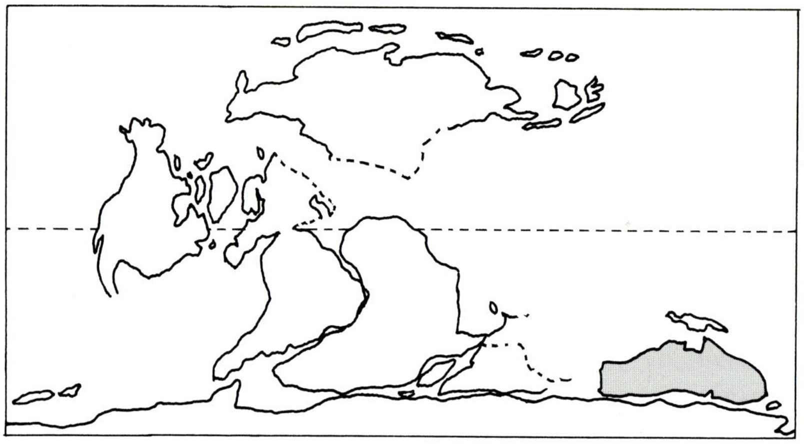

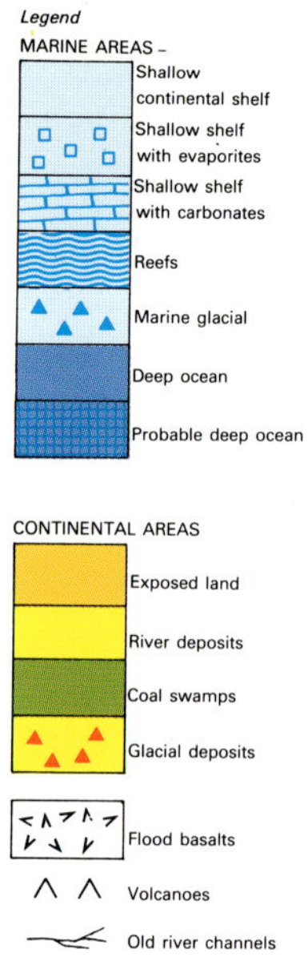

A	Adelaide	**D**	Darwin
AS	Alice Springs	**M**	Melbourne
B	Brisbane	**P**	Perth
C	Canberra	**S**	Sydney

LATE CARBONIFEROUS

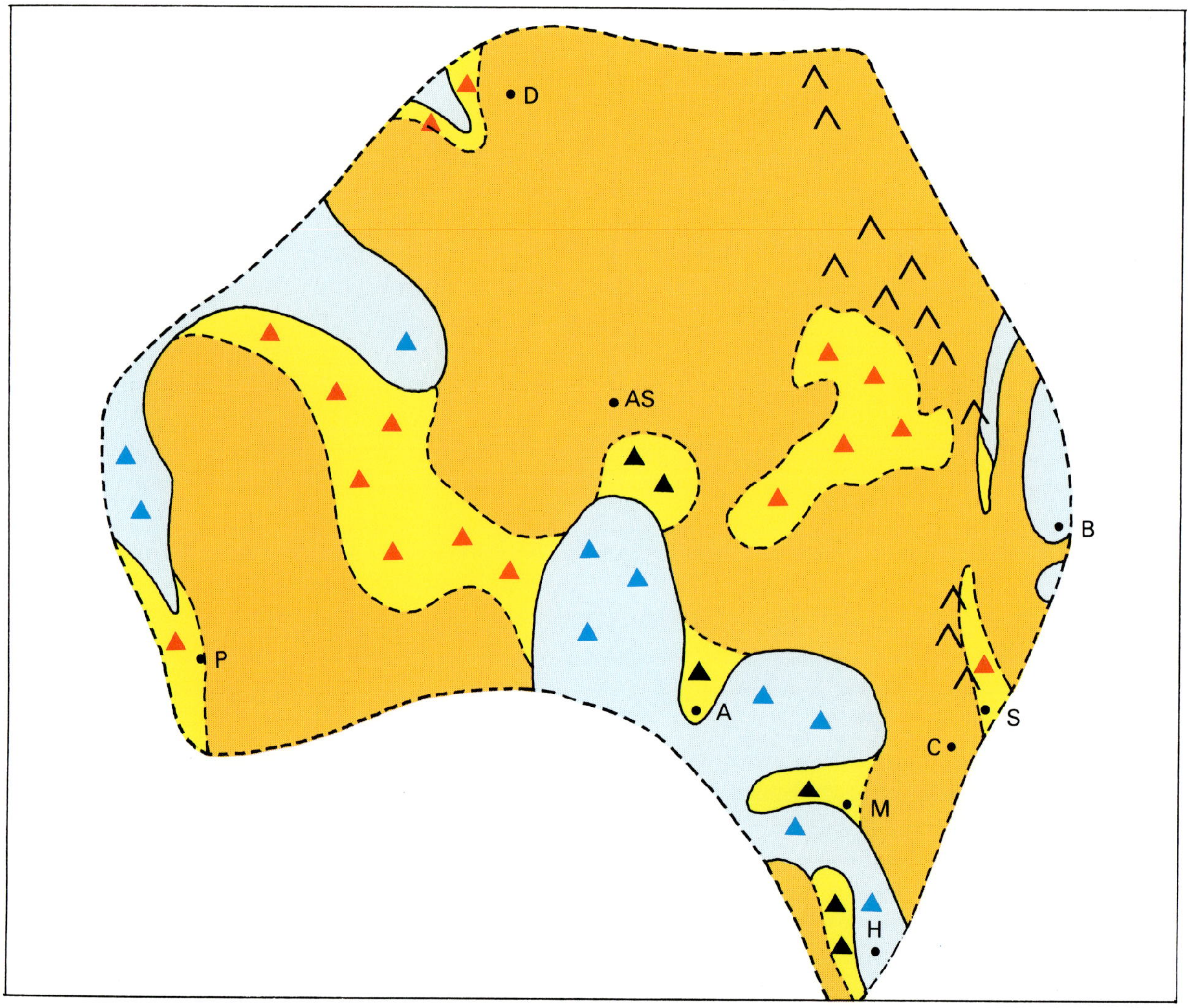

decimated the seed-fern forests, but at least one form survived — *Botrychiopsis*, which assumed the form of a sparse, low-growing, impoverished tundra vegetation. This plant colonized otherwise bare and stony regions of glacial outwash.

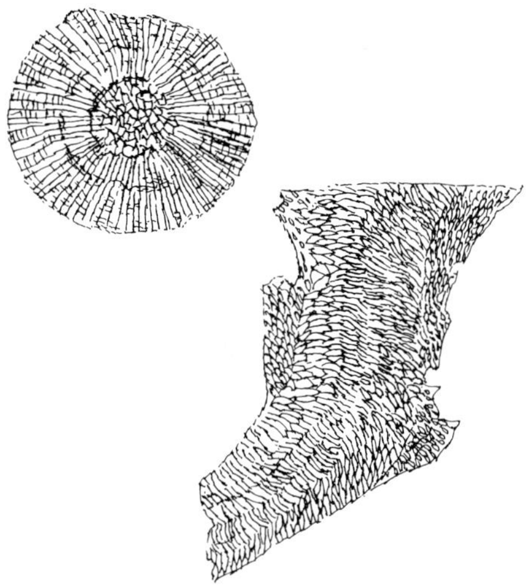

A Carboniferous tetracoral, also known as a rugose or horn coral because of its horn-like shape. Top view is on the left and a cross-sectional side view is on the right. This specimen is about 15 mm in width, but these corals came in a variety of sizes. The fossil represents the external, carbonate skeleton of a solitary rugose coral.

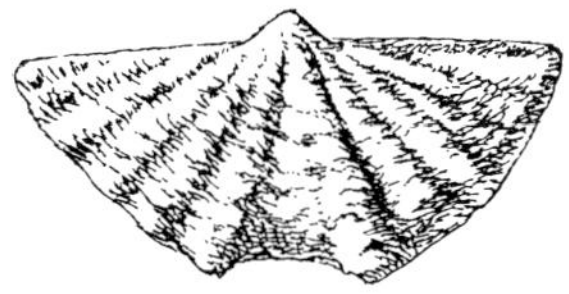

A speriferoid brachiopod from Late Palaeozoic rocks. Speriferoids were characterized by having a very wide hinge line; where the two shells articulated, and often a deep groove (or ridge) down the middle of the shell. This specimen is about 35 mm in its greatest dimension, but speriferoids came in a variety of sizes.

The head (or calyx) of *Nymphaeoblastus*, a blastoid from the Carboniferous. During the Palaeozoic stemmed echinoderms such as blastoids, crinoids (sea lilies) and their relatives were quite varied. This specimen is about 25 mm wide.

Permian (230-280 million years ago)

By the end of the Carboniferous, the effects of glaciation were felt over much of southern Australia. The form and extent of the ice sheet remains uncertain; whether, for instance, it was a massive continuous icecap comparable to that covering Antarctica today, or whether there were ice-free regions which served as refugia for the fauna and flora. The fact that some plants seem to have persisted through the glacial interval suggests that such refuges did in fact occur. Separate ice centres seem to have occupied southwestern, central, southeastern and northwestern Australia. Evidence for ice on the continent is to be found in polished rock surfaces with parallel grooves or striae, in 'chatter marks' formed as the rocks frozen in the bottom of the ice scraped the underlying rock, and in accumulations of morainal debris, great thicknesses of boulders intermixed with sediment of finer grain size. The melt waters which flowed beneath the ice masses or glaciers deposited rock flour and sands.

Sometime in the Early Permian, the ice began to shrink. It probably disappeared earlier from Western Australia than it did from the eastern part of the continent. Conditions over much of Western Australia have been described as temperate late in the Early Permian (Dickins, 1978); contemporaneously, southeastern Australia was experiencing a cool temperate climate. The melting of the icecaps resulted in a world-wide rise in sea-level. In the Early Permian, shallow seas flooded the Sydney Basin and the Bowen Basin of Queensland. Across southern Australia they covered a wide sweep from Tasmania into Victoria, western New South Wales and inland South Australia. The seas also formed deep bays on the Western Australian coastal margin.

In the later part of the Permian the seas retreated from much of eastern Australia and coal swamp conditions prevailed — with very brief marine interruptions — in Tasmania, the Sydney region, and much of Queensland.

It is this scenario, of an early chilling, followed by a gradual climatic amelioration, that the Australian biota had to contend with during the Permian. In the marine realm, the invertebrate faunas showed pronounced endemism at the beginning of the Period — they formed part of a distinct high-latitude realm. On rocky sea-bottoms in cool regions there flourished communities of large thick-shelled bivalves such as *Eurydesma*; sandy shore environments supported other bivalve groups such as the megadesmids (Runnegar, 1979). The endemic character of the marine invertebrate faunas decreased after the early chilling phase; faunas, especially brachiopod faunas from Western Australia, were first to show affinities with those of the Tethys — the great latitudinal seaway extending through the Mediterranean eastwards to the north of Arabia and India. Later this faunal link became evident in Queensland. The faunas also show that truly warm waters reached only the extreme northwest of Australia — the Bonaparte Gulf area — and then only very late in the Permian.

On land, the *Botrychiopsis* tundra was invaded by new plants, as the climate warmed after glaciation. Pteridosperms, or seed-ferns, continued to dominate the vegetation; certain taxa were shared by all the Gondwanaland continents, and their widespread occurrence provided some of the earliest evidence that continental movement had occurred. First to appear in the fossil record were the oval leaves of *Gangamopteris*. This probably grew first in stands in the gentler coastal climates, later penetrating inland; forests in which it dominated have been compared to the birch-dominated taiga of today's northern regions.

Slowly the taiga gave way to the *Glossopteris*-dominated swamp forests, but this could not occur until permafrost was no longer a part of the environment. *Glossopteris* has large aeration chambers in its roots, which would have split apart on freezing. It was probably a large tree, adapted for growth in waterlogged soils; it was clearly deciduous, and its leaves have been found in autumn bank deposits (Retallack, 1981). While *Glossopteris* probably dominated the coal swamps of the Permian, other trees were also present; conifer logs and stumps occur in some diversity at the top of Permian coal seams, many showing annual growth rings which clearly reflect the strongly seasonal climate. Ferns, club-mosses and horsetails formed the undergrowth in these forests. Insect faunas were diverse. Curiously, evidence for terrestrial vertebrates is very meagre, and is confined to a single labyrinthodont in the youngest coal measures. There is no evidence from Australia of the rich reptile and amphibian faunas which occurred in southern Africa, in another sector of Gondwanaland.

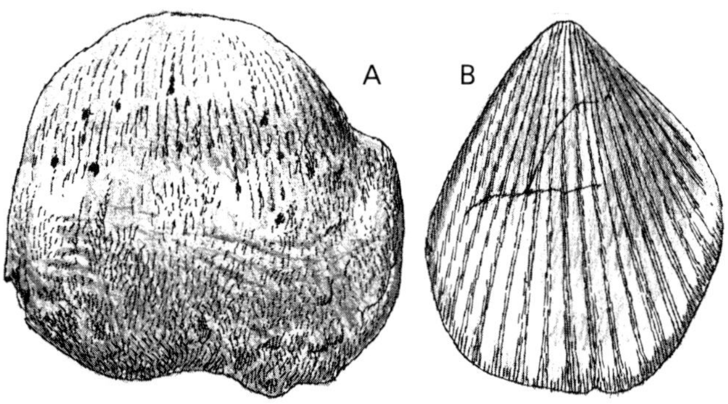

Terrakea solida, (**A**) a productid brachiopod (lamp-shell) from Permian marine rocks. Width across the shell is about 40 mm.
Deltopecten, (**B**) a bivalve from the Permian of Australia. Shell is about 70 mm in length.

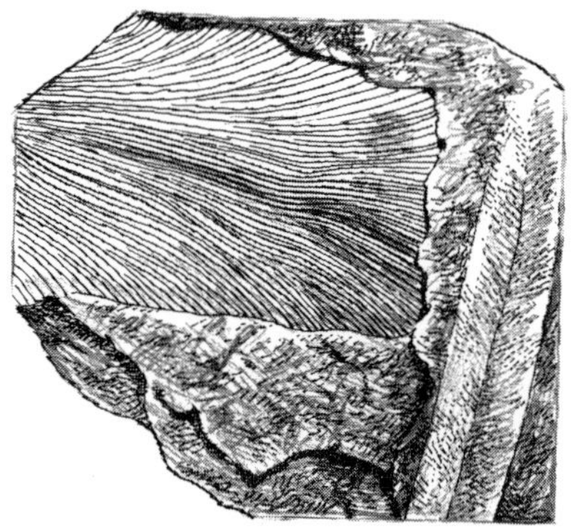

A leaf of *Gangamopteris*, a seed-fern from Permian terrestrial rocks. Many probably resembled modern tree ferns. The big leaf is about 80 mm long.

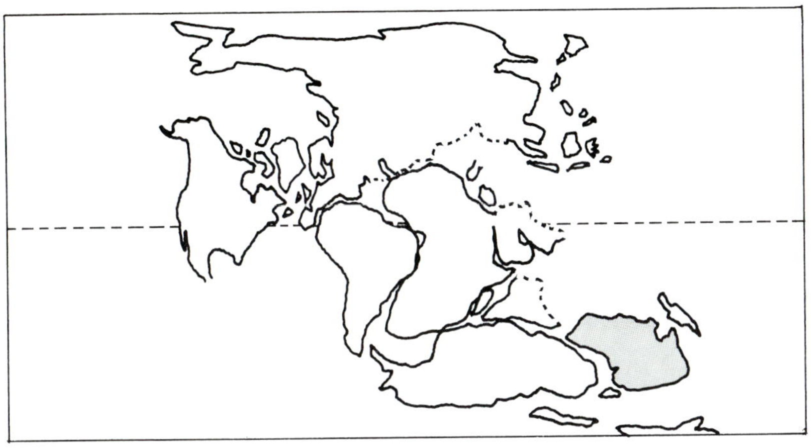

Legend

MARINE AREAS –

- Shallow continental shelf
- Shallow shelf with evaporites
- Shallow shelf with carbonates
- Reefs
- Marine glacial
- Deep ocean
- Probable deep ocean

CONTINENTAL AREAS

- Exposed land
- River deposits
- Coal swamps
- Glacial deposits
- Flood basalts
- Volcanoes
- Old river channels

A	Adelaide	**D**	Darwin
AS	Alice Springs	**M**	Melbourne
B	Brisbane	**P**	Perth
C	Canberra	**S**	Sydney

PERMIAN

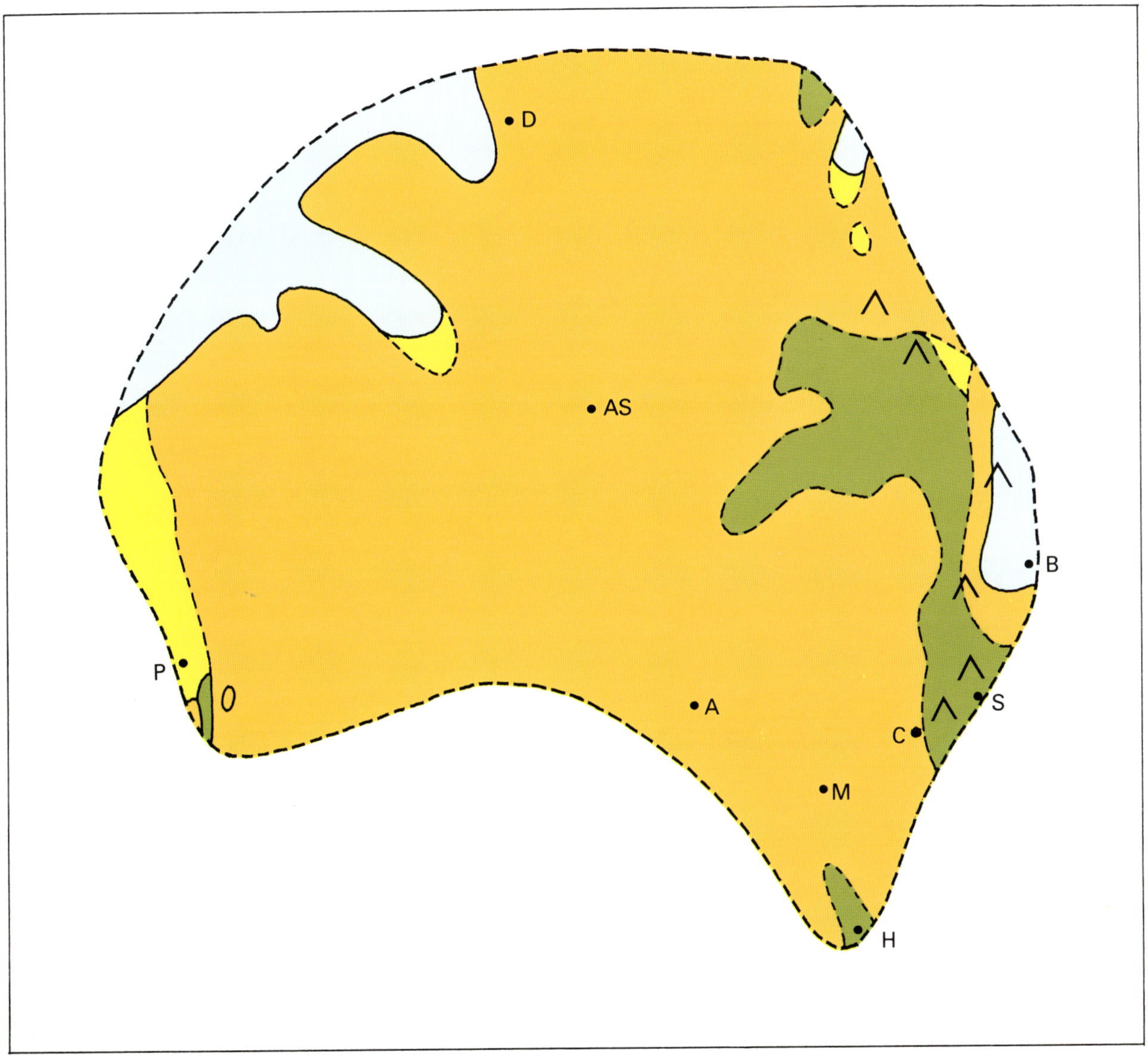

MESOZOIC ERA

Triassic (195-230 million years ago)

Towards the end of the Permian, the sea retreated from almost the whole of eastern Australia. The land was folded and uplifted in a broad band running north from Sydney through much of southeastern Queensland. In the early part of the Triassic shallow seas ingressed across the present margins of the continent in the Maryborough area of Queensland and in the Canning and Perth basins in Western Australia.

Australia lay probably in a high-latitude humid belt, outside the dry climates of mid-latitudes, latitudes in which great volumes of evaporitic deposits accumulated (Frakes, 1979). There is a great deal of evidence, in Triassic rocks, for abundant water-borne sediments. On the Northwest Shelf of Western Australia a large delta was built; in eastern Australia, the Hawkesbury Sandstone of the Sydney area may have been laid down in an environment having broad similarities to the floodplains of the modern Brahmaputra of India. Similar river-laid sediments abound in Queensland. There were also abundant lakes and swamps.

Although there is abundant evidence for a dense vegetation cover through most of the Triassic, the coal swamp habitats, so characteristic of the Permian, diminished markedly. Triassic coals occur only in the small Leigh Creek Basin in South Australia, in southeastern Queensland, northern New South Wales and Tasmania. The reasons for this are obscure, given the evidence for abundant moisture in the Triassic. The rarity of coals perhaps reflects periodic dry seasons, or better drainage than in the Permian, but this is conjecture only.

Fossil soil horizons in eastern Australia provide some clues to climatic conditions in the Triassic. The podzolic palaeosols of the Early Triassic in the Sydney Basin are compatible with a cool temperature climate — perhaps a climate less severely seasonal than that of the Permian, but still one with some of the characteristics of high latitudes. Red mudstones in the Rewan Formation of the Bowen Basin in Queensland represent, it has been suggested, the transported products of weathering under humid temperate conditions (Jensen, 1975); this contrasts with the conditions of tropical or near-tropical formation which have often been indicated for sediments of this colour.

The end of the Permian was marked by a crisis of considerable magnitude among marine faunas; the cause is unknown, but may be related to the widespread withdrawal of shallow seas from continental areas. The shallow seas which flooded across the western margin of Australia in the Early Triassic were rich in life forms very different from Permian ones. They contained an abundance of microplankton-algal forms or 'acritarchs', which suggest a phenomenon similar to the modern 'red tides'. These swarms of micro-organisms occurred on a global basis at that time and probably represent highly productive, immature ecosystems flourishing on nutrients provided by the newly-flooded land areas. Cephalopods, bryozoans and fish have been reported from these habitats.

On land, the Triassic saw major modification in the vegetation, although the seed-ferns remained dominant. Immediately above the youngest coal seams, with their *Glossopteris* flora, lies a fossil flora containing ferns, horsetails and conifers, but which is dominated by the seed-fern *'Thinnfeldia' callipteroides*; this flora probably grew on river flat alluvium, accumulating after the draining of the old *Glossopteris* swamps in the Late Permian (Retallack, 1981). Subsequently, conifer forests covered areas such as the vast floodplain of the Sydney Basin. The most typical of the Triassic floras displaced these forests — these were floras dominated by *Dicroidium*, especially by the broadleaf *D. zuberi*. This was a Gondwana-wide flora, assuming different forms

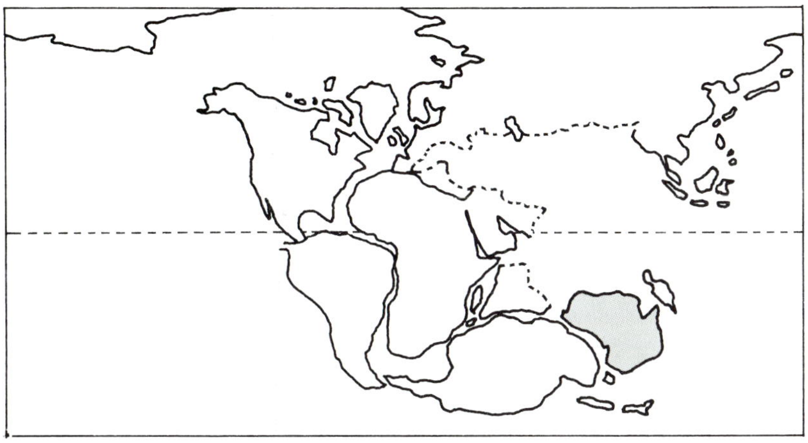

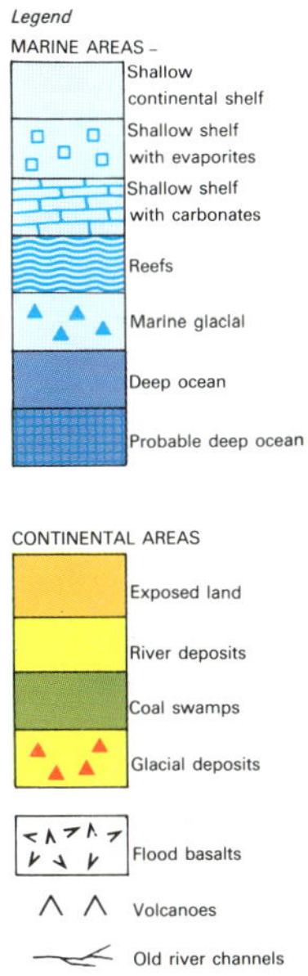

A	Adelaide	**D**	Darwin
AS	Alice Springs	**M**	Melbourne
B	Brisbane	**P**	Perth
C	Canberra	**S**	Sydney

TRIASSIC

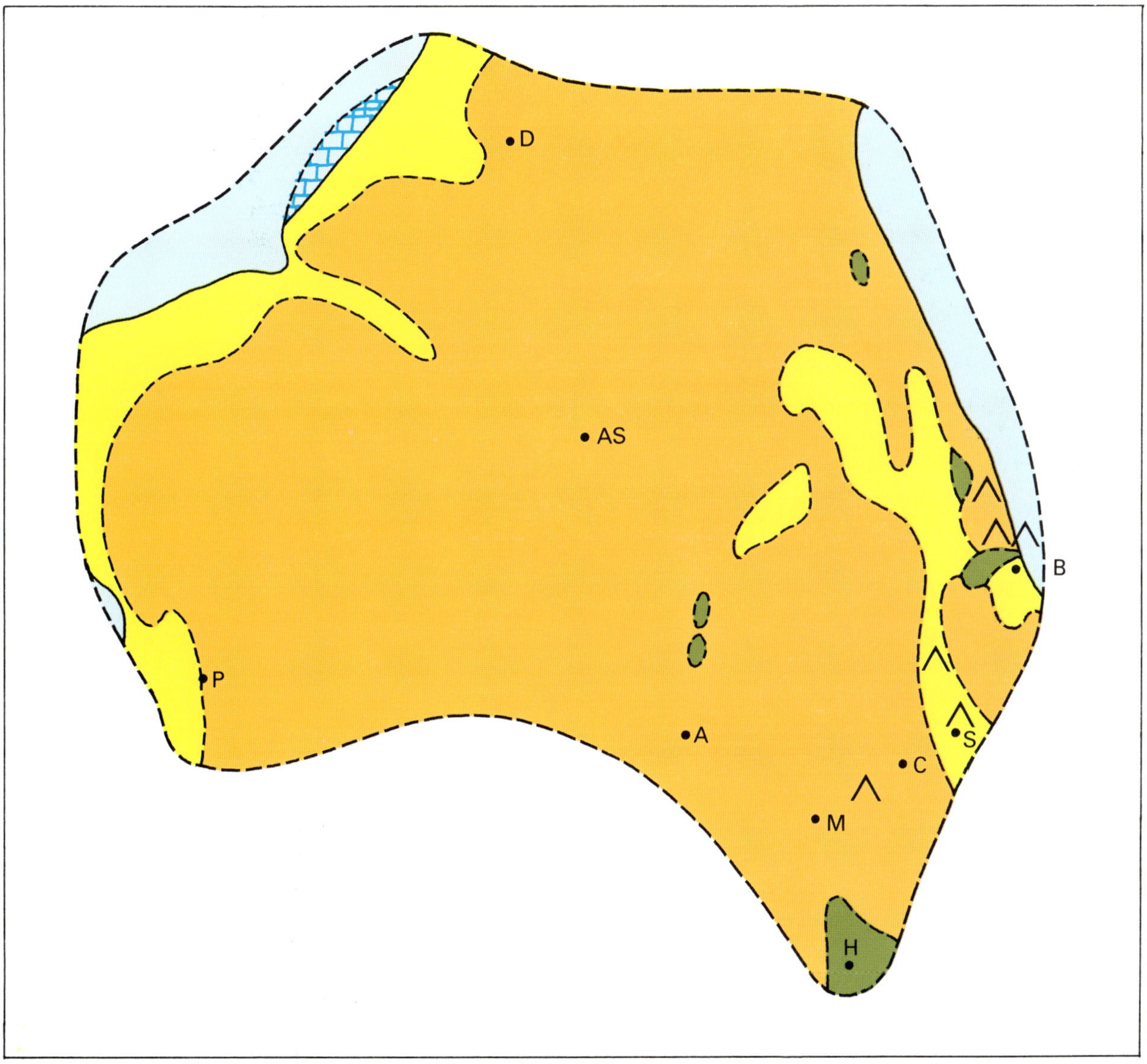

in different habitat conditions. In favoured habitats it formed forests, but on the sandy soils of the delta plain of the Sydney Basin it may have grown as a heath. In water-logged alluvial soils, there and in northwestern Australia, the lycopod *Pleuromeia*, its thick stem crowned with strap-like leaves, formed monospecific thickets occupying the modern mangrove habitat.

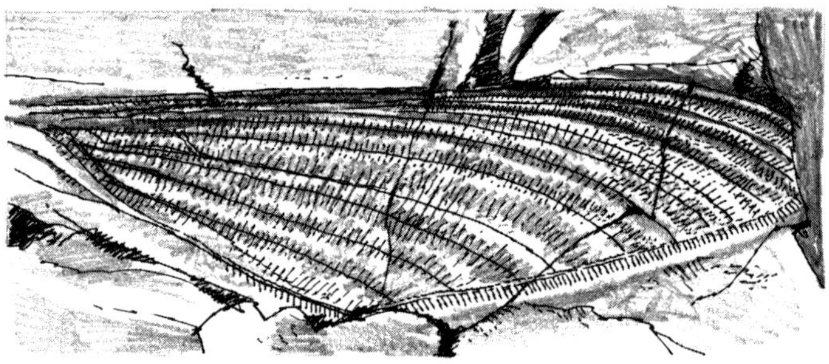

Wing of an insect, *Mesogereon* from the Triassic (Middle to Late) of Queensland. Wing length is about 45 mm.

The freshwater lagoons and swamps of the Triassic provided a habitat for a diversity of vertebrates. Labyrinthodont amphibian remains are known from the predominantly red-bed sequences of Queensland, from lagoonal deposits in the Sydney area, and from tidal flat habitats in northwestern Australia — habitats that were periodically inundated by streams during seasonally high rainfall. Lizard-like reptiles also inhabited a range of Triassic environments, as did the lungfish *Ceratodus*, whose descendants live today in eastern Australian rivers, and display temperature preferences in the 10°-30° range.

Jurassic (140-195 million years ago)

The Jurassic Period marked the beginning of dramatic changes in the earth's geography. Late in the period West Gondwanaland (Africa and South America) began to split from East Gondwanaland (India, Madagascar, Australia and Antarctica). Great outpourings of basaltic rock presaged this continental breakup — flows and sills of enormous thickness extend from southern Africa, through the Transantarctic Mountains of Antarctica, to Tasmania. The beginnings of the eventual rift between Australia and Antarctica can be seen in the extension of these basalts westwards from Tasmania to Kangaroo Island.

On the western edge of Australia there occurred considerable movements on the very large Darling Fault in the Perth Basin, and sediments, chiefly freshwater, accumulated there in a deep narrow trough. A short-lived invasion of the sea occurred in the area in the middle of the Jurassic; later the sea flowed across the continent's northwest margin.

In eastern Australia a major river system developed, its drainage basin spreading over much of inland Queensland, northern New South Wales and northeastern South Australia; the rivers may have entered the sea near the present position of Maryborough. Initially, they spread a great sand sheet across the Great Artesian Basin. Later, coal swamps and freshwater lagoons abounded in the same area, with the general diminution of river flow. Other rivers drained northwestern Australia, flowing to the sea through the Canning Basin. On Australia's southern margin the volcanic activity associated with the initial Australia-Antarctica break occurred in coastal basins where river and freshwater lake deposits also accumulated.

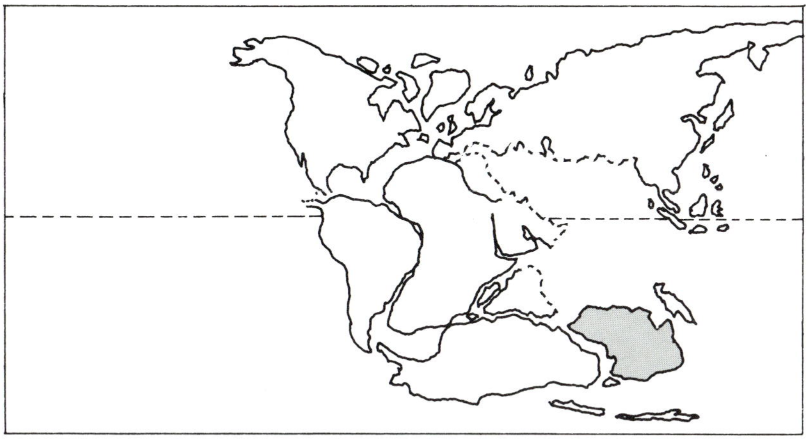

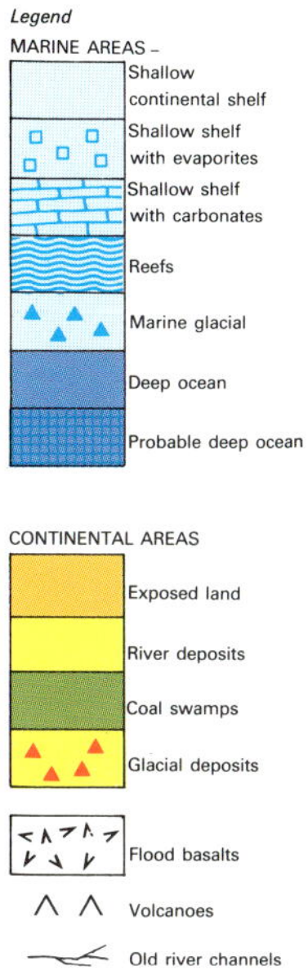

A	Adelaide	**D**	Darwin
AS	Alice Springs	**M**	Melbourne
B	Brisbane	**P**	Perth
C	Canberra	**S**	Sydney

JURASSIC

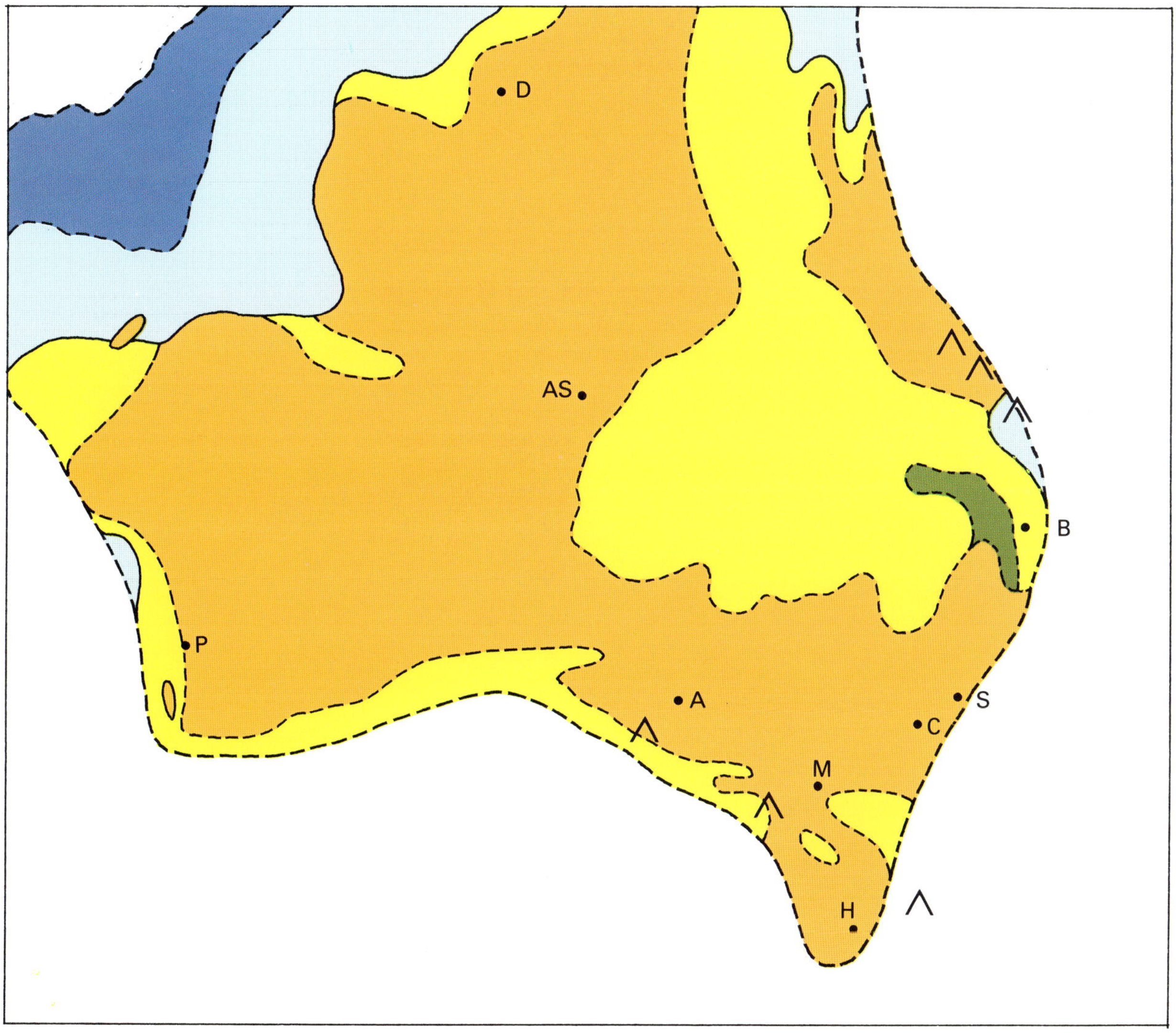

On a global basis, the Jurassic was a time when warm climates extended to high latitudes; in a very general way the warmth was accompanied by aridity in a belt extending to about 45° on each side of the equator. Poleward of this there seems to have been a rainy belt, as evidenced by the abundance of coals in high Jurassic latitudes. Australia lay within this belt, and the evidence for strong river flow and for the presence of coal swamps accords with a generally rainy climate. There is, however, as might be expected, evidence for patchiness of climatic types in time and space. Early Jurassic rocks north of Perth, for instance, contain red-beds and an abundance of the pollen type *Classopollis*, produced by an extinct family of conifers. Elsewhere in the world, sediments with abundant *Classopollis* have been used to imply aridity, and it has been suggested (Filatoff, 1975) that hot, arid conditions may have prevailed on Australia's west coast early in the Jurassic. Later, an abundance of *Araucaria*-type pollen (produced perhaps by ancestors of the Norfolk Island pine and Monkey-puzzle tree), and abundant ferns, suggests that precipitation increased and temperatures were lower.

The temperatures experienced in Australia in the Jurassic are difficult to estimate. Diverse ammonite faunas in the seas of Western Australia suggest reasonable warmth, but probably not as high as the 25°C suggested by oxygen-isotope analyses, which may be in error. In eastern Australia, the fossil floras provide the main clue; abundant araucarian conifers, and early forms of the Podocarpaceae, the great southern conifer family, suggest fairly cool conditions by analogy with their modern growth preferments.

The habitat suggested by beds containing the Talbragar fish fauna, in northern New South Wales, might be characteristic of the eastern Australian Jurassic. Here a lake lay in a region of high, probably uniform rainfall, surrounded by a forest dominated by *Agathis* (the Kauri pine), with scattered podocarpaceous conifers, and near the lake edge, a low heath with ferns and pteridosperms. Cicadas probably stridulated in the trees at certain seasons, and their remains are preserved in the Talbragar lake (White, 1981). Given the high latitudes of Australia it seems difficult to escape some seasonal changes of climate. However, the presence of giant sauropod dinosaurs such as *Rhaetosaurus* in freshwater deposits in Queensland, seems to indicate that winter temperatures remained well above freezing.

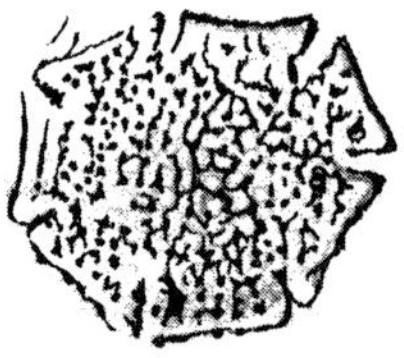

A pollen grain of *Nothofagus*, the Southern Beech. Forms related to this genus ranged from Late Mesozoic to present day. The grain is about 0.025 mm in width.

Cretaceous (65-140 million years ago)

In the Cretaceous the fragmentation of Gondwanaland continued, and the seas surrounding Australia began to assume their modern configurations. In the west, the Indian Ocean began to develop as India (with Greater India, its eastern extension) began to move northwards away from Australia/Antarctica. Off Australia's east coast, a new continental fracture developed in the Late Cretaceous, and a low ridge incor-

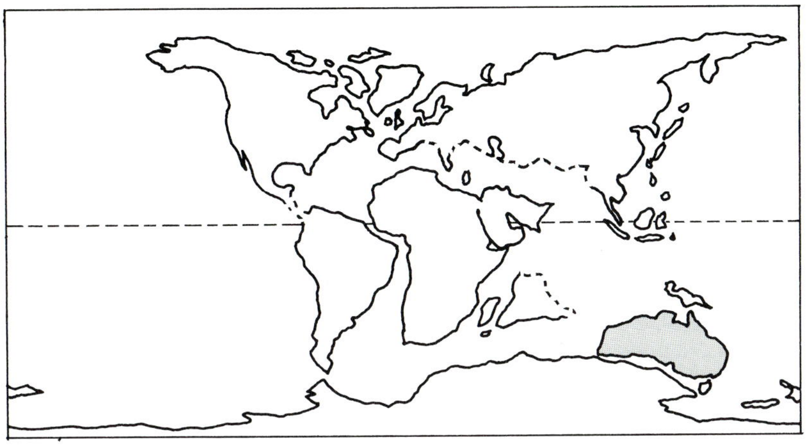

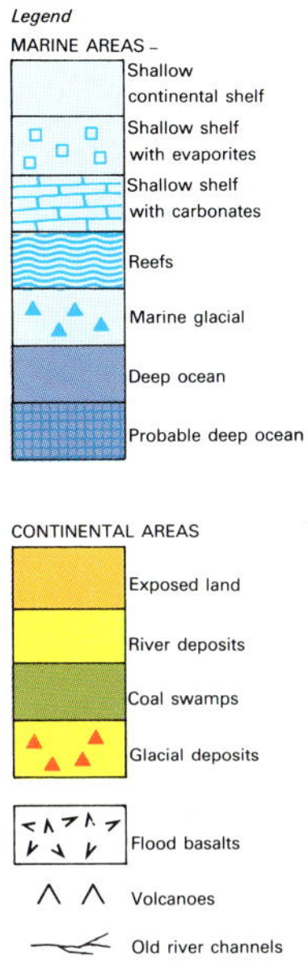

A	Adelaide	**D**	Darwin
AS	Alice Springs	**M**	Melbourne
B	Brisbane	**P**	Perth
C	Canberra	**S**	Sydney

CRETACEOUS

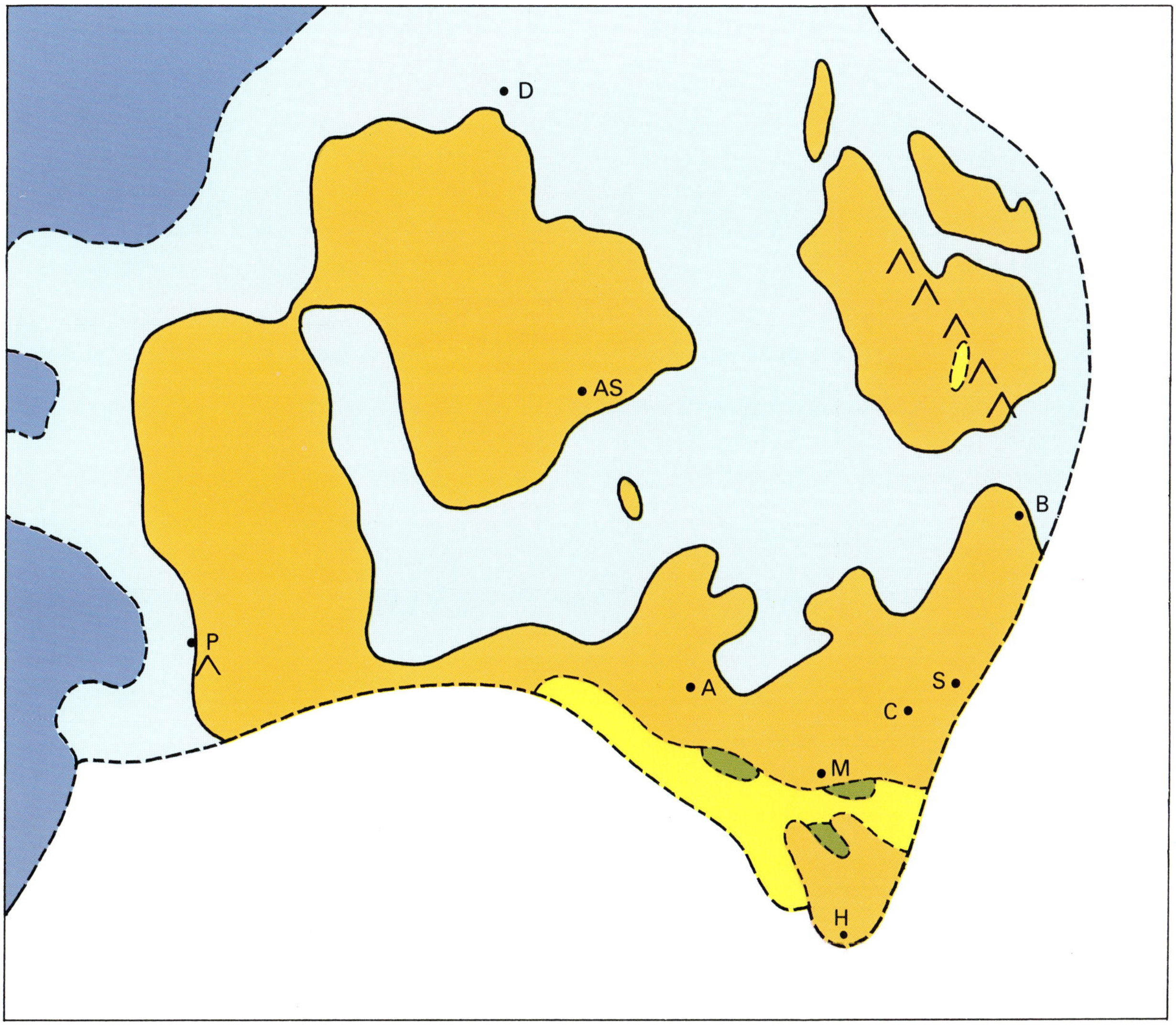

porating the Lord Howe Rise moved eastwards to form the Tasman Sea. Extensive volcanic activity occurred along the present Queensland coast. To the south, further rifting between Australia and Antarctica allowed the development of very narrow embryonic seas.

A substantial rise in global sea level in the early part of the Cretaceous resulted in the flooding of much of inland Australia by shallow seas. The ingress of these seas came probably from the north, as fossil molluscan faunas from the coastal belt of the Northern Territory are the oldest Cretaceous ones known in Australia. These shallow seas flooded most of inland Queensland and northern South Australia and extended to the Canning Basin in the northwest and the Nullarbor Plain in the south. They almost certainly had a profoundly moderating effect on the climate. Late in the Cretaceous the seas retreated again. In the inland Queensland-northern South Australia region, sediments were deposited from freshwater lakes and rivers in the early part of the Late Cretaceous. The muds of these deposits preserve a diversity of dinosaur footprints.

The Cretaceous was a time of great warmth over the globe. Palaeotemperature curves drawn on the basis of oxygen isotopes are available from the northwest Pacific, New Zealand and Europe, and these show that the Albian Stage, late in the Early Cretaceous, was perhaps the warmest time in the whole of the Phanerozoic. For much of the Cretaceous, a zone of tropical to sub-tropical conditions extended to 45°N and perhaps to 60°S. Beyond this zone, climates in high latitudes were temperate to cool temperate. The configuration of the continents then, as in much of the Mesozoic Era, was perhaps largely responsible for warm conditions at high latitudes. With the continents clustered into a single landmass, and an extraordinarily wide Pacific Ocean, equatorial currents flowed in the equatorial zone for long distances, warming progressively as they went. The western Pacific in the Cretaceous was considerably warmer than at present, and the deflection of these waters into south polar regions ensured warmth and wetness there. In the last phases of the Cretaceous, a cooling trend began. This was probably the beginning of the trend that led eventually, through zigzag fluctuations, to the most recent ice ages. At the end of the Cretaceous, the cooling may have been at least partly responsible for some of the extinctions — for the disappearance of the ammonites from the oceans, and the dinosaurs from the land.

For Australia, latitudes in the Cretaceous may have been even higher than shown in the figure. A recent estimate suggests that southern Victoria moved from 70°S at the beginning of the Cretaceous to about 85°S a short time later. At such latitudes, the polar nights must have been long and temperatures low, yet the Australian biota of the time showed no clear evidence of adaptation to such long, dark winters. Rather, the floras — conifers, pteridosperms, ferns and ginkgos in the Early Cretaceous — contain a large percentage of apparently evergreen types, suggesting warm to cool temperate conditions. Tree rings in woody conifers suggest that moderate seasonality prevailed.

The Cretaceous vegetation cover changed dramatically during the period. The Early Cretaceous is of 'Mesozoic' character, with conifers forming the dominant woody element, and pteridosperms and ferns the understorey. The flowering plants made their debut late in the Early Cretaceous. From a beginning, probably as small coastal shrubs, this now highly successful group of organisms diversified rapidly, until, by the end of the period, they challenged the conifers for the supreme position in Australian forests. Some of the modern Australian taxa — such as the Proteaceae and *Nothofagus*, the Southern Beech — appeared before the end of the Cretaceous.

The faunas from terrestrial rocks in Australia suggest, like the plants, that temperate, not cold, conditions prevailed during the Cretaceous. Although disagreement still exists as to whether dinosaurs were cold-or warm-blooded, it is nevertheless unlikely that they could tolerate prolonged periods of low temperature. The presence of lungfish remains in Queensland, and in high-latitude Cretaceous Victoria, offers further support for a picture of a warm to cool temperate, equable climate with a moderate rainfall. It has been suggested that to achieve the conditions under which apparently evergreen plants could flourish, the inclination of the earth's axis in the Mesozoic must have been more nearly vertical than it is now (Douglas and Williams, 1982). This suggestion remains largely untested.

Inoceramus, a bivalve or clam that inhabited Mesozoic seas in Australia as well as overseas. This shell, one of two valves, is about 100 mm in length.

Stems and leaves of *Cladophlebis australis*, a fern from terrestrial rocks ranging from Triassic to Cretaceous. Length of the stem is about 55 mm.

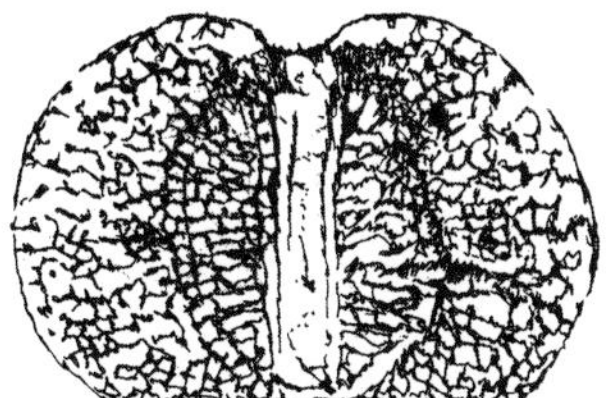

Podocarpidites, pollen grain of the gymnosperm tree, which ranges from Jurassic to present day. This grain is about 0.1 mm in diameter and must be viewed under a high magnification to be seen.

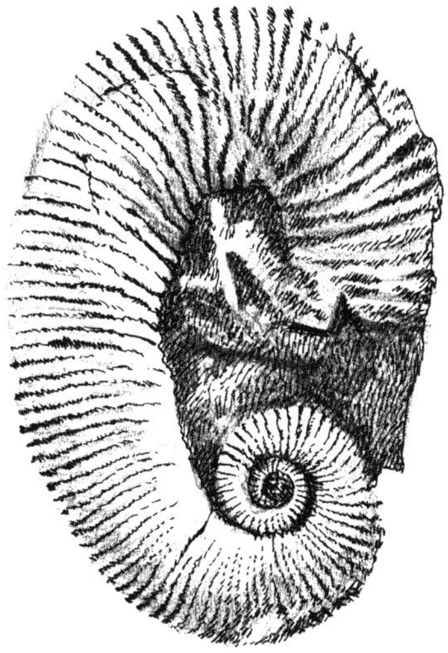

Labeceras, an ammonite from the Cretaceous of Australia. Ammonites are related distantly to octopi and squids, but, unlike these modern forms, had external shells like the living *Nautilus*. Ammonite shells were divided into several separate chambers, some air-filled; the walls of these chambers were often complex, producing a branching pattern on the outside of the shell, called sutures. Length across this shell is about 70 mm.

Leaves of *Ginkgo digitata*, from the Cretaceous of Australia, one of the maiden-hair trees. The only living relative of this group, which was once widely spread, was found growing in China. Since then it has been successfully reintroduced in many parts of the world. Length of the leaf in the middle is about 70 mm.

CAINOZOIC ERA

The Tertiary (65 to 2 million years) and Quaternary (2 million years — present) periods together make up the Cainozoic Era and they will be discussed together.

It was during the early part of the Tertiary that the Australian continent achieved essentially its present outline. On its southern margin, separation from Antarctica continued, and the Southern Ocean came into being. In the Eocene Epoch (55–38 million years ago), warm ocean waters transgressed from the Indian Ocean eastward along the southern margin of the continent. This early Southern Ocean formed a large deep gulf south of Australia through the Eocene. The South Tasman Rise, which is a block of continental crust, lying south of Tasmania, formed a kind of 'trailing edge' as the Australian continent moved north; this last connection with Antarctica was not severed until sometime in the Oligocene, about 30 million years ago. After the severing of this last link with the southern continent, oceanic waters south of Australia could flow freely between the two continents. This circum-Antarctic circulation probably accelerated the chilling of Antarctica, as it prevented the access of warm waters from the north.

Off the eastern margin of Australia, sea floor spreading movements that had formed the Tasman Sea in Cretaceous time migrated northward to form the Coral Sea Basin. On land, basaltic lavas were periodically erupted in a broad band down the eastern edge of the continent, related in some way to its northward drift. This movement appears to have proceeded steadily through the Cainozoic, and is evident from the successive positions of magnetic poles measured from the basalts. The eastern highlands, Australia's most prominent topographic feature, are probably the result of up-arching movements and faulting associated with the opening of the Tasman Sea and the seas marginal to the Queensland coast. The timing of their uplift remains a problem — it probably began in the Mesozoic, but later phases of uplift have occurred; at the beginning of the Tertiary then, it is likely that there was already a significant belt of highlands down the east coast.

The northward motion caused the eventual collision of the Australian Plate with the Pacific Plate in the region of New Guinea. This collision resulted in complex mountain-building in New Guinea from about the Middle Miocene, 12–15 million years ago. This history of Australia's development meant that the continent was isolated from the early Tertiary, after separation from Antarctica, until the mid-Miocene, when mountain-building and earth movements in the Indonesian region

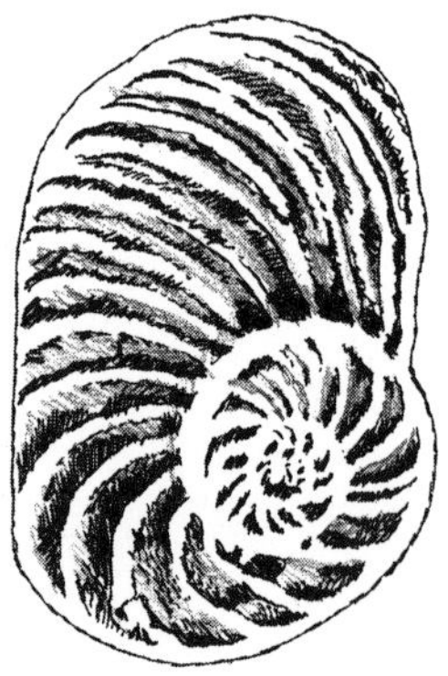

A single-celled foraminiferid, *Operculina*, about 6 mm in length. *O. bartschi* is Miocene-Recent in age.

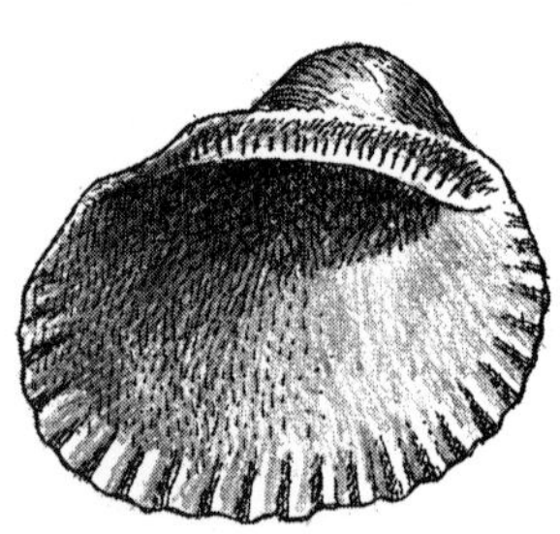

A bivalve, *Anadara*, an internal view of the left valve, about 45 mm in width. *A. trapezia* is Pleistocene to Recent in age.

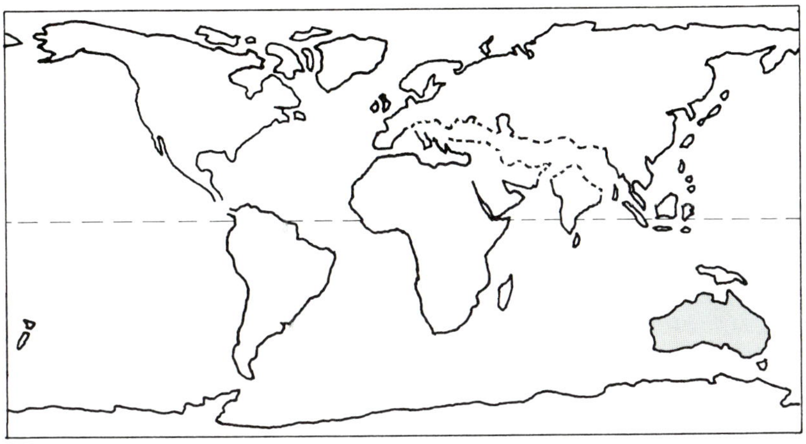

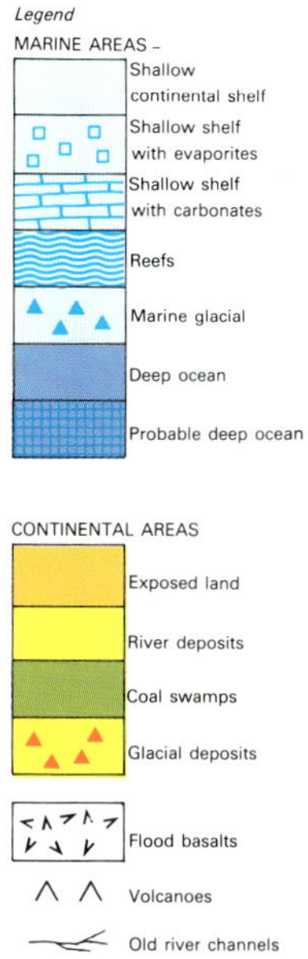

A	Adelaide	**D**	Darwin
AS	Alice Springs	**M**	Melbourne
B	Brisbane	**P**	Perth
C	Canberra	**S**	Sydney

EARLY TERTIARY

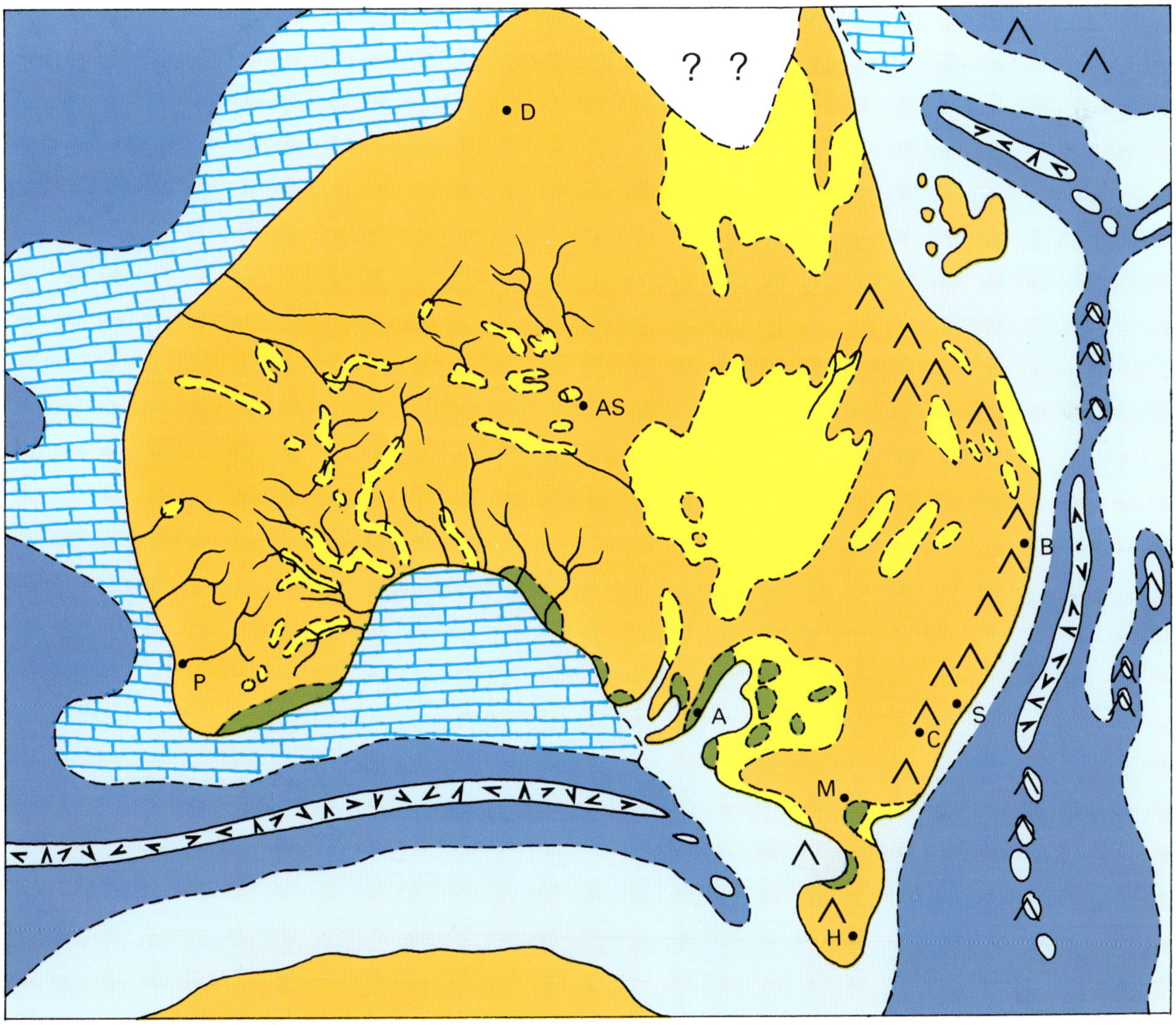

resulted in the creation of a biological stepping-stone route to and from Asia. This long phase of isolation is clearly reflected in Australia's modern biota.

The history of the marsupials, for instance, embodies the long isolation. These evolved in the Cretaceous perhaps in the northern hemisphere, and were in Australia at least by the Early Tertiary. Shortly after their first recording in Australia, the separation from Antarctica occurred, isolating the ancient populations and allowing much evolutionary radiation free of outside influences. For reasons not yet understood, placental mammals were excluded from Australia until late in its history, and so offered no competition to the evolving marsupials.

The Cainozoic was a time of pronounced climatic change. The general global cooling, which had begun in the Late Cretaceous, continued through the Tertiary, and eventually led to the formation of the Antarctic icecap. This probably reached maximum size in the Late Miocene, some 5–6 million years ago. Just how extensive it was in the Early Tertiary remains something of a mystery. Climates in Australia were profoundly influenced by developments in Antarctica — at times when there was little or no ice, the patterns of wind and water circulation would have been sluggish and irregular; at times of ice buildup, the pronounced difference in temperature between equator and pole would have intensified circulations, and produced wind patterns much like those of today.

In the Early Tertiary, with little Antarctic ice, conditions in Australia were generally warm and wet, and rain-bearing winds penetrated deep into the inland. The warm sea surfaces, as suggested by the oxygen isotope information, provided abundant moisture through evaporation. Wind patterns may have been such that there was a narrow band of westerlies in high latitudes, and north of these, a wide band of erratic winds which watered inland Australia (Kemp, 1978).

Evidence from fossil leaves and from the pollen record suggests that there was a wide spread of rainforest across southern Australia at this time. The southern beeches and the podocarps, which today form the temperate rainforests, were abundant then, and are reflected in the pollen story; the leaf fossils suggest rainforest of more tropical aspect. All of these fossils reflect plants that grew at or near wet habitats where they are most likely to be preserved. Between the swamps, lakes and river banks, drier habitats must have existed, but these have left a poor record behind. These pockets of dryness enabled many plants and animals to become pre-adapted to the aridity which became so widespread later.

A system of lakes, surrounded by rainforest, seems to have existed through the Tertiary in low-lying central Australia; they extended from Lake Frome, north through Lake Eyre, to regions north of Alice Springs. This is the area that is now the driest in the continent. The high rainfall of the Early Tertiary allowed the development of a complex drainage system over much of Australia. In Western Australia traces of the old river system now remain as strings of lakes which are linked only after intermittent heavy rains. In the Nullarbor Plain region the southern edges of the old system of rivers which once flowed to the Southern Ocean are truncated by marine carbonates of the Miocene seas.

General cooling through the Cainozoic led to drier conditions being established over much of Australia. The first major drying out occurred around the middle of the Miocene, some 10–15 million years ago, as the Antarctic icecap built up. Dryness before then had been patchy and irregular. The sediments which accumulated in the Lake Eyre and Lake Frome regions reflect this Miocene drying out. Dolomites and a variety of clay minerals in the Etadunna Formation reflect prolonged, but perhaps sporadic, dry phases starting then. The vegetation record suggests that grasslands became well established in the continental interior at about this time, a factor which

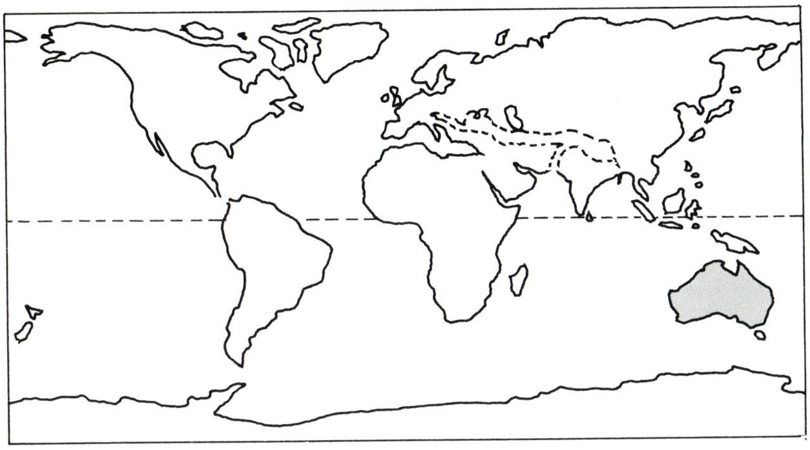

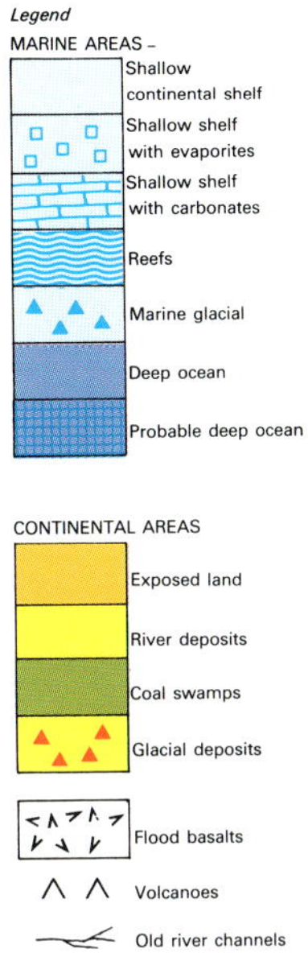

A	Adelaide	**D**	Darwin
AS	Alice Springs	**M**	Melbourne
B	Brisbane	**P**	Perth
C	Canberra	**S**	Sydney

MID TERTIARY

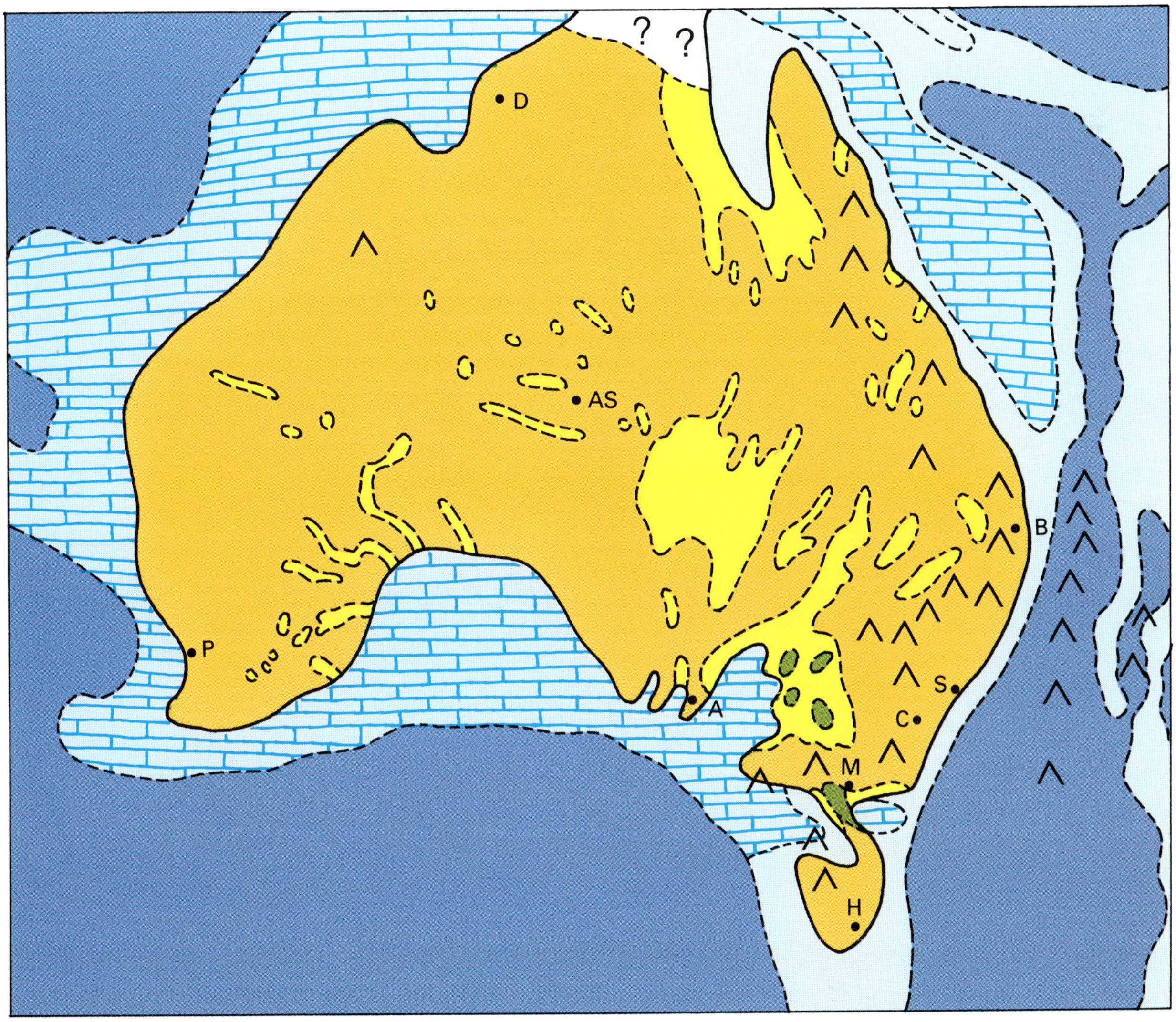

must have been intricately linked with the evolution of grazing mammals. Rainforest persisted along the inland watercourses.

The eucalypts, which provide the essential character of so much of the modern Australian landscape, had become part of the Miocene scene. Fossil evidence, from beautifully preserved casts of fruits in silcretes in now arid South Australia, suggests that they may have had a central Australian origin; from here they expanded coastwards, displacing the old rainforests, to coastal refuges. Eventually, late in the Cainozoic, they were themselves displaced from the driest regions of the continental interior, and replaced by a vegetation in which *Acacia* provided the dominant woody shrubs.

As the icecap in Antarctica grew, patterns of wind circulation approached those of today. A belt of anticyclonic circulation developed in high latitudes and migrated northwards with increasing cold, eventually to overtake Australia (Bowler, 1982). The aridity which became widespread in the Late Miocene/Pliocene (the period between 5 and 2.5 million years ago) thus may have developed first in southern Australia. There is, however, little fossil evidence to support this suggestion; indeed, fossils from continental deposits are rare altogether in this time interval, probably reflecting the general lack of wet habitats. Northern Australia at the same time began to experience a sub-tropical climate, with rain in the summer months. The present regime, in which the southwestern and southern margins receive winter rainfall as anticyclonic, high pressure belts move north, and the northern regions receive regular summer rain from monsoon-like systems, dates probably from the establishment of icecaps in the northern hemisphere, approximately 2 million years ago.

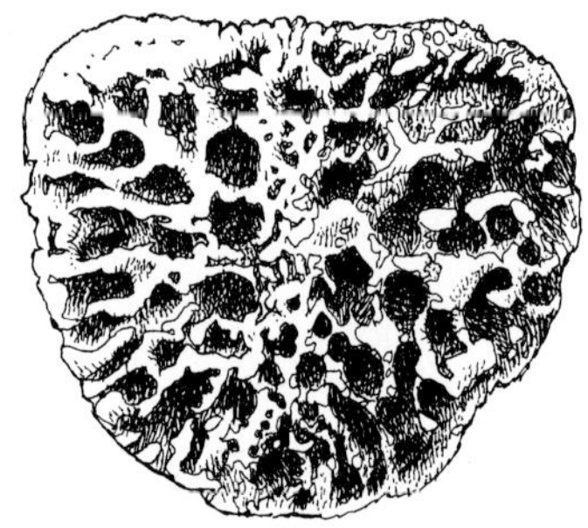

A bony scute from the skin (dermal) armour of a large, extinct crocodile, *Pallimnarchus pollens*, about 80 mm wide. Pliocene-Pleistocene in age.

Fluctuations in those icecaps determined Pleistocene climates. A period of maximum dryness some 20,000 years ago has left an imprint on the Australian landscape — conditions then were colder, drier and windier than at present. The longitudinal sand dunes of central and northwestern Australia formed at that time. The formation of northern icecaps, and probable extension of the southern during the Pleistocene, resulted in a sea level lowering of about 200 metres, exposing the continental shelves around Australia repeatedly. To the north of Australia this sea level lowering provided the opportunity for Aboriginal man to migrate through the Indonesian islands to Australia at least 40,000 years ago.

Kadimakara

'According to the traditions of some Australian Aborigines, the deserts of Central Australia were once fertile, well-watered plains. Instead of the present brazen sky, the heavens were covered by a vault of clouds, so dense that it appeared solid; where today the only vegetation is a thin scrub, there were once giant gum-trees, which formed pillars to support the sky; the air, now laden with blinding, salt coated dust, was washed by soft, cooling rains, and the present deserts around Lake Eyre were once a continuous garden.'

'The rich soil of the country, watered by abundant rain, supported a luxuriant vegetation, which spread from the lake shores and the river-banks far out across the plains. The trunks of lofty gum trees rose through the dense undergrowth, and upheld a canopy of vegetation, that protected the country beneath from the direct rays of the sun. In this roof of vegetation dwelt the strange monsters known as the "Kadimakara" or "Kadimerkera".'

'Now and again the scent of the succulent herbage rose to the roof-land, and tempted its inhabitants to climb down the gum-trees to the pastures below. Once, while many Kadimakara were revelling in the rich foods of the lower world, their retreat was cut off by the destruction of the three gum-trees, which were the pillars of the sky. They were thus obliged to roam on earth, and wallow in the marshes of Lake Eyre, till they died, and to this day their bones lie where they fell. After the destruction of the gum trees, the small holes in the forest-roof increased in number and size, until they touched one another, and all the sky became one continuous hole; wherefore the sky is called "Puri Wilpanina", which means the "Great Hole".' (J.W. Gregory, *The Dead Heart of Australia*, 1906: 3-4).

So goes the Dieri legend of how the bones of extinct, gargantuan animals came to lie on the surface of salt pans in the Lake Eyre Basin and along Cooper Creek. These were of the 'Kadimakara' or Dreamtime animals, which no longer exist.

The Dieri legends were not far from the theories we accept today about just what these bones signify. The Lake Eyre Basin was once more well-watered and supported a diverse array of medium-sized plant-eating marsupials, dolphins, primitive platypi, flamingoes and gigantic lungfish — in the Miocene Epoch, some 15 or so million years ago. The bones the Dieri probably saw along such saltpans as Lake Callabonna, however, were from a much younger time, the Pleistocene, perhaps no older than a few tens of thousands of years. *Diprotodon* was one of the most common marsupials of these times and probably lived under conditions not too different from those which exist today in this region.

The Kadimakara, after which this book was titled, have many 'tales of the dead' to tell about Australia's past — over the last 500 million years. Their bones give some idea of the evolutionary pathways followed by backboned animals in Australia and how such paths compared or differed from those of the vertebrates from the rest of the world. The history of marsupials on this continent, for example, has been very distinct from that on any other continent. It has provided a unique and varied, independent experiment in mammalian evolution, allowing comparison of, for instance, evolutionary rates with those of mammals on other larger and more interconnected continental masses such as North America and Europe. Some of these rates in Australia may, in fact, be slower.

The Kadimakara, in addition, allow the reconstruction of the peregrinations that vertebrate animals have, or have not taken, between different continents in times past. This gives an independent cross-check on the 'continental drift' maps provided by seismologists, geophysicists and geologists. The Late Devonian fossil fish record of Australia, for example, is more like that of China than that of any other part of the world, suggesting that freshwater fish may have moved back and forth between these areas at that time. This is in agreement with the maps of the world for 360-370 million years ago.

For the palaeontologist and geologist alike, the Kadimakara are also useful tools in determining the age of the rocks in which they occur. By studying rock sequence after rock sequence in which fossil vertebrates occur, the palaeontologist gradually pieces together a composite sequence, a patchwork of rocks and fossils, one on top of the other. Then when an isolated bone or tooth is found, a geologist can estimate the age of the rocks from which it came, fitting it in to the known sequence.

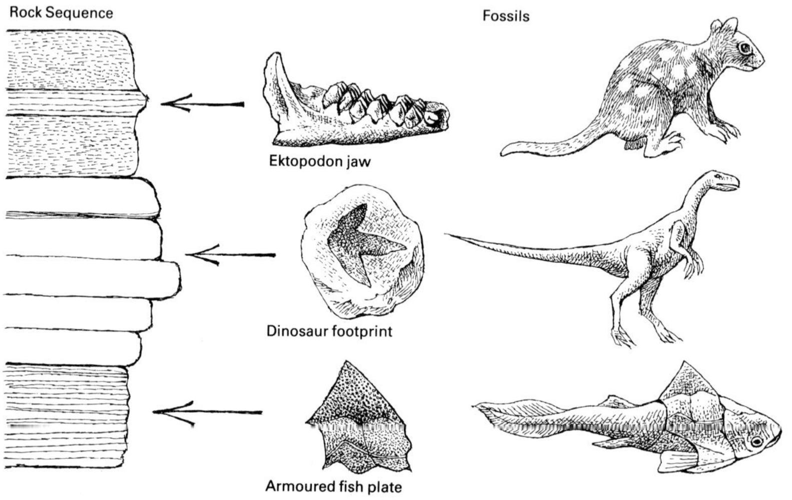

Rocks can be dated by the fossils that are preserved in them. Here the oldest beds contain primitive armoured fish (such as *Bothriolepis*), while younger ones contain dinosaurs, and the youngest contain mammals (such as *Ektopodon*).

Finally, the bones and teeth of the Kadimakara allow reconstructions of the animals themselves to be made. From the fossil fragments it is possible to determine if the animals were runners or diggers, grazers or browsers, and what part of the ecosystem they inhabited. Reconstruction of whole faunas can be made and then compared to the living ones of today, keeping in mind the biasing factors that are inherent in the fossil record.

The Kadimakara hold many secrets of the past that are being unlocked by vertebrate palaeontologists. Much has been discovered during the past 150 years since Australian vertebrate palaeontology had its beginnings — discoveries made by professional and amateur alike. With this continued cooperation and field and research support supplied by museums, the Australian Research Grants Scheme, the National Geographic Society, and many other private sector trusts, companies, and individuals, much more is bound to be discovered. So much is yet to be understood about the origin and evolution of vertebrates on this continent, the reasons why Australia has served as a refuge at the end of the world for numerous groups of backboned animals, the effects of changing latitude and climate on the fauna of a continent that has moved a drastic 20-25° in latitude during the last 40 million years — and is still moving. So much is yet to be discovered, which is what makes the search for more Kadimakara a fascinating endeavour!

Arandaspis prionotolepis

Ritchie and Gilbert-Tomlinson, 1977

The Southern Four-eyed Fish

CLASS: Heterostraci
ORDER: Arandaspidiformes
FAMILY: Arandaspididae
GENUS: *Arandaspis*
SPECIES: *Arandaspis prionotolepis*
AGE: Early to Middle Ordovician

The late Professor Alfred Sherwood Romer, one of the leading vertebrate palaeontologists of this century, often enlivened parties and lectures by singing an amusing little ditty to the tune of 'It's a long way to Tipperary.'

It's a long way from amphioxus,
It's a long way to us,
It's a long way from amphioxus
To the meanest human cuss.

Farewell fins and gill slits,
Welcome teeth and hair
It's a long way from amphioxus
But we all came from there!

All of us have a natural interest in knowing where we came from, although usually this knowledge extends only a few generations back on our father's or our mother's side of the family. Nearly every human culture has extended this range with traditions or legends of how their tribe, or country or even the world came into being. But it is only in the last 200 years, thanks to the painstaking researches of thousands of scientists all over the world, that we have pieced together an accurate picture, not only of the history of the earth but also of the myriads of life forms which live (and have lived) on its surface. We now know that the Earth is at least 4,500 million years old, that simple living organisms have existed on it for more than 3,400 million years, and that diverse and much more complex organisms have been abundant for the past 600 million years. We can even trace the development of the particular group to which we belong, the backboned animals, to Late Cambrian times about 500 million years ago.

We share a basic body plan with other backboned animals — fish, amphibians, reptiles, birds and other mammals — and, at different times in the distant geological past, we obviously shared a common ancestor with each of these groups. The picture is less clear when we look at evolution below the fish level, but even here some clues are available from living, yet primitive chordates such as the sea squirts (or tunicates) and an intriguing little marine animal known commonly as *Amphioxus*, the lancelet (*Branchiostoma*), which lives buried in the sand of the sea floor.

Amphioxus may not be our ancestor (as implied in the song above), but it provides a tantalizing glimpse of a stage through which our earliest fish-like ancestors may have passed in Cambrian or even late Pre-Cambrian times. It is a slender, leaf-shaped animal about 6 cm long, stiffened by a long, thin, flexible rod in its back. This rod, the notochord, is the forerunner of the backbone, or vertebral column, which forms around it in more advanced chordates (or vertebrates). In its organization, Amphioxus is clearly well below the level of even the simplest living fish-like animals, the jawless fish or cyclostomes. The first recognizable remains of armoured relatives of such jawless fish come from Late Cambrian rocks over 500 million years old (Repetski, 1978), but the oldest fossil of an amphioxus-like animal, *Pikaia*, comes from the Burgess Shale of western Canada (Conway-Morris and Whittington, 1979) of Middle Cambrian age, about 550 million years old.

We know a great deal about vertebrate evolution during the last 420 million years, since mid-Silurian times, but earlier than this the evidence is frustratingly rare. The first evidence came from North America where Charles Walcott, in 1892, discovered and described two early fish, *Astraspis* and *Eriptychius*, which occur abundantly as small bony fragments in certain layers of the Harding Sandstone Formation of Middle Ordovician age (about 460 million years ago) in Colorado, Wyoming and Montana. Despite intensive searching since the 1890s, only one or two partly articulated specimens have ever been found. They reveal that the head of *Astraspis* and *Eriptychius* was covered by a bony shield but tell us little about the size and position of the eyes, nostrils, gills, mouth and fins. The microstructure of the bone fragments resembles that of the heterostracans, a group of extinct armoured jawless fish (or ostracoderms) found in Siluro-Devonian rocks of 430-380 million years ago. Several distinctive groups of such ostracoderms (heterostracans, cephalaspids, anaspids, thelodonts, galeaspids) are locally abundant and often well-preserved in Siluro-Devonian rocks of North America, Europe and Asia but, until recently, have been virtually unknown from the southern hemisphere.

Now, from Ordovician rocks of central Australia, have come not only some of the oldest fossil vertebrates in the world but also some of the best preserved remains from this early period of vertebrate history. The circumstances in which they came to light illustrate the element of luck that attends many major fossil finds. In many instances they have to be 'discovered' twice, first in the field and later, only in the museum collection, is their significance recognized.

I joined the Australian Museum as Curator of Palaeontology in 1968, and visited Adelaide in 1969 to attend the Australia and New Zealand Association for the Advancement of Science conference. During a break in the lectures I took the opportunity to examine a small collection of fossil fish in the South Australian Museum and discovered a small block of iron-stained sandstone with the impression of a fish plate. It came from '? Devonian, Mt. Charlotte, Northern Territory, collected D.J. Taylor, 1959'. It appeared to be part of the bony armour of an ostracoderm headshield, thus from a fish group not previously recorded from Australia.

A check of geological maps revealed that the rocks of the Mt. Charlotte Range were not Devonian but Ordovician in age, raising the exciting possibility that the fragment might be one of the oldest of vertebrates and that there might be more

---

In early Ordovician times, about 480 million years ago, what is now the dry centre of Australia was covered by a widespread shallow sea inhabited by rich marine faunas. These were mainly invertebrates — trilobites and various molluscs, bivalves, gastropods, and straight-coned cephalopods. The trilobites are often represented as fossils only by the distinctive feeding burrows, called *Cruziana*, left behind in the sea floor sediments.

Arandaspis prionotolepis, one of the earliest vertebrates, also occurs in these rocks. It was a rather sluggish, bottom-living animal, a member of the primitive, armoured jawless fish known as ostracoderms. The preservation indicates that these fish lived and were buried near to where they died in a shallow-water, near-shore marine environment. A smaller related form is shown in the middle distance; others present in the same fauna are still too poorly known to reconstruct.

F. KNIGHT 81

where it came from. David Taylor, now a consultant palaeontologist in Sydney, provided more information on his 1959 find made while he was a field geologist working for Frome-Broken Hill Pty. Ltd. investigating the area south of Alice Springs in the Northern Territory. Taylor and his crew were working with the Bureau of Mineral Resources in Canberra, and it was agreed that many of the rock and fossil specimens collected would be sent back to Canberra for curation and storage.

Near the summit of Mt. Charlotte, an east-west trending ridge some 177 kilometres south of Alice Springs, Taylor found a slab of quartzite with a curious mixture of fossils. On one side were fragments of asaphid trilobites, extinct marine arthropods that are characteristic fossils in Ordovician rocks in many parts of the world. On the reverse were several impressions he interpreted as 'fish plates'. At the end of the field season he returned via Adelaide, where he deposited a small quantity of representative specimens with the South Australian Museum. Most of the material, however, went back to Canberra.

In 1963-4 Bureau of Mineral Resources field parties remapped the geology of the area. No more fish plates were found at Mt. Charlotte, but similar material was recovered from Mt. Watt, a small isolated hill 64 kilometres south of Mt. Charlotte. In both cases the fish-bearing unit was the Stairway Sandstone Formation of probable Early to Middle Ordovician age. The Bureau of Mineral Resources parties, led by A.T. Wells, also included a palaeontologist Joyce Gilbert-Tomlinson. Although the presence of the fish plates was noted at the time, they could not be studied immediately in the flood of diverse and stratigraphically more useful fossil finds and, like Taylor's original 1959 material, were deposited in the bulk store. How long they might have remained there is a matter for conjecture had not serendipity intervened in the form of the single small 'fish plate' deposited by Taylor in Adelaide and recognized ten years later!

Joyce Gilbert-Tomlinson, when contacted, confirmed Taylor's report and later forwarded the fossil fish to Sydney for preparation and study.

The fossils from the Stairway Sandstone are impressions, or natural moulds, of fish plates. Many of them display a delicate dermal ornament, but in no case has any original bony material been recovered. Most of the fish plates were fragmented, but one specimen, preserved in relief in a block of quartzite, was obviously a moderately complete, articulated individual at the time of burial.

When all the available material had been prepared, Ritchie and Gilbert-Tomlinson (1977) described the first Ordovician vertebrates from the southern hemisphere. Two quite distinctive forms were recognized. *Arandaspis prionotolepis*, the most abundant and best preserved form, probably reached 14-16 cm in length. The genus is named after the Aranda, the dominant Aboriginal tribe in the area where the fossil fish occur and the group to which the famous artist Albert Namatjira belonged; *aspis* is from the Greek word meaning 'shield'. The specific name comes from *prionotos* (Greek), meaning 'jagged' or 'serrate' and *lepis*, meaning 'scale'. This reflects the unusual nature of the trunk scales of *Arandaspis*.

The entire head and branchial (gill) region was encased in a bony shield formed by a deep, rounded ventral plate and a flatter, triangular dorsal plate. Between them lay a row of smaller, rhomboid plates probably associated with the excurrent ducts leading from the gill pouches. Behind this headshield lay a small, triangular patch of trunk scales, not enough to reveal the shape of the body and fins, but sufficient to show how the scales were arranged. Each scale was long, narrow and decorated with a single row of pointed tubercles. These projected beyond the rear margin, and the scale superficially resembles a portion of fretsaw blade. Although the only articulated specimen is tantalizingly incomplete, it is still by far the best-preserved fish yet

Outcrops of the Ordovician Stairway Sandstone Formation on the flanks of Mt. Watt, N.T. from which many specimens of some of the earliest known fossil vertebrates have been recovered.

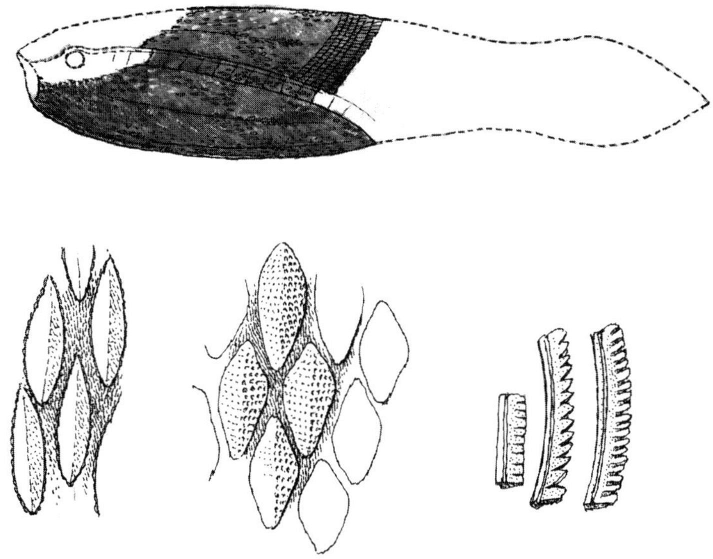

Arandaspis prionotolepis Ritchie and Gilbert-Tomlinson. Restoration of *Arandaspis* in left side view; most of trunk and tail are unknown at present. Outline of rostral (nose) area, restored from later finds, is currently awaiting description. Detailed view of ornament on the bony plates of the headshield and scales of trunk region. Headshield 6-7 cm long; the estimated total length was about 13 cm.

recovered from Ordovician rocks anywhere in the world.

From the material now available we can confidently reconstruct what *Arandaspis* probably looked like, as well as the environment in which it lived. It was a deep-bodied, rather rounded fish. The front half was encased in a headshield formed of large, thin sheets of bone. It probably did not have any paired fins and was probably a clumsy swimmer. The trunk was covered with oblique rows of long, narrow, serrated scales.

Exposures of the Stairway Sandstone Formation extend the length and breadth of the Amadeus Basin in central Australia, and the rocks range in thickness from a maximum of 560 metres in the north to a mere 8 metres at Mt. Watt on the extreme southeastern margin of the basin. The formation was deposited from a shallow sea that lay upon ancient folded and eroded beds of the Pre-Cambrian Winnall Formation.

Associated with *Arandaspis, Porophoraspis* and the two other ostracoderms in this near-shore marine environment were various marine invertebrates. Molluscs are represented by straight-coned (orthocone) cephalopods, distant relatives of the living, coiled shelled *Nautilus,* and by various types of bottom-living bivalves. Many of the bedding planes, which carry impressions of the fish plates, are criss-crossed by shallow burrows and feeding grooves of bottom-dwelling arthropods. These trails, called *Cruziana,* are thought to have been made by various types of trilobites; although no actual trilobite fossils have been found at Mt. Watt, such remains are locally abundant in the Stairway Sandstone elsewhere in the Amadeus Basin.

One of the most intriguing features of *Arandaspis* lies between the forward and laterally placed eyes, in the middle of the headshield. There, two small holes are clearly visible, lying side by side. Many living and fossil vertebrates have a single, centrally-placed opening which is occupied by a small outgrowth from the dorsal surface of the brain, known as the pineal complex. The functions of the pineal, the so-called 'third eye' of vertebrates, are still the subject of considerable research and discussion. It is undoubtedly involved in important hormonal control processes. By Triassic times such median eyes had apparently gone out of fashion amongst vertebrates, but in modern man and other advanced vertebrates although the pineal does not reach the skull roof it is still present, buried deeply within the brain itself. In those forms where it does still reach the surface — certain fish, amphibians and reptiles — it is a light-sensitive, but not an image-forming, organ.

The implications of the *Arandaspis* paired openings are instructive bearing in mind that they are based on the interpretation of impressions in rock, not on preserved remains of actual soft parts. The large, well-developed lateral eyes of all vertebrates, which receive and transmit visual images to the brain, arise as outgrowths of the mid-brain, but the pineal complex arises in the midline between the fore- and mid-brain sections. In many of the most primitive vertebrate brains available to us for study today, the pineal complex is a paired structure consisting of the pineal and parapineal bodies side by side. In living jawless fish, for instance, the lampreys, both organs form eye-like structures, but the pineal is dominant and more dorsally placed than the parapineal. By contrast, in various living lizards and in the New Zealand tuatara, *Sphenodon*, the dominant organ is the parapineal body. The obviously paired nature of the pineal complex led Professor Romer (Romer and Parsons, 1977), to speculate 'that possibly the remote ancestor of the vertebrates may have had paired dorsal eyes as well as paired lateral ones.' Rather surprisingly none of the Siluro-Devonian ostracoderms to which *Arandaspis* is most closely related show any evidence of a paired pineal condition. By serially sectioning beautifully preserved, uncrushed headshields of one group of ostracoderms, the cephalaspids, Professor Stensiö (1927) of Stockholm was able to reconstruct the shape of their brain cavity and demonstrate that it must have housed a brain very similar to that of the superficially dissimilar living lampreys; very little had changed in the 400 million years between the cephalaspids and modern lampreys. The discovery that the mid-Ordovician *Arandaspis* probably possessed an additional pair of equally developed light receptive organs on top of the skull confirms Romer's prediction. Such an interpretation, based on a few impressions in ancient rocks may seem far-fetched to the non-specialist, but it illustrates how knowledge of living animals may legitimately be used to interpret fossil remains of long-extinct, but related forms.

Porophoraspis, the second Stairway Sandstone fish, is easy to distinguish from *Arandaspis*. Although known only from a few small fragments of bone preserved as natural moulds, the surface sculpture consists of tiny, close-set tubercles with deeply crenulate margins and perforated by rows of minute pores.

The name *Porophoraspis crenulata* is derived from the Greek word *porophoros*, meaning 'pore-bearing' and *aspis*, meaning 'shield'. The specific name comes from the Latin word *crenulata*, reflecting the crenulate nature of the dermal ornament. The *Porophoraspis* fragments, unfortunately, were much too incomplete to give any indication of the size and shape of the animal.

It was not until 1977, when the first paper on these ancient fish was in press, that I was able to visit the Northern Territory Ordovician sites with Robert Jones (also from the Australian Museum's Palaeontology Department), Brian Chatterton (University of Alberta, Calgary) and Karl Roth, from the Alice Springs branch of the Northern Territory Museum. We concentrated our efforts on Mt. Watt, site of the 1963-4 Bureau of Mineral Resources discoveries. From the scree-covered flanks and by extensive excavations of the fish-bearing layers the team recovered many more large and well-preserved portions of *Arandaspis* dermal armour. These included the first nearly complete dorsal shields of this form. Several dorsal and ventral shields of a smaller ostracoderm related to *Arandaspis* were also found, but all efforts to locate well-preserved remains of *Porophoraspis* were unsuccessful, and we still know very little about this animal. A fourth ostracoderm, as yet unnamed, is represented by several large irregular fragments of bony plates bearing a distinctive surface of flat-crowned, polygonal tubercles but too incomplete to reveal what it may have looked like. The surprising variety of the Stairway Sandstone jawless fish, four quite distinct forms, illustrates clearly that vertebrates must have been in existence and diversifying long before Middle Ordovician times.

The Gogo Fish

CLASS: Elasmobranchiomorphi
SUPERORDER: Dolichothoracomorpha
ORDER: Euarthrodira
FAMILY: Unknown
GENUS: *Harrytoombsia*
SPECIES: *Harrytoombsia elegans* Miles and Dennis, 1979

GENUS: *Camuropiscis*
SPECIES: *Camuropiscis coneinnus* Dennis and Miles, 1979

GENUS: *Tubonasus*
SPECIES: *Tubonasus lennardensis* Dennis and Miles, 1979

CLASS: Elasmobranchiomorphi
SUPERORDER: Dolichothoracomorpha
ORDER: Euarthrodira
FAMILY: Holonematidae
GENUS: *Holonema*
SPECIES: *Holonema westolli* Miles, 1971.

CLASS: Elasmobranchiomorphi
SUPERORDER: Ptyctodontomorpha
FAMILY: Ptyctodontidae
GENUS: *Campbellodus*
SPECIES: *Campbellodus decipiens* Miles and Young, 1977

CLASS: Teleostomi
INFRACLASS: Chondrostei
ORDER: Palaeonisciformes
FAMILY: Stegotrachelidae
GENUS: *Moythomasia*

CLASS: Teleostomi
INFRACLASS: Crossopterygii
ORDER: Rhizodontia
FAMILY: Onychodontidae
GENUS: *Onychodus*

AGE: Late Devonian

One of the most remarkable fossil fish sites in the world occurs in Devonian rocks bearing the unlikely name of Gogo. This is not, as some might assume, named after a night club, but refers to Gogo Station, a large property in the southern Kimberley region of far northwestern Australia.

Gogo lies south of Fitzroy Crossing in an area with a fascinating geological history. In Late Devonian times, 360-370 million years ago, the Kimberley region was an upstanding landmass surrounded on three sides (south, west and north) by warm shallow seas. The southern margin was a fault belt bordering a large, subsiding sedimentary basin, the Canning Basin. Nearby lay a great reef more than 300 kilometres long and several kilometres wide. In many ways it must have resembled the Great Barrier Reef that lies off the east coast of Queensland except that the organisms living in and forming the 'Devonian Great Barrier Reef' were quite different from those alive today. Eventually, as conditions changed, the Devonian reef complex was buried under younger sediments and hidden for over 300 million years until recent earth movements re-exposed it by the natural processes of uplift followed by erosion. The resistant limestone reef masses now form low, rugged ranges cut through and flanked by black soil valleys and plains. The black soil areas represent the inter-reef channels and basins that were filled with the softer sediments of the Gogo Formation containing the remains of the Devonian reef fauna.

Geologists have been keenly interested in the Kimberley reef complexes for many years for very good economic reasons. Similar reefs elsewhere in the world have proven to be lucrative sources of natural gas and oil deposits. Most of the research work in the Kimberleys has focussed on the detailed structure and genesis of the reef deposits and on the fossil invertebrates found there, since these are especially useful in stratigraphic correlation (Playford and Lowry, 1967; Playford, 1970, 1976, 1980). As early as the 1940s field geologists had collected 'coccostean remains in concretions' in the Gogo Formation, but this brief record mentioned in a technical report passed unnoticed by vertebrate palaeontologists. Fortunately some of the specimens of fish remains in concretions found their way into the collections of the Western Australian Museum in Perth.

It was obviously only a matter of time before their significance was recognized, but it happened in a rather interesting way. For most of this century the British Museum (Natural History) in London has been a centre of research on fossil fish, first under Sir Arthur Smith Woodward, then from the 1930s to the 1960s under Errol Ivor White, and carried on to the present by his colleagues and successors.

In the 1950s Dr. White, aware of the enormous untapped potential of Australia for Devonian vertebrate fossil fish discoveries decided to send his chief assistant, Harry Toombs, to Australia to prospect likely sites. Toombs made three trips in all, in 1955, 1963 and 1967, the results of which were to provide a remarkable stimulus to research work on Devonian fish not only in Britain, but also in Australia.

Toombs' first visit, between March and May 1955, concentrated entirely on a search of Early Devonian marine limestone formations in the Taemas - Wee Jasper area, southwest of Yass, New South Wales. A few exceptionally interesting fossil fish specimens had been recovered from here previously (White, 1952). Toombs succeeded in increasing the number of known specimens from about 8 to over 170, many of them new to science. The fish remains were embedded in solid limestone, from which they had to be painstakingly extracted by new techniques with acetic acid, perfected by the staff at the British Museum (Natural History) (Toombs and Rixon, 1959).

When the 1955 material had been prepared, Toombs returned to Australia in March 1963 to resume the search of the Taemas - Wee Jasper sites. On the outward trip by sea the S.S. Himalaya stopped at Fremantle for a day enabling Toombs to visit the Western Australian Museum, where the Director, David Ride, showed him some of the Devonian fish specimens in limestone nodules from the Kimberley sites. Ride had been trying to interest some palaeoichthyologist in the material for years since Basil Balme and Brian Glenister had found it in the 1950s. To many other researchers they would have appeared interesting but virtually impossible to prepare by manual techniques. To Harry Toombs, one of the pioneers of the acetic acid technique, they appeared ideal material for the new preparation method. In fact the little acanthodian skull had already been partially acid-etched by Glenister. Ride encouraged the involvement of the British Museum and thoroughly concurred with acid preparation. It was arranged that at the end of his fieldwork in New South Wales, Toombs would return to Western Australia, flying directly to Fitzroy Crossing. There he would be joined by a support team provided by the Western Australian Museum. In July 1963 Toombs combined forces with Ken Buller, Senior Preparator at the Western Australian Museum and his assistant, Eric Car. Both were experienced field workers (Brunton, *et al.*, 1969; Miles, 1970, 1971a, 1971b; Gardener and Miles, 1975) and incidentally had collected at Gogo before with another expedition led by Charles Camp and John Cosgriff from the University of California in 1960.

Fortunately the geological structure and history of the reef complex area was being intensively surveyed for the Western Australian Geological Survey by Philip Playford. At the request of David Ride he met the museum team, explained the complex geology and indicated the sites he believed were most likely for 'fossil-fishing'. As a result, after only one day traversing the area in a Land Rover, the party was able to set up its base-camp in the middle of a richly fossiliferous area on Gogo Station, a cattle property of some 1.2 million hectares.

The team did not have to look far for the fossils they had come so far to seek. As far as the eye could see the surface of the ground was littered with rounded limestone nodules, which had weathered out of the shales of the Gogo Formation that formed the valley floor — it was a palaeoichthyological goldmine! Surrounding the valley some distance away were the algal and stromatoporoid reef masses forming low ranges.

Although only a small proportion of the nodules was fossiliferous, the number of nodules present was so large that a rich haul of fine specimens was soon obtained. Most abundant were those containing marine invertebrates, mainly arthropods in the form of the distinctive bivalved carapaces of phyllocarid crustaceans. Cephalopod molluscs were reasonably common and included straight-shelled (orthocone) nautiloids and coiled-shelled goniatites; the latter are particularly useful in stratigraphic correlation even between continents. Other molluscs such as gastropods and bivalves were present but not abundant. Fish remains were relatively uncommon in the concretions, but this was offset by the remarkable preservation of the specimens recovered. The fish skeletons were virtually uncrushed, often almost complete and showed no signs of transportation of the fish after they had died. The excellent preservation indicates that the concretions formed around the bodies soon after death, in still-water conditions and in the absence of scavengers. The enclosing rock had hardened quickly and protected the fragile skeletons from the strong compaction, which did affect the surrounding sediment.

In three weeks, working in 40°C temperatures, the team collected half a tonne of

---

A scene in Late Devonian times in part of a reef complex which fringed the Kimberley landmass in northwestern Australia about 360-370 million years ago. These animals have been restored from superbly preserved complete skeletons found in the Gogo Formation near Fitzroy Crossing, Western Australia. Most common were various armoured fish, or placoderms (arthrodires and antiarchs) but also present were early representatives of the ray-fin fish, or actinopterygians and dipnoans, or lungfish.

The arthrodire *Harrytoombsia* (centre left) is pursuing several ray-fins, called *Moythomasia*. *Holonema*, a more massively built but slower arthrodire cruises (lower left) towards an elegant, long-snouted lungfish, *Griphognathus* (lower right). A common, but unexpected, element in the marine Gogo fauna is *Bothriolepis*, an antiarch with strange, armoured pectoral fins, which is commonly found in Devonian rocks of several continents but almost always in continental freshwater deposits. *Rolfosteus*, an arthrodire with a long, pointed, tubular snout, and *Ctenurella* (top right) illustrate the diversity achieved by Devonian placoderm fish.

F. KNIGHT 84

fine specimens, which were shipped back to London via Derby and Perth for preparation and study. It did not take long to confirm that they included some of the best-preserved and most scientifically important Devonian fish remains from anywhere in the world. There are many different kinds of armoured placoderm fish; three different types of lungfish; ray-finned palaeoniscids; and rare remains of lobe-finned crossopterygians, a group which probably gave rise to amphibians.

When the quality and diversity of the Gogo fauna was fully realized, Harry Toombs returned once more to Australia accompanied by a team of palaeontologists and fieldworkers: Howard Brunton, British Museum (Natural History); Ian Rolfe, Hunterian Museum, University of Glasgow; George Kendrick and Ken Buller, Western Australian Museum. Hundreds more nodules were recovered from the initial fossil fields on Gogo Station and from new and equally rich exposures located in other outcrops of the Gogo Formation. Although, again, all the material was sent back to Britain there was an agreement that the type material of new taxa would be returned to Australia after it had been described. Australian teams have visited the area since then and recovered more material.

The Gogo fauna is rich both in numbers of individuals and in diversity of species, and it is likely that in the many nodules awaiting preparation there are other new forms still to be discovered. The fish skeletons are so well-preserved that even the most delicate of bones are virtually uncrushed and not distorted. The technique by which they are extracted is long term and tedious but has amazing results, (Miles, 1970 for details). The fish remains are often discovered when the enclosing nodule is cracked open in the field, and in this case the pieces have to be cemented together firmly before the acid treatment. The nodule is immersed in dilute acetic acid, which dissolves limestone (calcium carbonate) but not bone (calcium phosphate). After several days in acid, sufficient rock has usually been dissolved to leave some portions of fish plates projecting from the surface. The nodule is then soaked in fresh water to remove any salts formed and then carefully washed and dried. The bones, which are extremely fragile, are protected and strengthened by impregnating them with a weak solution of a plastic compound in a solvent, after which the specimen is re-immersed in the acid for a few more days. This cycle is repeated over and over until all of the enclosing rock has been removed and the bones extracted. Although the acetic acid technique is basically quite simple, it requires enormous patience, skill and time and the procedure has to be modified frequently as problems are encountered.

Once extracted, the individual bony plates can easily be reassembled after they are strengthened to form an absolutely accurate, three dimensional skeleton of a long-extinct fish, not merely a replica, but in its original bony form. Laymen find it difficult to believe that such delicate specimens have actually been extracted from a 360-370 million year old lump of solid limestone. The end results can be quite breathtaking even to fossil fish specialists, who are more accustomed to working with crushed, broken and weathered skeletal remains in rock that is resistant to acid treatment.

In the Gogo collections the most common fish were the armoured placoderms; forms in which the head and front part of the trunk were encased in a box-like shell of bony plates. They ranged in length from about 10 cm to 1 or 2 metres. At least 16 placoderm species are known to have been present and others are steadily coming to light as the unprepared nodules are processed. The two main groups of placoderms (the name means 'platy-skin') present are the arthrodires ('jointed-neck') and the antiarchs, the latter represented by only one species.

Some of the largest arthrodires such as *Eastmanosteus* and *Coccosteus* were carnivorous forms, which have long been known from Devonian rocks in the northern hemisphere. Many of the other arthrodires represent new genera and species, some of them with distinctive specializations in the head and trunk shield or in the jaws.

One form, *Harrytoombsia elegans* Miles and Dennis (1979), an arthrodire, must have been about 35 cm long in life and an efficient predator, probably eating phyllocarid crustaceans and some of the smaller fish. The 'teeth' in arthrodires are not like those of most other fish, however. There are no real 'teeth' in the jaw; instead the cutting edge of the upper and lower toothplates, made of bones, has a serrated and self-sharpening surface not unlike the crimping scissors of dressmakers. *Harrytoombsia* is a relatively rare, but distinctive, element in the Gogo collections and it seems quite appropriate that it is named after the man who did so much to reveal the fascinating fish fauna of the Gogo Formation.

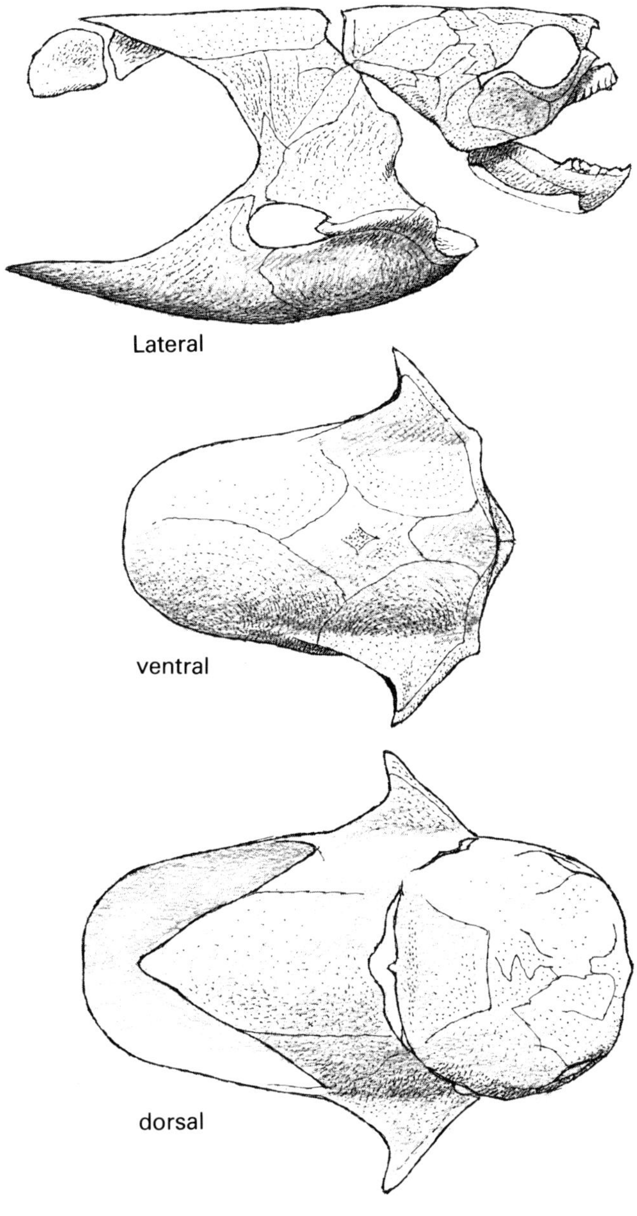

Harrytoombsia elegans, an arthrodire from the Late Devonian Gogo Formation, Western Australia. Head and trunk armour about 12 cm in length.

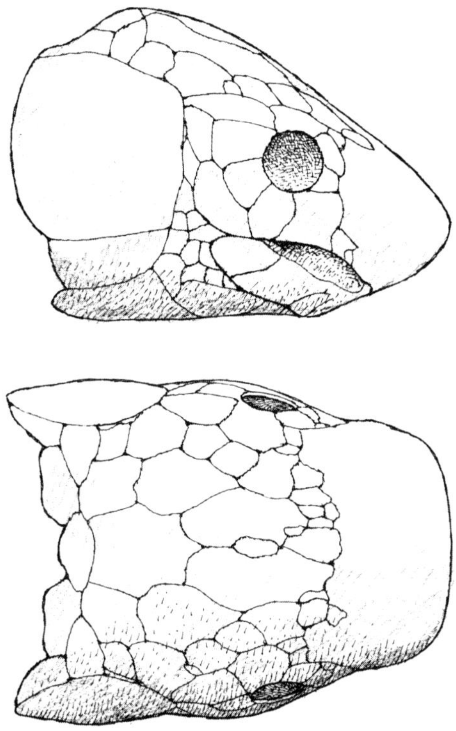

Chirodipterus australis, a lungfish. Lateral (top) and dorsal views.

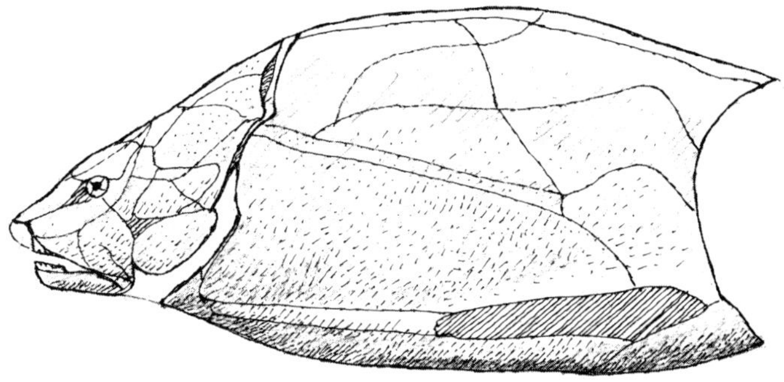

Holonema westolli, a solidly built arthrodire, one of the larger placoderm fish in the Late Devonian Gogo fauna. The armoured head and trunk shield was about 40 cm long; the unarmoured trunk and tail was not preserved in the fossils.

Smaller arthrodires such as *Camuropiscis* and *Incisoscutum* are much more abundant elements. Others, such as *Rolfosteus* and *Tubonasus,* had specialized snouts formed by long tubular rostral plates. Three genera, *Bruntonosteus, Bullerichthys* and *Kendrickichthys,* had specialized crushing toothplates instead of the usual shearing dentition.

One of the largest of the Gogo arthrodires was *Holonema,* a solidly built round-bodied animal with a distinctive pattern on the bony plates. Its remains have long been known from North American, European and more recently, from Asiatic sites but mainly from fragmentary material. Many of the Gogo *Holonema* are virtually complete skeletons, and they have been critical in determining the relationships of this rather unusual animal.

Much rarer placoderm elements in the Gogo fauna are ptyctodonts, smallish fish with greatly reduced head and trunk armour. These show a remarkable resemblance to some modern cartilaginous fish, the chimaeras or rat-fish, although it is not certain if there is any direct connection. Ptyctodont remains are widely distributed in Middle to Late Devonian sediments, mostly in marine deposits, but are normally represented only by their very specialized upper and lower shearing toothplates. One of the two Gogo ptyctodonts, *Ctenurella,* is well-known from a different species found in German Devonian deposits but the other, *Campbellodus,* is only known, so far, from Gogo.

Bothriolepis, the only other common placoderm in the Gogo fauna, is the sole representative here of a distinctive group, the antiarchs, which are discussed in this book by John Long. The presence of *Bothriolepis* in the purely marine Gogo fauna is a surprise, because although it is well known from about 40 species found in Devonian rocks of North America, Greenland, Europe, Asia, Australia and Antarctica, elsewhere it is always found in strictly freshwater deposits, either of rivers or lakes. The marine Gogo species of *Bothriolepis* may thus provide a clue as to how its freshwater relatives achieved such a world-wide distribution.

Of the other elements in the Gogo fauna the lungfish or dipnoans, whose relatives are discussed by Anne Kemp in this book, were relatively abundant and often remarkably well-preserved. Three different genera are present, *Griphognathus, Chirodipterus* and *Holodipterus,* all of which are known from Devonian deposits elsewhere but are not as well-preserved as at Gogo, with their delicate brain-cases completely uncrushed and even the gill-arch supports still in place.

The smaller ray-finned fish, or actinopterygians, are represented by several palaeoniscids, some of which (*Moythomasia*) are well-known from northern hemisphere localities, but are not as well-preserved as at Gogo.

The last group of fish, the lobe-finned crossopterygians, are relatively rare at Gogo, being represented by one genus, *Onychodus,* known from elsewhere and an osteolepid, as yet undescribed. Detailed accounts of many of these different fish have been appearing regularly over the years: by Dennis and Miles (1979a, 1979b, 1980, 1981, 1982), Gardiner (1970, 1973), Gardiner and Bartram (1977), Miles (1971a, 1971b, 1977), Miles and Dennis (1979) and Miles and Young (1977).

The results so far have been spectacular and there is more to come. Harry Toombs, now living in retirement in the south of England, can look back with satisfaction on the results of his fieldwork in Australia and his contribution to our knowledge of Australian Devonian fossil vertebrates.

Bothriolepis cullodenensis

Long, 1983

The Crested Box-fish

CLASS: Elasmobranchiomorphi
SUBCLASS: Placodermi
ORDER: Antiarcha
FAMILY: Bothriolepididae
GENUS: *Bothriolepis*
SPECIES: *Bothriolepis cullodenensis*
AGE: Late Devonian

Bothriolepis belongs to a group of fish called placoderms ('plate skin'), which bear an armour of bony plates covering the head and trunk region. Although the genus *Bothriolepis* (*bothros*, meaning 'pit'; *lepis*, meaning 'scale') has been known since 1840 (Eichwald, 1840), it was not discovered in Australia until 1929, when Professor Edwin Hills, then a keen young student of geology, found pieces of *Bothriolepis* armour in the rugged mountains north of Marysville, Victoria (Hills, 1929, 1931). Since Hills' discovery, *Bothriolepis* has turned up commonly in rocks of Late Devonian age throughout Australia (Hills, 1932; Gilbert-Tomlinson, 1968; Long, 1982) and surprisingly in much older rocks near Canberra (Young and Gorter, 1981). Despite the numerous and often well-preserved fossils of *Bothriolepis* from Australia, little work has been done on the material. In 1980 I began working on some well-preserved *Bothriolepis* fossils from near Mt. Howitt, Victoria, and later decided that it was necessary to collect more *Bothriolepis* from other fossil sites in order to clarify comparisons. It was during a visit to the scenic Freestone Creek region, north of Briagolong, that the plates of a very peculiar *Bothriolepis* were discovered.

I can never forget that day. The rain was coming down in buckets on a miserable day in May. My future wife, then an aspiring assistant, and I had found an outcrop of mudstone on the banks of Freestone Creek near Blue Pool, which looked like it could contain fossils. We were enthusiastically banging at it with hammers. One of the local lads, wearing a farmers' raincoat and a broad-brimmed hat, had heard the noise and came to investigate with the help of his dogs.

'Lookin fer gold are ya, mate?', he asked us.

'Nah, fossils — you know, the remains of old animals', I replied.

'We think there's fish here'.

He thought we were either lying or crazy, but his curiosity got the better of him.

'Oh! Ah well, come up to the house fer a cuppa when youse are through, and show us what ya find!'

It was only minutes after the puzzled local had left us that we found the first good fossil, which left no doubt that our suspicions were correct. We found a very large trunkplate with quite distinctive features from those of the previously described species of Hills, *Bothriolepis gippslandiensis*, and we knew that we had found an unusual new species. The afternoon's booty was soon displayed back at the farmhouse in front of an open fire.

Several seasons collecting at Freestone Creek and other sites provided enough material to describe the new species. Aside from this material, some well-preserved fossils had been collected by geologists from the Victorian Mines Department in the mid 1950s from Freestone Creek, and these gave the first clues that the *Bothriolepis* from there was unusual. A couple of articulated specimens of the same species were later identified from the Mt. Howitt collection, one of which includes the outline of the whole tail. *Bothriolepis cullodenensis* (Long, 1983) takes its name from the now deserted town of Culloden, which is north of Briagolong, Victoria; not to be confused with Bonny Prince Charlie's battlefield in Scotland!

The material of *B. cullodenensis* from Freestone Creek consists mostly of isolated bony plates, with rarer articulated parts of armour. From the line drawings it can be seen that the shield is divided into two sections. The headshield bears a large central window (foramen) for the eyes and nostrils, and is hinged to the box-like trunkshield. The pectoral fins have been modified into bone-covered props, which are connected to the trunkshield by ball and socket joints. Hundreds of isolated plates from Freestone Creek have been collected, enabling the armour to be fitted together like a three-dimensional jigsaw puzzle. The fossils are usually studied by dissolving the bone from the rock to enable a clean rubber latex cast of the impression to be made. Well-preserved material from Mt. Howitt shows the crushed armour with the scales of the tail trailing from behind the trunkshield. The specimens from Mt. Howitt differ slightly from the Freestone Creek material in having a lower crest on the back, but this is not significant enough to warrant the naming of a new species.

The characteristic features of *B. cullodenensis* which make it distinctive from other species are the long, narrow headshield and short, high trunkshield with a high median crest on the back (Long, 1983). Anatomical details on the internal surfaces of the armour such as the large embayments inside the headshield, lateral pits, are also distinguishing features of this species. The closest relative of *B. cullodenensis* is *Bothriolepis gippslandiensis*, also from Victoria. These species both have short, high trunkshields and unusually large pits inside the headshield but clearly differ by the shape and proportions of the component plates. These two species, and possibly a third Victorian species, which is less well known, constitute a group of bothriolepids which are more primitive than species outside Australia (Long, 1983). *Bothriolepis verrucosa* from the Middle Devonian near Canberra (Young and Gorter, 1981) is the only species believed to be more primitive than the Victorian species.

So far approximately sixty species of *Bothriolepis* have been described worldwide. *Bothriolepis* has been reported from Europe, North America, Canada, Russia, Greenland, Scotland, England, Turkey, China, Australia and Antarctica (Denison, 1978). The most detailed study of this fish was done by Stensiö (1948) in which the anatomy of every species known at that time was reviewed. Well-preserved *Bothriolepis* remains from Canada were thinly sliced to study the sediment infilling the body cavities. This revealed that the fish ingested mud rich in organic material (Denison, 1941) and may have possessed paired air-sacs. Undoubtedly the best material of *Bothriolepis* anywhere in the world is from the Gogo Formation, Western Australia (Gardiner and Miles, 1975). The bones from this locality are freed from the limestone by acetic acid and can be put together to restore the actual armour in three dimensions. Delicate cartilages, which had a thin lining of bone, such as some of the tiny bones of the mouth are also preserved in this material. The Gogo material demonstrated that the jaws were weak bones in *Bothriolepis* as befitted a mud-grubber (Young, 1984).

Perhaps one of the reasons why *Bothriolepis*, such a bizarre little fish, has been the subject of much worldwide interest is its use in correlating rock sequences. In China the Devonian period has been subdivided into nine time zones using the entry and disappearance of *Bothriolepis*-like fossil fish (Pan Kiang, 1981). *B. cullodenensis* and *B.*

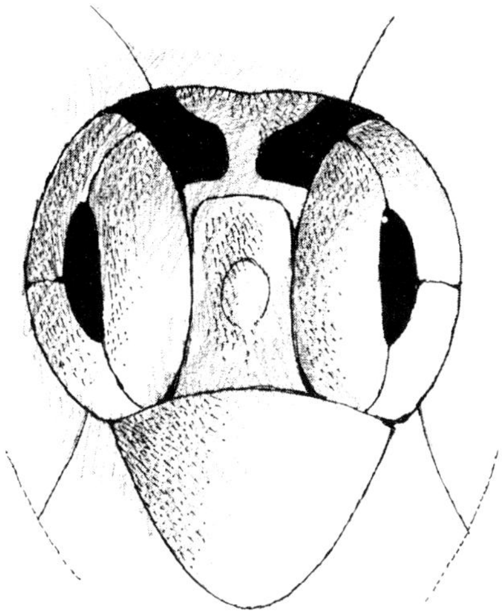

Detailed restoration of the eyes and nostrils of *B. cullodenensis*. The plate separating the eyes has a raised pineal 'boss' — location of the third 'eye'; x 3.

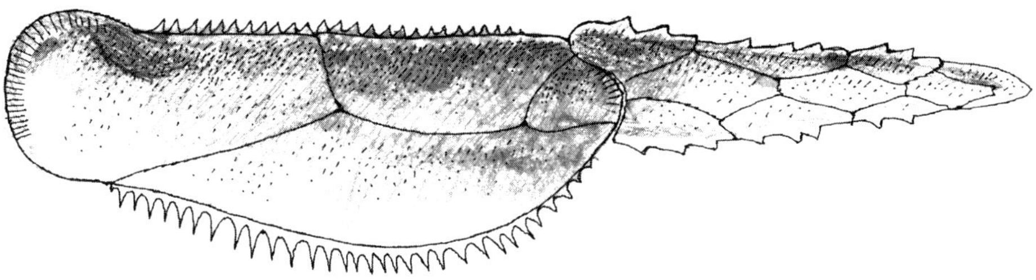

Pectoral appendage of *B. cullodenensis* in dorsal view. The rounded area on the left is the ball that fits into the socket in the trunkshield. The spines along the front and back of this fin-like structure are for defence; x1.

gippslandiensis have been used as time zone fossils for subdividing the Late Devonian rock sequences in Victoria (Long, 1983; Long, Turner and Kemp, 1982). The economic potential of fossil fish studies lies in this application. If fish are the most useful fossils for correlation of otherwise barren rock sequences, then faunas must be described in order to make accurate relative age determinations. In Victoria this is done with the help of radiometric dating on volcanic rocks, which occur in the same successions as the fish-bearing rocks (Williams *et al.*, 1982).

B. cullodenensis lived in freshwater rivers and lakes, and probably used the high crest on its back for stability in swift-flowing currents. The long bony arms were either used to fend off attackers or perhaps assist in turning manoeuvres or even for locking the fish inside crevices in rocks as some armoured catfish do today. Alongside *Bothriolepis* lived a flat placoderm with characteristic wavy lines of bone on its armour. This was *Phyllolepis* (meaning 'leaf scale'), which is shown lying on the bottom sediment in the colour plate. Well-preserved remains of this placoderm from Mt. Howitt reveal that it possessed many small pointed teeth on its jaws and was probably a voracious predator. It may have waited on the bottom for unsuspecting fish to come along then lunged up at them with a quick flash of its long tail.

Another placoderm which frequently occurs with *Bothriolepis* and *Phyllolepis* at Mt. Howitt and Freestone Creek is *Groenlandaspis*. *Groenlandaspis* was first discovered in Greenland, hence the name, and until recently was not known from anywhere else. Ritchie (1974, 1975) discovered *Groenlandaspis* in Antarctica and Australia. Soon afterwards, he toured Europe and found *Groenlandaspis* turning up all over the place. The isolated plates of *Groenlandaspis* bear a distinctive ornament of rounded tubercles which distinguish them from *Bothriolepis* and *Phyllolepis*.

The true bony fish (osteichthyans) were also common in the rivers and lakes of the Devonian age. Fast swimming ray-finned fish such as those seen in the top of the colour plate, were to survive and dominate long after the extinction of the placoderms. Indeed, these fish thrive today, being the most successful and diverse of all vertebrates.

The reconstruction of *B. cullodenensis* was based upon the assemblage of many individual plates from Freestone Creek and entire specimens with the tail from Mt. Howitt. The plates of similar size were measured, and the angles at the edges of the armour were used to make a rough cardboard model. Once the final adjustments were made, a refined model of a wire frame, covered in latex rubber, was constructed. This model was the basis for accurate sketches of the restored armour. There is no way that the colour of the fossil fish could be determined from their remains so this was left to the discretion of the artist. Colours depicted here are based upon typical colour schemes found in living freshwater fish.

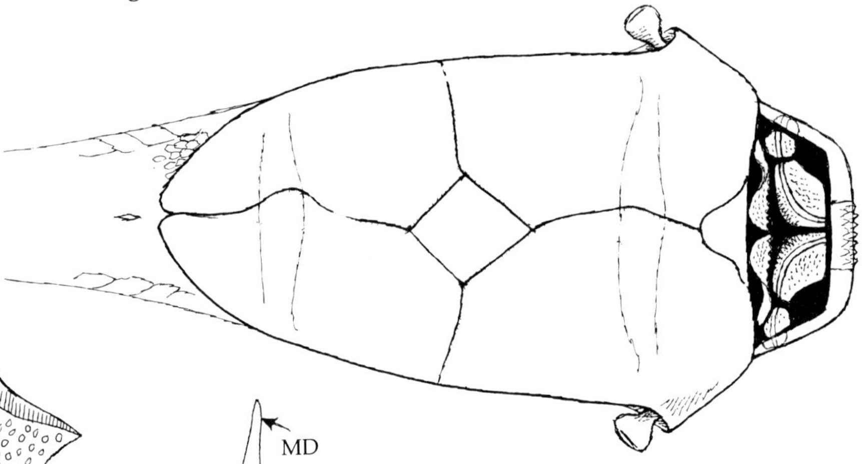

Ventral (underside) view of the armour of *B. cullodenensis*; x 1.

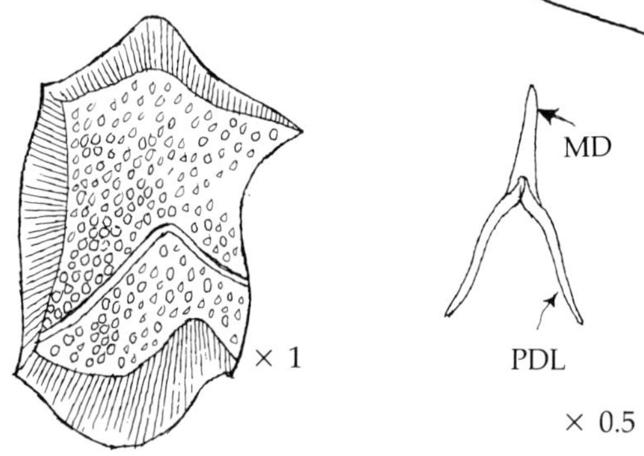

Posterior dorsolateral plate (PDL) of *Groenlandaspis* sp. Fish in this genus have tubercular ornament on their plates and an inverted V-shaped sensory canal. These plates are unusual in that they are in contact below the median dorsal plate (MD), which sits on top of the trunk armour like a narrow hat.

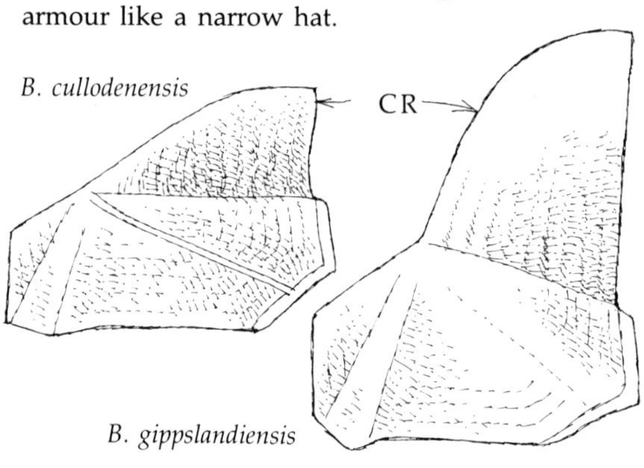

Anterior median dorsal plate (AMD) of *Bothriolepis gippslandiensis*, in side view. This species has a much lower median crest (CR) than that of *B. cullodenensis*; x 1.

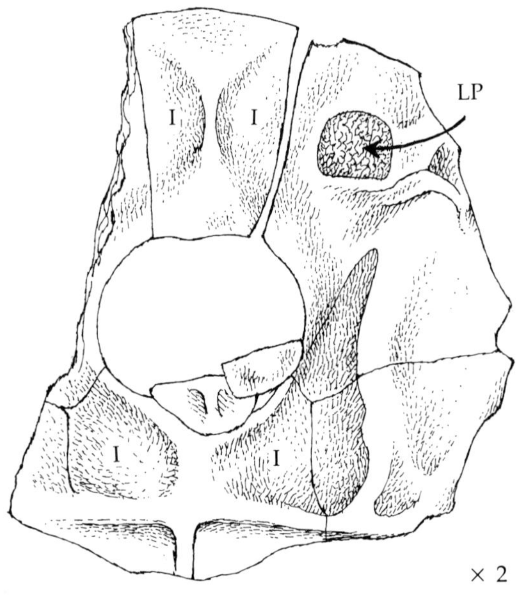

Ventral (underside) view of the headshield of *B. cullodenensis*. The indented region (I) formed a brace for the cartilaginous braincase, whereas the lateral pit (LP) was probably an attachment area for the jaw musculature.

Fossil fish from the Late Devonian highland streams and lakes of central Victoria. In the centre are two individuals of the unusual antiarch *Bothriolepis cullodenensis*. Another placoderm, a flattened phyllolepid, lies on the sandy stream bed below them waiting for prey to pass above. In the top- to mid- water range are the swift swimming palaeoniscoid fish, precursors to the modern bony fish, which today predominate in seas, rivers and lakes.

KNIGHT

As mentioned before, the environment of these fish was freshwater. This is determined from study of the rock sequences which contain the fossils. Thinly layered black shales are typical of lake deposits in which sediment settles slowly out of the still water. The Mt. Howitt locality, which yields entire articulated fossil fish, was such an environment. In more turbulent environments like rivers, the dead fish are soon disarticulated and isolated bones, often damaged or worn, are the characteristic fossils.

Despite the abundance of Devonian fish fossils found in Australia, there is much work to be done before the significance of the faunas can be fully evaluated. Fish such as *B. cullodenensis* tell us something about the isolated nature of the faunas in Victoria. *Bothriolepis* faunas in New South Wales have been found near Canowindra (Ritchie, 1969), Jemalong Gap (Campbell and Bell, 1977), Eden and Pambula (Ferguson *et al.*, 1979) and many other places. However, all of these species of *Bothriolepis* possess a regular type of armour, similar to other species worldwide, but very unlike *B. cullodenensis* and *B. gippslandiensis* from Victoria. This raises the question of whether the environments may have differed. Recent studies on the geology of southeastern Australia indicate that the area now covering Mt. Howitt and Freestone Creek was a highland mountain region in the Late Devonian (Cas, 1983). To the north the topography was flatter with large meandering streams covering the coastal plains. Perhaps the Victorian bothriolepids developed their crests to stabilize themselves in the currents in the mountain streams, and the New South Wales bothriolepids, like most others from Scotland and elsewhere, inhabitated large slow-flowing rivers and lake systems close to the sea where stabilizing crests would be of no advantage. At present this concept needs much detailed work to be proven, yet it shows what the study of fossil fish can tell about the continent of Australia many, many years ago.

Ceratodonts
Lungfish

CLASS: Teleostomi
ORDER: Ceratodontida
FAMILY: Ceratodontidae
GENUS: *Ceratodus, Neoceratodus*
SPECIES: *Ceratodus wollastoni* Chapman, 1914
Neoceratodus eyrensis White, 1925
Neoceratodus gregoryi White, 1925
AGE: Cretaceous, Miocene and Pleistocene.

Known distribution of *Ceratodus wollastoni*

Known distribution of *Neoceratodus eyrensis*

Known distribution of *Neoceratodus gregoryi*

For thousands of years, the centre of Australia has been an arid region where most water stands in stagnant pools — the remnants of the rare floods from wetter country often thousands of kilometres away or from intermittent and unpredictable rain storms. This was not always so. Throughout the Tertiary period, central Australia must have had a higher rainfall, and most of the land was covered by a system of lagoons and rivers supporting a variety of animals and plants no longer able to survive in this region. Remains of large primitive algae (charophytes), snails, small shrimps, fish and other vertebrates have been found in the rock sequences in this area. Among them were at least three species of lungfish, members of a conservative group of vertebrates which has changed very little since it first appeared in the Triassic, 200 million years ago.

The lungfish belong to the order Ceratodontida, a name derived from two words, *ceras* meaning 'horn' and *dens* meaning 'tooth'. The fossils of this group are mainly teeth, or really toothplates, and are usually the only part of the fish to be preserved. Agassiz (1833-1843), in describing them, was reminded of horns or antlers. Others have described similar fossils as 'duck feet', a more apt description of their shape.

In action, the toothplates are rather like the crushing molars of a horse but have the horns, now known as 'ridges', facing outwards. Most ceratodont lungfish simply crushed their food between the two undulating opposing surfaces. Some used the ridges like many-bladed scissors.

Since *Neoceratodus forsteri*, the living Australian lungfish, was first described by Krefft in 1870, lungfish have been variously classified as fish or as amphibians. As ancestors of the tetrapods (four legged animals), they have been regarded as important by some, or insignificant by others (Rosen *et al.*, 1981; Jarvik, 1964). The literature on this subject is extensive, complex, indigestible and sometimes even rude, and perusal thereof should not be undertaken lightly. It is true that lungfish share many features with amphibians, particularly in their anatomy and development, but they also have much in common with other primitive fish. My view is that the lungfish, or Dipnoi, are an order of fish within the subclass Choanichthyes, which includes many fossil species as well as the coelacanth, a 'living fossil', Old Four Legs (Smith, 1958), discovered in deep water off the coast of Madagascar in 1938. Lungfish are not important as the ancestors of higher vertebrates, but represent an evolutionary side-line left behind by the mainstream long ago. This is not to deny that fossil lungfish provide us with a fascinating glimpse of past eras.

Fossil lungfish had a worldwide distribution, far wider than the present ranges of the living lungfish, *Lepidosiren paradoxa* (South America), three species of *Protopterus* (Africa) and *N. forsteri* (southeast Queensland). Fossils of the latter species were much more widely distributed in the past, and have been found at Lightning Ridge, New South Wales (Cretaceous), in South Australia (Miocene) and at Chinchilla, Queensland (Pleistocene), (Kemp and Molnar, 1981).

Records of living lungfish over the past seventy odd years show that *N. forsteri* is presently found in rivers and lakes in southeastern Queensland. Its distribution at the time of its discovery in 1870 is not necessarily as small as has often been stated in the past, i.e. confined to the Mary and Burnett River systems in southeastern Queensland. There is a record of lungfish from 'Rockhampton Market', many kilometres away from the so-called 'ancestral' localities.

The collection date of the specimen was 1876, nearly twenty years before O'Connor decided that the lungfish was in danger of extinction. He caught 98 fish from a small area in the Mary River and transplanted the 69 survivors to a number of other river systems, thus denying us the chance to find out whether the distribution was originally restricted to the Mary and Burnett Rivers! *N. forsteri* has done very well in most of its new localities, and has shown that it can cope with a fair variety of freshwater environments (Welsby, 1905). If the distribution was really restricted to such a small area in recent times, it certainly was not because other rivers were unsuitable as lungfish habitats.

Many different kinds of fossil ceratodont toothplates have been found all over the world, and a large number of species have been described. About twelve belong to this continent, and three of these lived at times in the past in central Australia.

One, *Ceratodus wollastoni*, named after T.C. Wollaston, a South Australian opal buyer, is known only from its toothplates. The species was not entirely confined to central Australia, as specimens have been found in Cretaceous deposits at White Cliffs, at Lightning Ridge, and in the Grawan opal fields (all in New South Wales) (Chapman, 1914a), near Winton in Queensland, and in the Miocene Etadunna Formation of South Australia (Callen and Tedford, 1976). In the South Australian locales the lungfish was contemporaneous with teleost fish, porpoises, turtles, crocodiles, flamingoes and one species of platypus that had enamelled teeth, as well as lizards, possums, diprotodonts and kangaroos. The diet of this lungfish can be inferred with reasonable certainty from the wide, heavy toothplates, with rounded contours that show little wear. The teeth are not adapted for shearing and were not used for rough or abrasive food like snail shells. They are so long and heavy that one could not expect to find them in anything but a slow and bucolic animal, placidly chewing soft water

---

Three common species of Cainozoic lungfish from central Australia: *Neoceratodus gregoryi* in the foreground, *N. eyrensis* in the open water, and *Ceratodus wollastoni* in the water weeds.

KNIGHT

plants in the warm shallow lake. The breadth of the toothplates suggested a broad, flat head and a thick, heavy body. Also, one can predict that *C. wollastoni*, like other ceratodonts, had a diphycercal tail (a tail fin that comes to a point, like an eel's), paddle-shaped fins and large scales. It probably had cryptic colouring to match the weed bank where it lived.

The toothplates of *C. wollastoni* contrast sharply with those belonging to the second and smallest of the three species, *Neoceratodus eyrensis*. This was one of the two species of lungfish collected by J.W. Gregory on his journey to the Central Desert in the early twentieth century, and referred to rather casually as 'teeth of Mudfish (*Ceratodus*)' in his book *The Dead Heart of Australia* (1906). *N. eyrensis* was later described in great detail by E.H. White (1925), who named it 'eyrensis' after Lake Eyre. *Neo* meaning 'new' was originally added to the generic name to distinguish the living Australian lungfish from fossil forms, but *N. eyrensis* is closely related to *N. forsteri* (Kemp and Molnar, 1981). *N. eyrensis* is represented by numerous toothplates of all sizes up to 40 mm (representing an animal about 2 metres in length), by a solitary vomerine (incisor) tooth, and by many fragments, some reasonably complete, of the jaw bones to which the teeth were attached. The ridges on the toothplates, which are narrow and sharp, fit between each other as the opposing teeth move, like the blades of scissors. The jaw articulations of modern lungfish allow mostly up and down movement, but little front to back movement, and the wear pattern on the toothplates of *N. eyrensis* shows that the jaw action was similar. Possibly, *N. eyrensis* sliced its food, as well as crushed it, and like the living South American and African lungfish, was mainly carnivorous. However, the vomerine tooth is blade-shaped, like the corresponding tooth of *N. forsteri*, and unlike the sharp, pointed vomer of the African lungfish. In relation to the length, the jaws are narrower than in *C. wollastoni*, and this suggests a narrow head. So the reconstruction in this book depicts *N. eyrensis* with a long narrow body and slender fins, hunting small fish in the deeper water clear of the weeds.

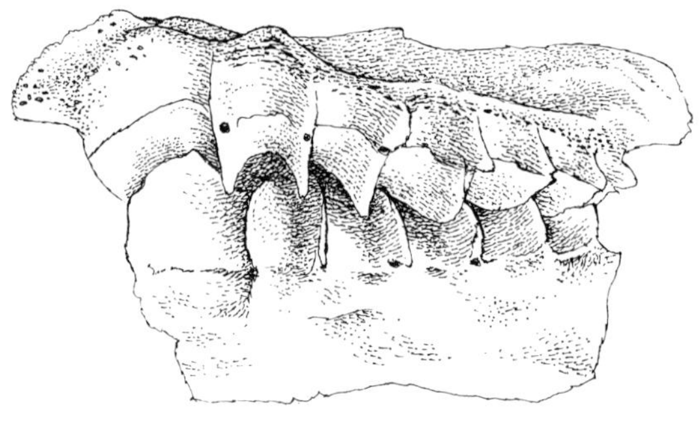

Upper and lower toothplates of *Neoceratodus eyrensis* in occlusion, viewed from the side; x1.

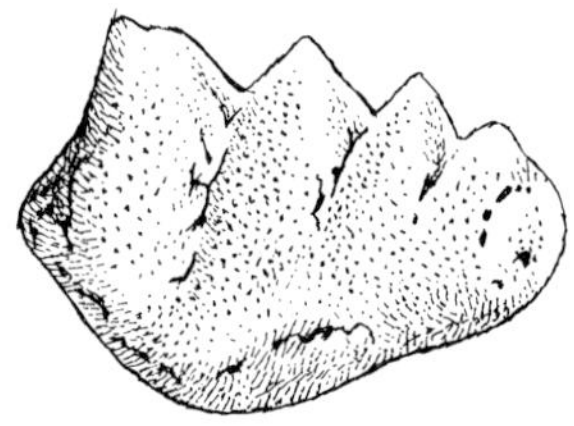

Upper toothplate of *Ceratodus wollastoni*; x1.

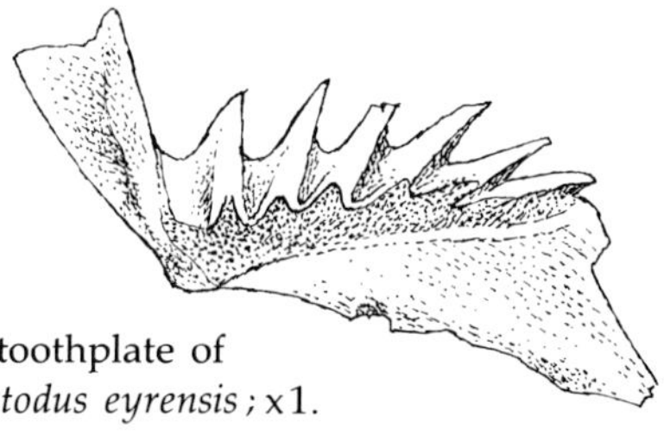

Upper toothplate of *Neoceratodus eyrensis*; x1.

N. eyrensis has so far been found only in the centre of Australia, in the Katapiri Sands and in the Etadunna, Namba, and Wipijiri formations. The third central Australian species, *Neoceratodus gregoryi*, also occurred in these localities, as well as in the Carl Creek Limestone and at Redbank Plains, both in Queensland and of Tertiary age.

N. gregoryi, first collected by Gregory, was ultimately named after him by White. Several recent expeditions have discovered much more material, and *N. gregoryi* is now the most completely known of the Australian fossil ceratodonts. Enough remains of the skull structure, the teeth, the jaws and even the ribs to allow reconstruction of a whole animal in some detail. This species closely resembles the living lungfish *N. forsteri* in structure, except that it grew to a very considerable size. A fully grown adult of *N. forsteri* is about 1.1 metres long, 0.5 metres in girth, and has a rostral bone (or bone of the snout) 20 mm wide, with toothplates 30 mm long. A specimen of *N. gregoryi* with these dimensions would have been a 'minnow'. The biggest toothplates available belonging to *N. gregoryi* have a length of 75 mm, and the widest rostral bone a breadth of 85 mm. Assuming that these are related to the body length in a similar way in both species, the largest *N. gregoryi* known would have been between 3 and 4 metres long, quite an impressive lungfish.

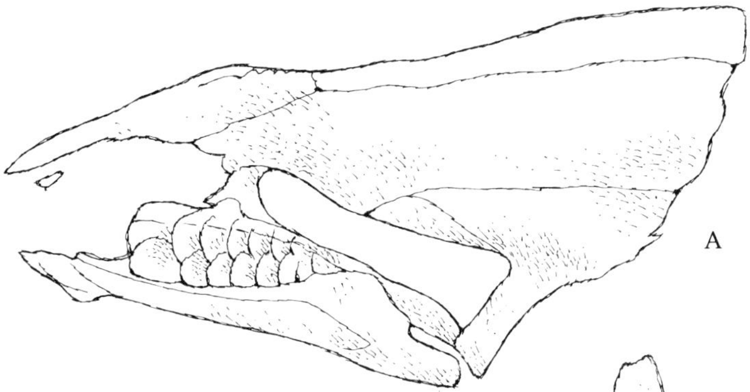

A

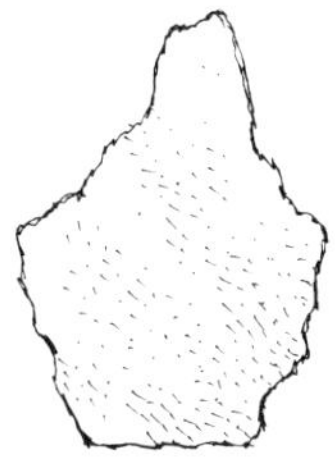

B

A A reconstruction of the skull of *Neoceratodus gregoryi*; x0.5.

B The rostral bone of *Neoceratodus gregoryi*; x0.5.

The skulls of the two species are similar. Aged specimens of both have a peculiar dished effect in the bone above the orbit, not mirrored in the soft structures at least in *N. forsteri*, which has smooth rounded contours to the head. The major bones in the skull of *N. forsteri* are also found in *N. gregoryi*. These bones are traditionally labelled with letters, not names, to emphasize the peculiarity of their shapes and positions, and their lack of comparison with the skull bones of other vertebrates. They are not, of course, complete, and the reconstruction of the *N. gregoryi* skull needs a little imagination to fill in the details! Much of the skull and lower jaw would have been cartilaginous, not bony, including the support for the vomerine tooth and the jaw articulation. The bones, which formed the major part of the skull roof, are thickened at the suture lines where the bones meet and taper to thin sheets at the back of the skull, as in *N. forsteri*. The small post-orbital bone (behind and above the eye socket) is stout and roughly oval as in the recent species. The rostral bone, which protects the snout, has some unusual features. It is pentagonal in shape with a relatively short anterior process, while the corresponding bone in *N. forsteri* is oblong with a long anterior projection. In old specimens of *N. gregoryi* this bone is very thick, and the hind part of the outside surface is covered with sharply curved projections, rather like a lunar landscape. We do not know what these structures are for, but possibly they developed as the animal grew older to support the sensory pits of the snout. There is no parallel in the structure of the rostral bone in the living lungfish, although it does have a sensory line system in the snout. The exact function of this sense organ in lungfish is not known, but a similar system in teleost fish and sharks is used to detect mechanical, electrical and chemical stimuli.

While the herbivorous mammals were evolving a grinding dentition with enamel, dentine, and cementum, all materials that wore at different rates, some ceratodont lungfish had developed a system along similar lines, equally efficient but more refined. *N. gregoryi* was one of these. The toothplates have a combination of several types of dentine, producing opposing surfaces with the consistency of fine sandpaper. Wear patterns on some of the teeth indicate a limited degree of fore and aft movement, as well as the usual dorso-ventral crushing. *N. gregoryi* must have been able to reduce its food to a fine paste. Some of the older toothplates are severely worn, suggesting a rough diet, probably omnivorous and consisting of aquatic animals and plants with a quantity of mud or grit. This species may also have indulged in a little scavenging on such dead animals as possums and turtles, in much the same way that the living lungfish will eat dead animals.

Based on the close structural similarity of the dentition and the skull of these two species it is quite possible that they may have lived in similar places, ate similar diets, and had comparable life cycles and breeding habits. *N. gregoryi* is shown in a habitat similar to the one chosen by *N. forsteri*, in deeper water beside weed banks, which provided some of its food and shelter for its eggs and juveniles.

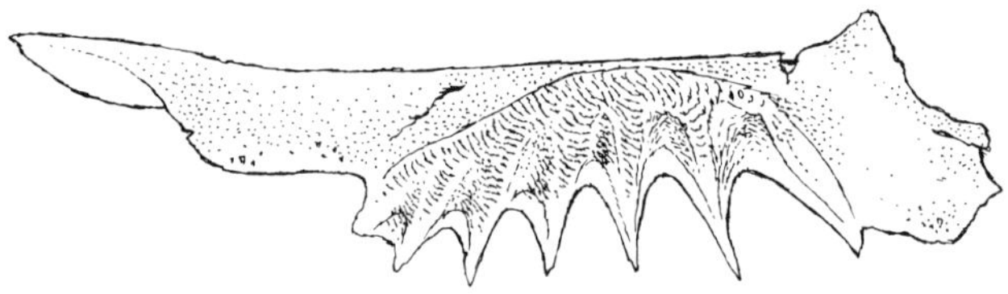

Lower jaw toothplate of *Neoceratodus gregoryi* × 1

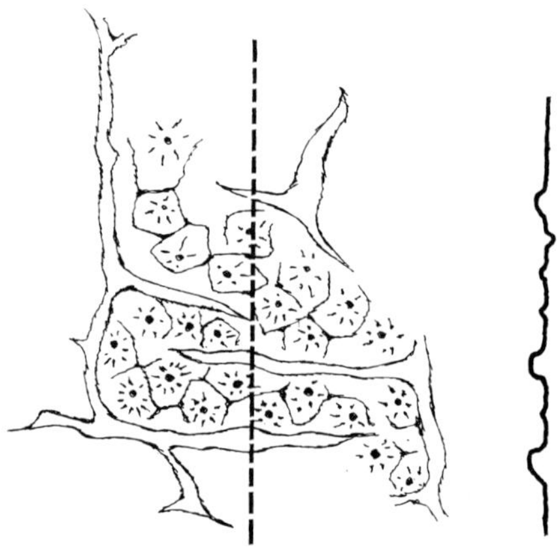

Detail of the occlusal or grinding surface of a toothplate of *Neoceratodus gregoryi* ; × 10

Siderops kehli

Warren and Hutchinson, 1983

The Kolane Labyrinthodont

CLASS: Amphibia
SUBCLASS: Labyrinthodontia
ORDER: Temnospondyli
SUPERFAMILY: Brachyopoidea
GENUS: *Siderops*
SPECIES: *Siderops kehli*
AGE: Jurassic

Labyrinthodont is the name used for most fossil members of the Class Amphibia found in Australia. Unlike their descendents, the frogs and salamanders, the labyrinthodonts were often large crocodile-shaped animals. They first appear in the geological record in the Late Devonian sediments of East Greenland and Australia, and later members are found in all continents including Antarctica. Until recently the last recorded labyrinthodont was from the latest Triassic of Germany.

On 26 June, 1940, a bone was exhibited at a meeting of the Royal Society of Queensland. This was not an unusual event in the early days of Australian palaeontology, but few exhibits have raised such controversy in subsequent years. The bone had been found by Mr. John Wadley in the sandstones of the Brisbane River, nearly 2 kilometres south of Lowood railway station. It was described by Longman (1941) as a piece of labyrinthodont amphibian upper jaw. He named it *Austropelor wadleyi* as a tribute to Wadley whose enthusiasm produced the find in an area in which Longman himself had searched without success.

The rocks in which *Austropelor* was discovered belong to the Marburg Sandstone, a formation of Jurassic age. Here lies the controversy, as no labyrinthodont amphibian had then been found in sediments younger than Triassic. At first it was thought that the bone had been misidentified, but when others established that it was indeed labyrinthodont, Whitehouse, the geologist who originally aged the Marburg, re-dated it as Triassic (Whitehouse, 1955). Subsequent evidence from fossil pollen indicates that the formation is undoubtedly Jurassic, and this led De Jersey (1963), to suggest that *Austropelor* was reworked from rocks of Triassic age. In the most recent publication Colbert (1967) established that the specimen was certainly labyrinthodont but was from the lower jaw. He stated however, that 'there will have to be more definite evidence than is afforded by the single specimen of *Austropelor wadleyi* to justify the extension of the labyrinthodont amphibians into Jurassic sediments'.

Another discovery, however, by Colin Kehl on his cattle station, Kolane, some 300 kilometres northwest of Lowood has provided the first unequivocal evidence of Jurassic labyrinthodonts (Warren, 1977; Warren and Hutchinson, 1983). This single, almost complete specimen was recovered over several years from the Westgrove Ironstone Member of the Upper Evergreen Formation. The Evergreen has been firmly dated as late Early Jurassic, and is a lateral equivalent of the lower part of the Marburg Sandstone.

While the field collection of this specimen was relatively easy, its preparation in the laboratory presented tremendous problems. As the fossilized bone was softer than the ironstone, any exposed bone was first impregnated with reinforcing plastic. Next, the ironstone blocks were trimmed with a diamond saw to within a few millimetres of the bone. This last layer of extremely hard matrix had to be removed by means of a small pneumatic drill (airscribe), a noisy, dirty process which took several years to complete. A further problem was that some blocks had broken so that the underside of the skull roof was exposed. In order to prepare the upper surface of these blocks, their lower surface was backed with silicone rubber before being encased in plaster of Paris. These casings absorbed the shock of preparation so that even millimetre thick bone did not shatter. Eventually, after as much of the ironstone as possible had been removed, the skeletal remains were assembled, and the animal was reconstructed.

The animal is known therefore, from a single specimen, which is fairly complete. The body would have measured 2.5 metres long, while the skull was 57 cm long and 63 cm in maximum width, making it one of the largest known labyrinthodonts. Compared with the skull size, the body was extremely small. The pectoral (shoulder) area is only half as wide as the skull compared with the one to one ratio found in other labyrinthodonts. In addition, the pectoral area lies well behind the skull so that there seems to be a developed neck region, a situation which could have provided manoeuverability essential for defence in an animal living alongside advanced dinosaurs. On the other hand, the front view of the skull of the Kolane amphibian suggests that it may have been an aggressive carnivore itself. Its battery of teeth is unequalled in Mesozoic labyrinthodonts. On the palate a row of approximately 100 teeth are small towards the back, but rapidly increase in size to 30 mm daggers near the front of the skull. Inside these is a continuous row of approximately 80 teeth, which are a little larger than the outer row and project beneath them. In this inner row of teeth are three sets of tusks on either side, the largest of which extends almost 70 mm below the jaw line. In addition to this, each tooth, from the tiniest to the largest tusk, is shaped like those found in animals known to be aggressive carnivores such as goannas, some mammal-like reptiles, and many mammalian carnivores. On each side of the tooth a sharp ridge extends towards the tooth tip resulting, when upper and lower jaws are brought together, in a series of scissor-like blades slashing past one another. To complete the 'dental armoury' the surface of the palate, at least in the area of the palatal bones, seems to have been covered by a shagreen of tiny 1-2 mm denticles. This strong-toothed, weak-limbed creature most probably fed in water, preying on fish and small tetrapods. However, with its awesome dentition, the Kolane amphibian would appear to have been quite capable of ambushing and killing land animals up to the size of a small dinosaur.

What relationships has this lone survivor with other labyrinthodonts? *Siderops* has a skull that is short-faced and parabolic in outline with the jaw articulation well below the level of the occiput. This type of skull is found in two groups of Mesozoic labyrinthodonts, the families Brachyopidae and the Chigutisauridae, which are united in the superfamily Brachyopoidea. The little that is known of *Austropelor* indicates that it, too, may belong to the Brachyopoidea. While the brachyopid labyrinthodonts have a wide distribution, particularly on the Gondwana continents, where they are absent only from South America, the chigutisaurids were restricted to Argentina until their recent discovery in Queensland (Warren, 1981). The Kolane labyrinthodont is certainly more closely related to the chigutisaurids than to the brachyopids, and it does share one specialized character with the Australian chigutisaurid, *Keratobrachyops australis;* they both have a uniquely expanded process of one of the skull bones (pterygoid). This feature is not illustrated in the literature on the Argentinian chigutisaurids, and perhaps was not preserved in them. When the nature of this characteristic in the Argentinian chigutisaurs becomes known, it may be necessary to separate the two Australian forms from them, but until that time the Kolane labyrinthodont should be regarded as a member of the Family Chigutisauridae.

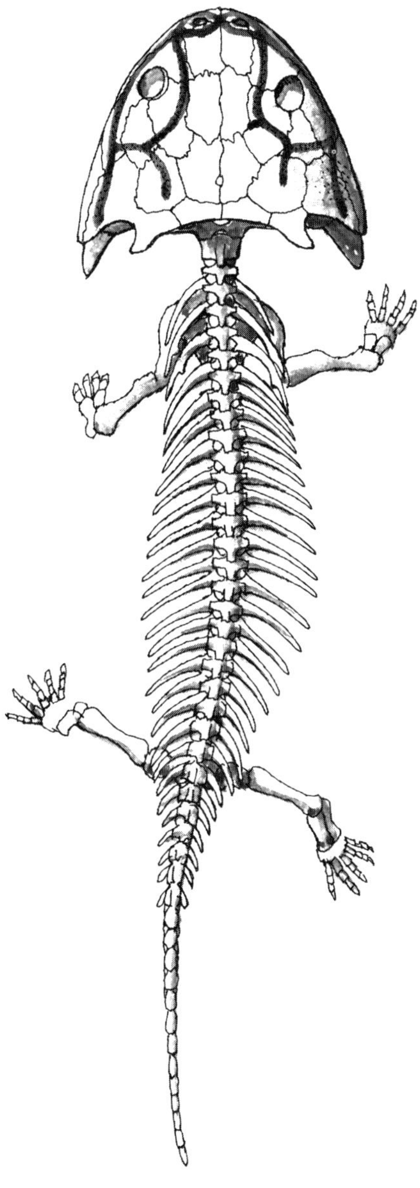

Siderops kehli from above, about 1/15 natural size.

In the reconstruction, two animals are shown, one on land and one in the water, emphasizing the amphibious nature of this first tetrapod ('four-footed') group. Their proportions are inferred directly from the reconstructed skeleton. The limbs are small, with front limbs well separated from hind limbs. Webbing on the toes is purely speculative, but put there by analogy with modern amphibians. The head is very broad, being twice as wide relative to the body as it is in other labyrinthodonts. On the skull surface, grooves for detecting water-borne sound are well developed in places, and these and the coarse surface sculpturing can be seen through the skin. The eyes face directly forward, perhaps giving some degree of stereoscopic vision. On land the animal could probably hold its head high.

Colouring and skin pattern behind the head are based on camouflage patterns found in some modern amphibians. The fish shown being pursued were about 20 cm in length, giving an idea of the great size of this very distant relative of the frogs.

Although the Kolane labyrinthodont is interesting in itself and is important for extending the range of labyrinthodonts into the Jurassic, perhaps its chief importance will be in the nature of the animals found with it. Early Jurassic sediments of terrestrial origin have rarely been preserved, a notable exception being the Kota Formation of India. Thus, we know little of this important time in the earth's history. The last two fieldtrips to the Evergreen Formation have yielded two plesiosaurs, the earliest adequate plesiosaurian remains from the Gondwana continents (Thulborn and Warren, 1980), as well as freshwater shells and quantities of fossil wood. Other evidence of life in the Early Jurassic comes from within the skeleton of the labyrinthodont itself. Firstly, a single vertebral fragment, found inside the animal's mouth, appears to belong to a smaller, less robust labyrinthodont. Secondly, fine preparation of the palate of the amphibian just inside the palatal tooth row revealed a number of small, segmented invertebrates. These have been determined (Jell, 1983) as remains of a single species of millipede, the only one yet found between the Carboniferous and the Oligocene, a time space of 250 million years.

Amphibians in a Jurassic landscape. On the bank, from a clump of horsetails and club-mosses *Siderops* keeps a wary eye on a pterosaur. His companion in the water pursues some air breathing lungfish.

F. KNIGHT '82

Notochelone costata

Owen, 1882

The Southern Turtle

CLASS: Reptilia
ORDER: Chelonia
FAMILY: Cheloniidae?
GENUS: *Notochelone*
SPECIES: *Notochelone costata*
AGE: Early Cretaceous

?*Ornithocheirus*

The Western Flying Dragon

CLASS: Reptilia
ORDER: Pterosauria
FAMILY: Ornithocheiridae
GENUS: ?*Ornithocheirus*
SPECIES: not identified
AGE: Early Cretaceous

Placid beneath the clear blue sky, with a scattered flock of white cumulus, lies a turquoise sea: to the west, a low, heavily-wooded shore, bounded by salt marsh, while to the east, north and south only calm, crystalline water. The water of this shallow sea is unclouded by sand, silt or mud, and trunks of trees and tree ferns washed out from the western shore are seen clearly, strewn water-logged, on the bottom among seaweeds and a glittering array of small silvery fish. Enormous bivalves (*Inoceramus* — kin to today's oysters) lie on the bottom, their shells reaching almost a metre in length, among shoals of much smaller brachiopods. Lithe-limbed, armour-plated brittle stars writhe slowly over the bottom in search of food, while coiled-shelled ammonites with piercing eyes and a serpentine multitude of trailing tentacles swim through the sea. Long, massive fish, *Belonostomus*, the size of modern gropers, drift below the gentle waves sharing the sea with reptiles: short-necked ichthyosaurs (*Platypterygius*) and long-necked plesiosaurs. Above, slender-winged pterosaurs fly through the sky, while mirrored below, small turtles 'fly' through the sea. Across the soft lime ooze of the bottom are cast a myriad tiny bones of the small fish. Here and there is a bone of a pterosaur or turtle, or even a miniature bird-like dinosaur. About a hundred million years later, long after the sea has vanished, the bed of the Hamilton River, and the houses of Boulia, Queensland, lie near this spot.

Most probably the western part of the Australian inland sea looked like this at the time that the Toolebuc Limestone was being deposited near the Hamilton River in the Early Cretaceous. We know that the water was clear, without sand or mud, because the limestone when dissolved in acid leaves behind only bones, and not a trace of mud or sand. The fossils of the brachiopods, ammonites and logs are still there to be seen today, but the ammonites are exceedingly rare. Bivalves are represented only by fragments. Brittle stars have yet to be found near the Hamilton, but they do occur in the Toolebuc Formation further to the northeast. Bones of *Belonostomus, Platypterygius,* plesiosaurs and most of all small sardine-size fish litter the limestone. A single bone from the hind leg of a very small (but not juvenile) theropod dinosaur was found among the thousands of fish vertebrae. But most importantly, bones of pterosaurs are also present; the first known from Australia. Also abundant (for reptiles) are bones of the turtle, *Notochelone*.

The small sea turtle *Notochelone* was among the first fossil reptiles found in Australia, and judging from its abundance in western Queensland this is no surprise. It was described by Sir Richard Owen (1882a) under the name *Notochelys*. This name was already in use for another, living turtle from Southeast Asia, and thus it was replaced by the name *Notochelone*. Both mean 'southern turtle'.

Although first described a century ago, little is yet known of *Notochelone* — this more because of a lack of interest in turtles, than because of a lack of material for study. *Notochelone* looked, and probably behaved, much like modern sea turtles. It was small, only about half a metre in length. Along the back of the shell (carapace) was a low ridge, a feature still seen today in very young sea turtles but lost in the adults. Its skull was lower and broader than those of modern sea turtles and more nearly triangular when viewed from above. The orbits for the eyes were large, and the skull was deeply embayed from behind, for the eardrums, with a blunt beak at the front.

The flippers of *Notochelone* are yet unstudied, but look similar to those of the living green turtle, *Chelonia mydas*. *Notochelone* looked like a modern sea turtle, and probably lived like one as well, spending most of its time at sea and coming ashore only to lay eggs. What it ate cannot be determined, as tortoises use basically similar beaks to eat a range of items from sponges and jellyfish to apples and oranges. But describing it as flying through the sea was not mere fancy — turtles use the flippers to generate lift and thrust hydrodynamically in the same fashion that birds and pterosaurs use their wings to generate lift and thrust aerodynamically.

Notochelone has been found in a rough crescent in Queensland, from near Hughenden in the east, west to near Julia Creek, and then south beyond Boulia. A few partial bones of a larger turtle, *Cratochelone*, with a shell about 1.5 metres long, have been found near Hughenden, but there is no sign of such a large turtle elsewhere. Although marine turtles are widely distributed today, *Notochelone* has been reported only from the inland sea of Australia and so perhaps was a localized (endemic) form. On the other hand when finally studied, *Notochelone* may turn out to be related to, or even the same as, some form found overseas.

Pterosaurs, also recovered from the western Queensland Cretaceous, were flying animals and thus have very thin-walled, hollow bones, which makes the entire skeleton extremely light-weight. With the exception of a single specimen of *Rhamphorhynchus*, now in the museum at Copenhagen, pterosaur bones have been distorted after burial, usually being crushed flat. The pterosaur fossils from Queensland are neither crushed nor distorted, and hence are important not only for what they may tell us of ancient Australia, but also for what they tell of pterosaurs, which may not be learned elsewhere in the world.

Pterosaurs are creatures that have long fascinated those who knew of them. The story of their study is itself interesting, largely in the shortcomings of scientists in their striving for an objective interpretation of the past — and an example of how difficult it is to learn what afterwards seems obvious. Because pterosaurs were extinct and birds are still alive, pterosaurs were once believed to have been somehow inferior to birds. No one actually thought this out, but it was accepted tacitly and without awareness. As Mark Twain said, 'it isn't what we don't know that causes trouble, its what we "know" that isn't so'. Pterosaurs were long cited as poor fliers because they lacked the keel on the sternum (breastbone), to which major flight muscles are attached in birds. It was reasoned that since pterosaurs lacked such a keel they also lacked the well-developed flight muscles and therefore were poor fliers. As a kind of consolation prize they were conceded to have been good gliders. The argument was wrong at two points. Pterosaur sterna do have well-developed keels just as do those of birds. When a sternum is crushed flat it can be difficult to tell if it had a keel. Some sterna were not badly crushed, but were simply ignored. Secondly, bats have no keel on their sterna but are still good fliers, as anyone can see on evenings when the flying foxes are abroad.

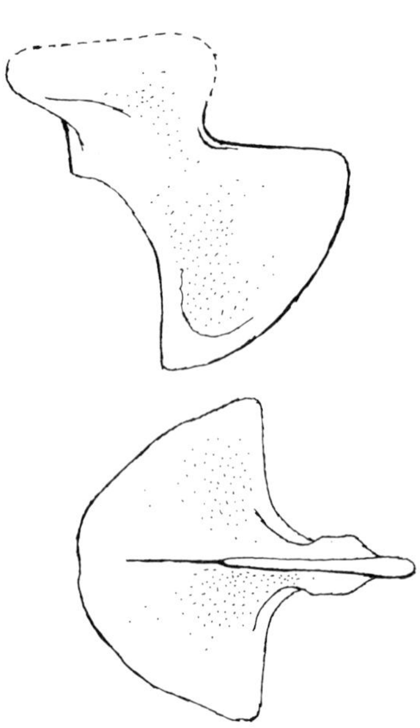

The sternum (breastbone) of the pterosaur, *Nesodactylus*, showing the strongly developed keel. The sternum is viewed from the side (top) and from below (bottom); x1.

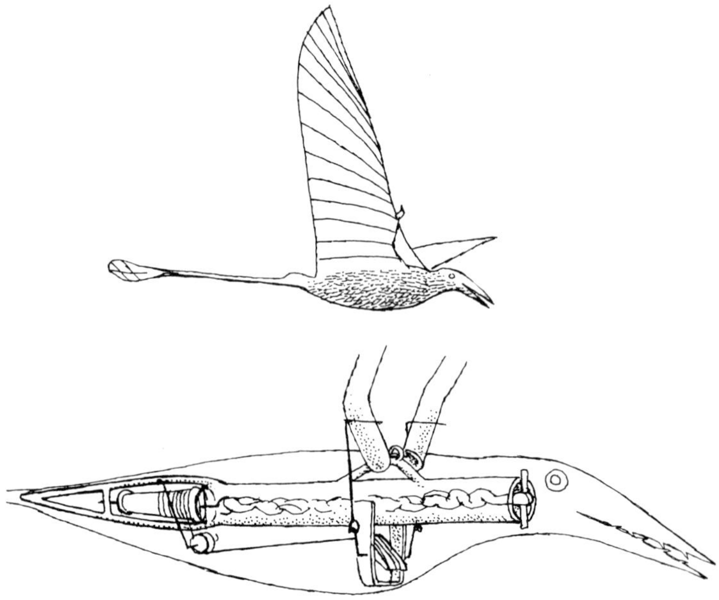

Flying model of pterosaur, constructed at the Max-Planck Institute for Comparative Physiology in Germany. This model was used in studying pterosaur flight capabilities, and was propelled by an elastic band.

This kind of unconscious bias was also involved in the story of the fur of pterosaurs. The skin of modern reptiles is usually covered with scales, just as that of mammals is usually covered with fur, and that of birds with feathers. Thus, when Goldfuss reported in 1831 that pterosaurs were covered with fur, he was ignored. Pterosaurs were known for less than fifty years at this time and had just been attributed to reptiles — in fact the whole idea of reptiles was developed at this time. Similar reports were made at the end of the nineteenth century, and were ignored.

Notochelone costata, a marine turtle and ?*Ornithocheirus*, a flying reptile, inhabitants of the great inland sea that covered much of Australia in Early Cretaceous times.

F.KNIGHT 82

Ferdinand Broili (1927), another German, again reported pterosaur fur and again was quietly ignored. Much of the reason for claiming a furry coat for pterosaurs came from microscopic examination of the skin, which was covered with small projections (Desmond, 1975) like those found in mammals at the base of hairs — although there were also reports of actually seeing hairs.

The balance was between interpreting these structures in terms of their analogy to hair in modern mammals, or in terms of their classification as reptiles, which do not have hair. The choice fell with the latter, although a long chain of deductions was involved: pterosaurs are reptiles, so even though they differ from modern forms in their skeletal structure (and hence probably their behaviour), they must have been the same in their soft structures. Since modern reptiles do not have fur, pterosaurs could not have fur, and the small projections must have some other, unknown explanation. Modern reptiles do not have wings either, but the relevance of this was not appreciated. Only when the Russian Sharov (1971) found two magnificent pterosaurs in Fergana (Soviet Central Asia) each preserved in a matting of fur, did scientific opinion finally accept that the long chain of deductions was wrong.

German work with flying models of pterosaurs indicates that the small forms may have been good fliers, while computer simulations of the large forms show that, depending upon the assumptions used, they were either good at powered flight or excellent at gliding. Other work has indicated that the aspect ratio (ratio of wing width to span) is lower than expected, i.e. the wings were narrower and hence more efficient aerodynamically. There is no evidence that pterosaurs flew any less well than birds. Pterosaurs appear to have had very low wing loadings (i.e. the amount of weight borne by the wings) and hence could remain airborne at lower speeds than comparably-sized birds. Low wing loadings also increase manoeuverability over that of comparable birds. Much recent work on pterosaurs, in English, has been done by Bramwell and Whitfield, (1974); Stein, (1975); Brower and Veinus, (1981) and Frey and Reiss, (1981).

Some pterosaurs (*Rhamphorhynchus*) had webbed feet and could doubtless swim on the surface. Some had long spear-like beaks, lined with sharp teeth (*Dorygnathus*) and likely fed on fish. Others had deep short snouts (*Criorhynchus*) like puffins. Still others (*Gnathosaurus*) had long, fine, closely-set teeth that formed a 'spoon' almost like the bill of the spoonbill, and most interesting, was one form from Argentina (*Pterodaustro*) that had multitudes of thin baleen-like teeth and probably fed on plankton at the surface (Wellnhofer, 1978).

The brain of pterosaurs, insofar as can be determined from studies of the internal structure and contours of the braincase, was very similar to that of birds. This development of the brain, together with the loss of the tail in many forms, implies great manoeuverability in flight. The necessity of continuous muscular action in flapping flight, as well as the coat of fur has suggested that pterosaurs were homoiothermic; they could keep their body temperature relatively constant physiologically, as in living mammals and birds.

Having been good fliers, pterosaurs might well be expected to have occurred over the whole globe, so their discovery in Australia was not a great surprise. The Queensland pterosaur appears to have been closely related to a moderately large, maritime pterosaur known from contemporaneous beds near Cambridge, in England (Molnar and Thulborn, 1980). This form, *Ornithocheirus* (meaning 'bird hand' — which it really is not), is unfortunately known so far only from scattered individual bones or pieces of bones, and thus has been difficult to compare in detail with the Queensland form. The similarity, however, is great enough to refer the Queensland form to ?*Ornithocheirus*, the question mark before the name signifying that although the bone fragments known match up satisfactorily, not enough is known of either form for confidence in the identification. The Queensland ?*Ornithocheirus* had an estimated wingspan of two metres, and a long beak, liberally supplied with sharp blade-like teeth, very likely used for catching fish.

The pectoral or (shoulder) girdle is basically V-shaped, with the humerus (wing) articulation at the apex and the two 'struts' contacting the trunk skeleton. The lower 'strut' (coracoid) abutted on the sternum and the upper 'strut' (scapula) on a structure formed of fused vertebrae, the notarium, not to be confused with the rather different notarium in some birds. A notarium is a structure known only from one group of pterosaurs, the pterodactyls, and served to support the trunk on the wings.

Like the shoulder girdle, the single vertebra known is hollow, with a strong, but thin, surface of lamellar bone. Internally both structures are buttressed by cross-struts of bone, very much like the structure of a bird bone, both similar to certain modes of aircraft construction. The pelvic construction likewise, is similar, with the position and form of the hip socket (acetabulum) suggesting considerable mobility of the lower limb. In most pterosaurs the lower bones of the pelvis (pubis and ischium) do not meet together at the midline of the animal (although pterosaurs have prepubic bones, which do meet on the midline), while in the large, crested *Pteranodon* both pubis and ischium do meet below to give a roughly conical form to the pelvis. The pelvis of the Queensland ?*Ornithocheirus* has the pubes meeting together below, but apparently not the ischia, although the ischia are not completely preserved. Final conclusions must await discovery of further material. This structure thus indicates a stage between the usual pterosaur condition and that of *Pteranodon*. Peter Wellnhofer of Munich, probably the most expert worker on pterosaurs, considered *Ornithocheirus* to be related to *Pteranodon*. Perhaps it is near the ancestral form.

The Queensland ?*Ornithocheirus* was probably a moderately large, fish-eating maritime form, closely related to similar forms spread across the world. Considering the different arrangement of the continents during the Early Cretaceous, this pterosaur is the most southerly known.

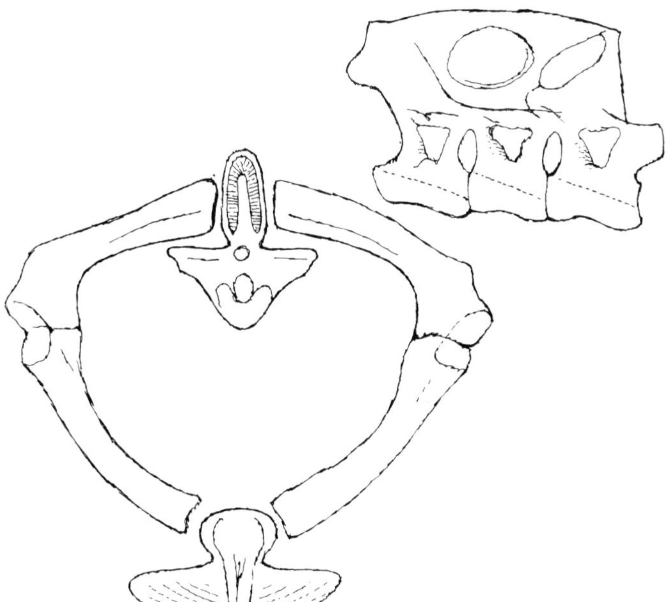

The pterosaur notarium viewed from the side (top). The lower diagram shows the relation of the notarium (at top) to the shoulder girdle (either side) and the sternum (at bottom).

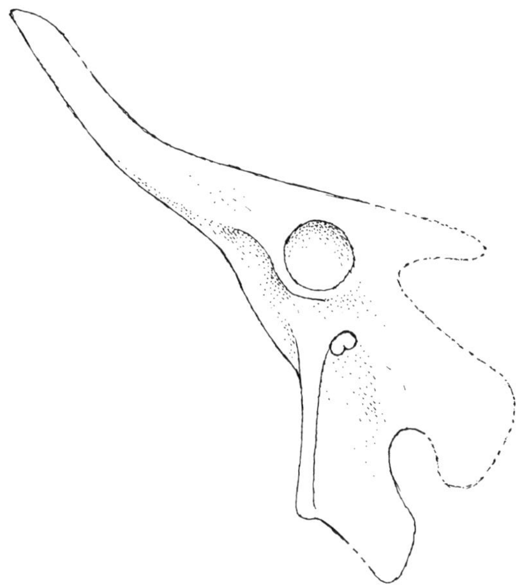

Pterosaur pelvis, viewed from the side, found in Cretaceous rocks near the Hamilton River, western Queensland; × 2

Meiolania platyceps

Owen, 1886

The Lord Howe Island Horned Turtle

CLASS: Reptilia
ORDER: Testudines
FAMILY: Meiolaniidae
GENUS: *Meiolania*
SPECIES: *Meiolania platyceps*
AGE: Pleistocene

Most vertebrate palaeontologists look for fossils in hot, dry parts of the world where the bones are easier to find weathering out of the rocks. It was therefore remarkable to find ourselves (a joint American Museum and Australian Museum crew) amid palm trees, broad white beaches, and cool ocean breezes looking for the remains of extinct turtles. The anomaly was bearable! The palm trees and beaches were on Lord Howe Island, a crescent-shaped volcanic remnant lying in the Tasman Sea about 670 kilometres ENE of Sydney, New South Wales. The island is about 13 kilometres long, rarely more than 2 kilometres wide, with a coral reef stretching across the horns of the crescent to form an idyllic lagoon.

But it is the mountains at the southern end that dominate the scene, they rise almost perpendicularly out of the sea to heights greater than 800 metres and lent an air of unreality to our searches. The 40,000 year old beach and sand dune deposits of Lord Howe Island have yielded hundreds of bones of the large, extinct turtle, *Meiolania platyceps*, which bore horns on its head, a club on its tail, and a penchant for stirring controversy among scientists for the past one hundred years. *Meiolania* means the 'lesser ripper' as compared with *Megalania*, the 'great ripper'; and *platyceps* means 'flat-headed'. The grotesque features of *Meiolania* have spawned arguments about its relationships to other turtles, while the distribution of *Meiolania* and its near relatives in Australia, Lord Howe Island, New Caledonia and South America has caused speculation about possible connections among these areas.

A horned turtle first made its debut upon the palaeontological scene in 1880 disguised as a giant horned lizard. Earlier, in 1858, the great British anatomist, Sir Richard Owen, had described some gigantic fossil lizard vertebrae from the Darling Downs, a rich agricultural region in southeastern Queensland. When, in 1879, more fossil reptile material was sent to him from the same region, he accepted the association of the lizard vertebrae with a newly collected skull (Owen, 1881, 1882b). In Owen's defence one must say that the skull was quite unlike any then known turtle

---

Fossils of *Meiolania platyceps* have been found in Pleistocene beach and lagoonal sediments on Lord Howe Island, a Miocene volcanic remnant in the Tasman Sea. This reconstruction of *Meiolania* is based on the first mounted skeleton, produced in 1981 at the American Museum of Natural History. The colour is speculative and derived from that of the large land tortoises of the Galapagos Islands. The environment was similar to that on the island at present (without the introduced species), but the Lord Howe Island pigeon and the Tasman booby are now gone.

F. KNIGHT 82

(or anything else for that matter), and he naturally looked at the Recent Australian fauna, which contained a fierce-looking but small lizard, *Moloch horridus*, which did have various horn-like excrescences. These were formed by epidermal scales, not bones. Nonetheless, he would have done well to pay a bit more attention to the person who sent him the specimen, Dr. George F. Bennett, who wrote:

> 'On examining the fossil skull. . . . . . I consider it Reptilian, and at first sight to be a Turtle; but on further inspection there are some points which are against that opinion.'

Just like a scientist; always looking at something in detail and confusing the issue.

The Darling Downs skull was followed in 1880 by a tail club from the same locality. The tail club, although matching the skull in its departure from normalcy, had no close parallel in other vertebrate groups and Owen added it to his giant horned lizard. There is some suggestion that during his later years Sir Richard (Owen, 1886) had lost some of his usual anatomical acumen, because he added the foot bones of a giant marsupial, *Diprotodon*, to the lizard-turtle mix, a few years after the tail club.

Up to this time (January 1886) there was still no notion of a horned turtle, but in April of that year Owen announced the discovery on Lord Howe Island of another horned lizard, very similar to the Queensland specimen. In retrospect this discovery was of great significance to an understanding of *Meiolania*, because another hundred years of fossil collecting in the Darling Downs has not turned up another scrap of a horned turtle, but Lord Howe Island has yielded hundreds of bones, including three partial skeletons and seven skulls.

In 1887, Darwin's 'bulldog', Sir Thomas Henry Huxley, took a bite out of Sir Richard by using the new material from Lord Howe Island to show that *Meiolania* was a turtle with similarities to the Recent snapping turtles. The new material included two skulls and, even with its grotesque horns and lumps, it was hard for Owen to deny that *Meiolania* had something to do with turtles.

Owen compromised and created an intermediate group, a favorite tactic of the evolutionists he was combatting, which had features of both lizards and turtles. His peers just ignored it. Owen died a few years later, in 1892, and although his composite lizard-turtle-marsupial is often the butt of old-time fossil hunter's stories, it should not mislead one into doubting Owen's formidable intellect. Owen had gone down in infamy for being the bad guy in that Victorian social phenomenon, the evolution debate. It can, however, be justifiably argued that Owen's contribution to vertebrate zoology was at least as significant as Darwin's; it was the kind of work that Owen did, meticulous description and wide-ranging comparisons, that advanced the search for natural groups.

With Huxley's identification of *Meiolania* as a turtle it was not long before the next round of argument brewed up, this time a bitter fracas between two herpetologists, George Baur and George Boulenger. The dispute was about the relationships of *Meiolania* within the turtles, a dispute that is still with us today and is the reason for my own entry into the '*Meiolania* stakes.' One of the more apparent characteristics of Recent turtles is their ability to withdraw their heads, at least to some extent, into their shells. Turtles can be divided into two groups based on the two types of neck retraction. In one group, which has come to be known as cryptodires ('hidden necks'), the neck is folded into a vertical plane, while in the other group, pleurodires (or 'side-necked'), the neck is folded horizontally. In the cryptodires the folded neck and usually the head are actually withdrawn inside the body wall, but in pleurodires the neck and head lie alongside the body wall beneath the shell overhang. Other features, particularly in the skull and pelvis, are coincident with the distribution of the neck retraction types and are further tests of the 'naturalness' or reality of the two groups.

Baur argued that *Meiolania* was a cryptodire while Boulenger argued it was a pleurodire. This would seem to be a problem that could be solved without too much trouble by looking at a neck of *Meiolania*. Unfortunately, in the late 1880s there were no specimens of a *Meiolania* neck and, more importantly, even a cursory examination of the horns on a *Meiolania* skull would show that either form of neck retraction would result in a kind of chelonian 'hara-kiri.' However, the Baur-Boulenger debate produced only insults, and no new information on the animal.

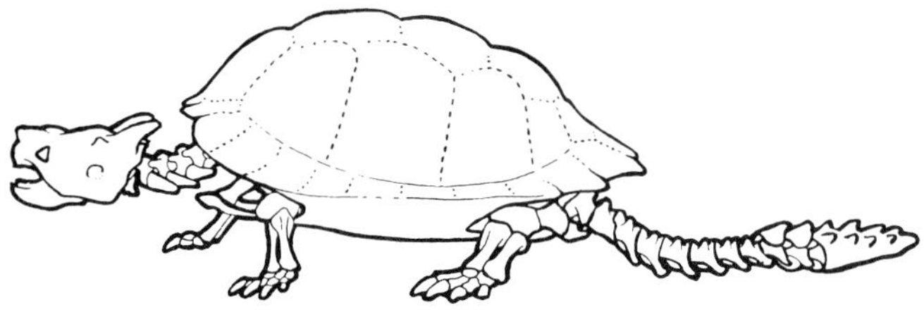

Side view of the restored skeleton of *Meiolania platyceps*, consisting mainly of Australian Museum specimen No. 57984, found in a swimming pool at Ocean View Guest House, Lord Howe Island. Length about 2 metres.

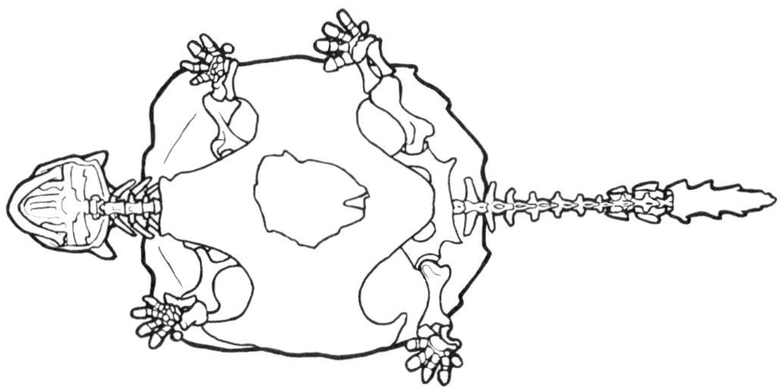

Bottom view of same specimen.

The early part of the twentieth century was characterized by a steady influx of *Meiolania* specimens from the Lord Howe Island residents, who have always had an interested and enlightened attitude toward their fossil turtle. This material, logically housed at the Australian Museum in Sydney, was described in 1925 and 1930 by C.A. Anderson, the Director of that museum. Anderson, and later in 1938 G.G. Simpson of the American Museum of Natural History, concluded that *Meiolania* was neither pleurodire nor cryptodire, rather it was a long surviving 'relic' of the ancestral turtle. Unfortunately, 'relics' often come about from ignorance, and in my own phylogenetic work on turtles (Gaffney, 1975) I have argued that *Meiolania* is a cryptodire that has retained some primitive features, but it is not a near relative of the ancestral turtle.

In addition to the neck retraction mechanism, cryptodires and pleurodires can be characterized by the pattern of arterial circulation in the head and by the jaw muscle mechanism (Gaffney, 1975). In fact, these cranial features evolved before the typical neck retraction features but after the cryptodire and pleurodire lineages split. *Meiolania* is now known from more than half a dozen skulls and many skull fragments, so that the condition of the arterial pattern and the jaw muscle mechanism is readily determinable. These features in *Meiolania* are the same as in the cryptodires and allow identification of *Meiolania* as a member of this group. The issue of *Meiolania* relationships has been confused by the retention in *Meiolania* of a number of primitive features and by the addition of unusual characters. The presence of well developed cervical or neck ribs is one of the more striking primitive characters. The earliest turtles from the Triassic have cervical ribs, but nearly all other turtles have lost them. Similarly, only the Triassic turtles and *Meiolania* have a tail club. But the grotesque development of cranial horns and similar excrescenses is a feature peculiar to the meiolaniids.

All of these characters combine to make a very bizarre looking animal. The shell is not particularly unusual. There are lumpy, ill-defined scute areas much like modern turtles, and the back edge has serrations. The skull has paired, cow-like horns and other variably developed lumps also covered by horny scutes. The neck is short and apparently was not covered with dermal armour. The legs are short with elephant-like feet rather than flippers or webs between the toes, and are covered with small bony plates and spines that form an extensive armour. The tail is relatively long and ends in a large, bony club, while the rest of the tail is covered with rings of bone that provide protection but allow movement.

What was this formidable armoured creature protecting itself against? The only other large vertebrates on Lord Howe Island were birds, none of which could have threatened a *Meiolania*. The mainland form, *Meiolania oweni*, could certainly have used protection against the giant lizard, *Megalania*, as well as crocodiles, but the Lord Howe Island species could only have been endangered by its own kind, and one can speculate on fanciful mating battles.

The reconstruction here is based on the first mounted skeleton of *Meiolania* produced in 1981 at the American Museum of Natural History in New York. Casts were made of the skeleton to make mounting easier and to provide more than one museum with a skeleton. The skeleton chosen for the cast was collected in 1970 from a swimming pool excavation at the Ocean View Guest House on Lord Howe Island. Large parts of the shell had to be restored from other specimens, but other than the exact number of tail vertebrae, virtually all the features of the *Meiolania platyceps* skeleton are known. No other species of meiolaniid is well enough known to make a reconstruction, and so I have chosen *Meiolania platyceps*.

Although work on dating the beach deposits that contain fossils of *Meiolania platyceps* on Lord Howe Island is very incomplete, it is likely that the species was living there when sea level was very close to or below its present position, perhaps 40,000-80,000 years ago. (McDougall, Embleton and Stone, 1981). The island would have looked much like it does now, but perhaps with more sandy beaches generating sand dunes behind the beaches. The ecological niche of *Meiolania* is unknown, but it seems to have been a terrestrial turtle very similar to the tortoises of the Galapagos and Aldabra Islands. The vegetation was probably not much different from the present Lord Howe Island vegetation, only lacking introduced forms, such as the Norfolk Island pine.

Platypterygius australis

(McCoy) 1867

A Cretaceous Marine Reptile

CLASS: Reptilia
ORDER: Ichthyosauria
SUBORDER: Longipinnatina
GENUS: *Platypterygius*
SPECIES: *Platypterygius australis*
AGE: Cretaceous

The name ichthyosaur was compounded from the ancient Greek words for 'fish' and 'lizard'; it records the excitement with which the first author discovered that he was looking at a 'saurian', or reptile, the shape of a shark, except for a snout like a long-nosed dolphin. Later workers have been more excited by the dolphin-like head. It often seems no elementary biology text is thought to be complete without a sketch of the three marine animals that show the convergence in external form between fish, reptile (the ichthyosaur) and mammal that lead somewhat similar lives. In defiance of that, the suggestion has repeatedly been made that ichthyosaurs were slow swimmers and inept divers. That seems most unlikely as the gut contents found in these reptiles are remains of fish, belemnites, squid and ammonites. The first three were agile and rapid swimmers, and ammonites were not surface dwellers; it seems that ineptness, like beauty, is in the eye of the beholder.

Reptiles of the ichthyosaur genus *Platypterygius* were distributed almost world-wide during the Aptian, Albian and Cenomanian stages of the Cretaceous Period, about 110-90 million years ago. One species, *P. australis,* lived in the shallow sea that occupied a quarter of Australia, the Great Artesian Basin, during the late Aptian and Albian time. *P. australis* was particularly common in the Albian, the final stage of Early Cretaceous, but the basin was filled to above sea level with sediment, and became non-marine a little less than 100 million years ago; just as the Late Cretaceous Cenomanian stage commenced.

The first Australian ichthyosaur was discovered in 1865 when James Sutherland was on a collecting trip in North Queensland. He wrote to Professor McCoy, Director of the National Museum of Victoria, from Marathon Station that he had despatched a box of fossils, the most significant of which was a series of 'several vertebrae' about 3 inches (75 mm) by 4 inches (100 mm), which he considered 'saurian'. In the leisurely haste of scientific publication, aggravated by sailing ship mail runs, McCoy's publication of the name '*Ichthyosaurus australis*' reached the English readers in 1867, about the time he received the fruits of Sutherland's follow-up search. Sutherland wrote again to McCoy from Marathon Station, 5 Nov. 1866:

'In Augt. last I sent a box of geological specimens for the National Museum; in this Collection was a specimen numbered 48 (five vertebral joints).

I this week send another case and in it a great many of the Specimens are also numbered 48; all of the same No. were found together and probably belong to one organisation.

The specimens numbered 60 were likewise found together.'

As in 1867, McCoy (1869) was worse than casual in his description of the new material; it is possible to judge that he had material from more than one animal, but not that there were two skulls. Perhaps that is to be expected of a man who spelt his own name two ways. He and Sutherland used M'Coy at this period, but later workers standardized on his other usage, 'McCoy'.

It was the skull fragment numbered 60,which Frederick Chapman figured in 1914 (1914b), together with a forefin fragment, which has no collection number now. The skull numbered 48 has never been illustrated, but now is being prepared. The original five vertebrae numbered 48, according to McCoy (1867) each about '4 inches wide, 3 inches deep and 1½ inch long', cannot be recognized among the 36 vertebrae numbered 48 at the National Museum of Victoria (now the Museum of Victoria), because all 36 are laterally squashed and higher than wide. Only two pairs are as big as 4 inches high (deep) and 3 inches wide. They are upper tail vertebrae with one process on each side to which the single-headed tail ribs attached. The smaller chest and neck vertebrae had paired knobs to which were attached double-headed ribs. Sutherland's use of 'probably' becomes apparent — however small the chance that the larger, scattered vertebrae came from a different animal, there is an indisputable break in the sequence. The loose material that led him to the sequentially arranged vertebrae and head cannot be shown to have come from the same animal.

The best fossil skeleton of an Australian ichthyosaur is the famous Telemon specimen, the most complete Cretaceous ichthyosaur the world has on view. This was collected and given a first round of preparation by J. Edgar Young, at Telemon Station in 1934, and presented to the Queensland Museum. There it has undergone further preparation. It is the skeleton of a young adult about 5.6 metres long. Its bones are rather large and strongly built for a moderate-sized ichthyosaur. They are not very distorted, although the skull is squashed obliquely and has lost about 10 cm from its snout tip. We are fortunate in having a better skull (Longman, 1922), which was

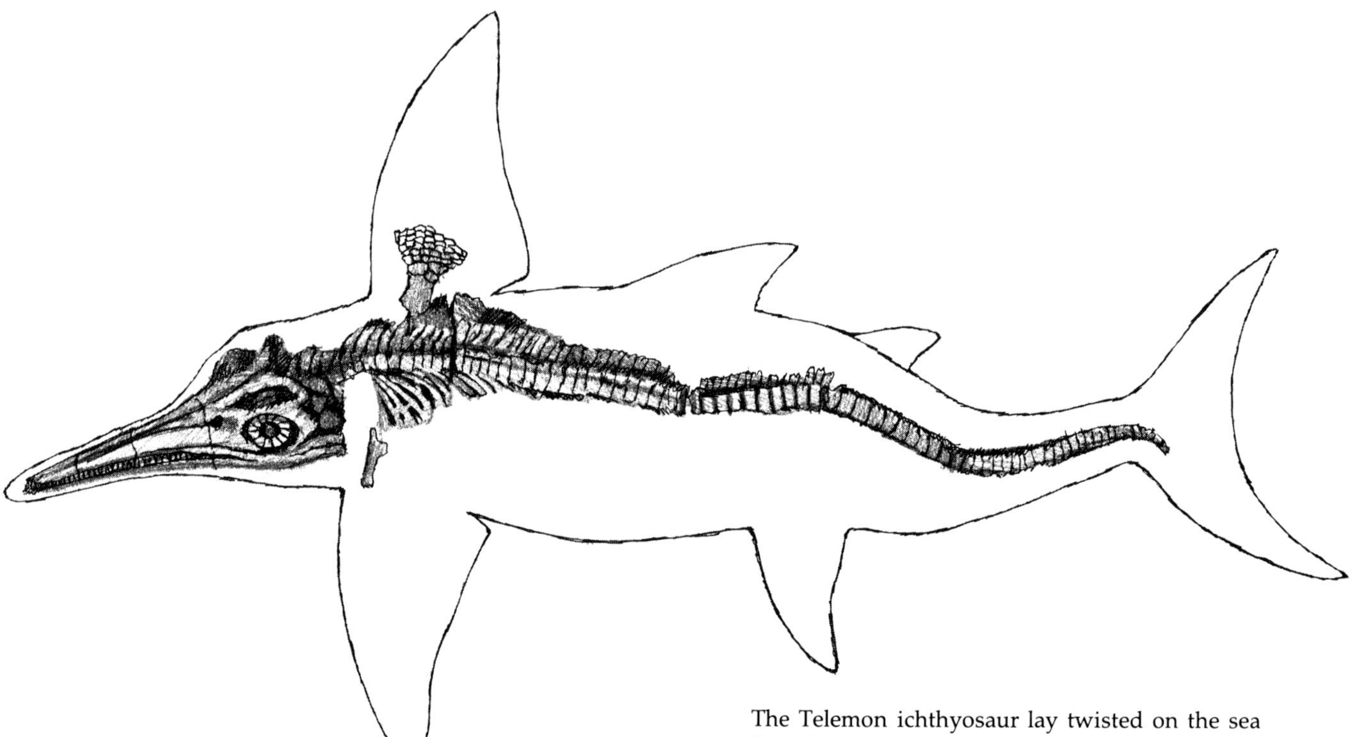

The Telemon ichthyosaur lay twisted on the sea floor before burial. There it was torn in two by scavengers; length 5.6m.

slightly compressed from top to bottom. By combining the knowledge derived from both these heads, the Telemon body, three more fragmentary skeletons, and several more fragments, a very good picture of the animal emerges. It was beautifully streamlined besides being powerful, and so appears to have been a rapid swimmer. Its head is about one fourth of its total length, and the skull bones are relatively heavy. The forefins were large and fitted in a near-horizontal position at the base of the chest; like most shark forefins. The hind fins are very poorly known as yet, for only dissociated elements have been found. The tail, exclusive of fin, was about one fourth of the body-length. The tail fin size and shape has to be estimated, for only the European *Platypterygius platydactylus* had a significant amount of its skeleton preserved. This slightly older specimen had a string of 38 tail vertebrae preserved, but both ends of it were missing. As a result we do not know the exact size, or the angle at which the fin diverged from the tail, although it is customarily shown oriented at 50°, as in the first reconstruction (Broili, 1907), and at minimum size, with 60-70 cm of tail fin vertebrae, i.e. a lower lobe 70 cm long, and upper lobe nearly as large. The overall length of *P. platydactylus* was about 5.2 metres; that it was only a large juvenile is shown by the rather narrow-chested shoulder girdle.

The Richmond partial skeleton was collected by scientists who recorded its relationship to the rocks in which it was deposited, and thus much is known of its history from corpse to fossil. As often happens with large vertebrate corpses at sea, it apparently sank, floated again as gas accumulated in the body, and then rotted until it de-gassed, so rapidly in fact, that it probably exploded. The remains plunged to the sea floor head first because of the weighty skull bones. It rammed almost the full-length of the snout into a sediment of shelly lime mud, and anchored there. As it hung in the water it continued to disintegrate, and the forefins fell to the sea floor to each side of the head. The right fin had lost its arm and wrist at some stage, and lay flat. It has a well-preserved fin-blade, with 9-10 tightly appressed digits, and the number of finger bones (or phalanges) in the longer digits exceeds 30. The left forelimb retained its arm and lay substantially on edge. Some of its digits overlie others and this makes interpretation difficult. All but one of the neck and chest vertebrae are missing, and so is one entire shoulder. These were heavy bones, and if any bones continued to lie on the sea floor, they should have. There are various possible scenarios for their removal. One is that they fell out of the floating carcass, before it sank; a second is that some large fish, perhaps a shark, tore off a great lump of shoulder, backbone and skull as the carcass hung anchored; or possibly a combination of both occurred. The remainder of the bones fell to the sea floor in an eccentric scatter around the snout, as the body decayed. Preservational history does not stop at that. The bones were covered in more of the shelly lime mud, and as that mud consolidated, and the mud below the bones consolidated further, compaction flattened all the scattered bones. The rear part of the lower jaws, all that remained of the head except for the buried snout, stood up in the slowly compacting sediment and shattered instead of distorting. The splinters still stand up at an angle. The tip of the snout drove against harder layers below and, probably because it was damaged in the initial penetration, cracked and expanded like a bullet while the more solid central parts retained their initial shape. A total length of 60 cm of mid-snout was preserved vertical in the compressed rock.

The plasticity of ichthyosaur bones under pressure makes 'absolute' characters the valuable ones when a scientist tries to classify the fossil form, e.g. the number of bones that articulate with the humerus (the upper arm bone) and which bones they are; the number of digits in a *complete* fin blade, and which bones they are attached to at the wrist (so that partial fin blades can be compared).

Ichthyosaurs have a long history prior to the occurrence of *Platypterygius*, and as a group are varied. The mixosaurs, the most primitive ichthyosaurs yet found, are set apart by their tail vertebrae and limb girdles (skeletal supports) as well as fin bones and details of the skull. The oldest Triassic ichthyosaur fossils yet collected belong amongst the well-established group of longipinnate ichthyosaurs. They are much more specialized than the next group to appear in the fossil record (the mixosaurs) and much larger. This is the probable reason for the order of discovery: not only are big objects more eye-catching, but smaller skeletons are constructed of smaller and weaker bones; and more open to attack by scavengers, to scattering by water movement, or dissolution after burial. The oldest mixosaurs recorded were already 'living fossils' of Triassic time, and they had died out by the end of the period.

Besides the other skeletal differences, mixosaur limb bones were still proportioned somewhat like ordinary reptiles, and the forelimbs still had five fingers or digits, although the number of bones in the digits had started to multiply, indicating fins instead of feet. All other known ichthyosaurs had not only the wrist but also the forearm elements shortened and broadened into typical fin elements. The newly-developed 'ichthyosaurian' bones are shown in black in the figure.

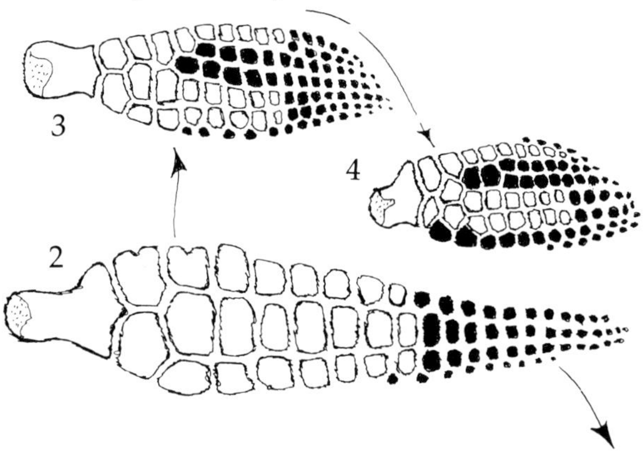

Large Longipinnatina (2) are the oldest known Ichthyosaurida. They gave rise to Heteropinnatina (3) in the Late Triassic, and these gave rise to Latipinnatina (4) in the Early Jurassic. Later, the line leading to *Platypterygius* (5) also originated from the Longipinnatina (2). All had flippers or fins in which only the bones shown in outline equate to those of ancestral reptiles. Strictly ichthyosaurian finger bones are shown in black. Intermediate forms between 2-3-4 occur, and between 2 and 5, but there are none between 3 to 4 and 5.

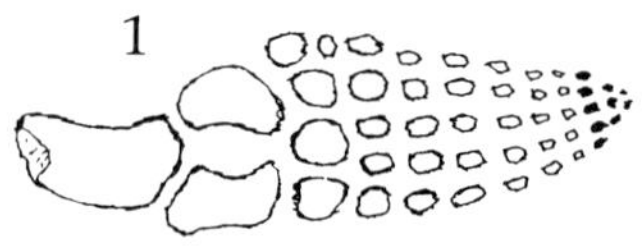

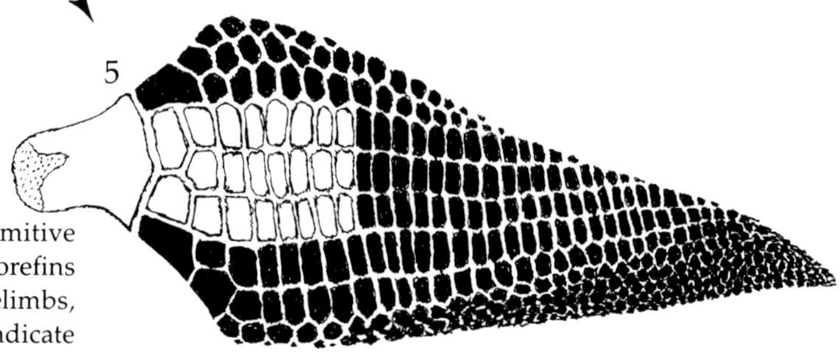

Tiny Mixosaurida (1) were the most primitive ichthyosaurs known. The bones in their forefins were reminiscent of ordinary reptilian forelimbs, although new bones (shown in black) indicate flipper tips, not toes.

The long-ranging longipinnate group differed from the mixosaurs by elongating the fingers and toes (digits) with extra bones while they narrowed the finblades, wrists and ankles, by the loss of other bones. Near the end of the Triassic and the beginning of the Jurassic (Appleby, 1979), some of the longipinnates gave rise to forms with inner digits that branched at various distances from the wrist, thus broadening the fins. Appleby called these heteropinnates. They were the forerunners of the latipinnates, broad-finned ichthyosaurs, some of which have left fossilized soft parts. All these and the modified longipinnate *Stenopterygius* had a broad attachment of the fins to the body, like most sharks.

Platypterygius australis, an icthyosaur from the Cretaceous shallow sea that flooded onto the Australian continent. An agile, rapid swimmer, this species grew to 6 or 7 metres. Here *P. australis* has caught a squid-like belemnite, a group that became extinct in the Eocene period.

F. KNIGHT

Platypterygius was derived from another longipinnate ancestor after the hetero-pinnates were extinct, also by broadening the fins. Fins were broadened by the addition of accessory digits on the front and back edges of the fins. Simultaneously, supporting accessory bones were formed or moved into position on each side of the wrist and forearm bones (radius and ulna). The result was a 4-bone wide forearm. As the accessory bones reached the humerus, those that could fit became inserted into newly-developed sockets on the end of the humerus, or to put it more simply, articulated with it. *P. australis* had the ulnar accessory bone on the trailing edge of the fin articulated to the humerus, a characteristic that was at least foreshadowed in *P. platydactylus*. Instead of this, *P. americanus* had the radial accessory bone on the leading edge of the fin articulated to the humerus. Exceptionally hefty humeri of *P. australis*, juvenile or adult, were proportionately large enough to afford a small attachment for the leading edge accessory bone as well, but this is classed as 'individual variation', while the strong preference for one bone or the other to have been articulated is a 'species difference' that justifies the names *australis* and *americanus*.

In the original, and often reproduced, diagram of *P. platydactylus*, Broili (1907) indicated 8 digits by very small stippled arcs; at the time this was far-sighted and courageous. Although ichthyosaur literature is full of amazement about the 8 digits preserved there, 9 digits would better finish the blade in normal-sized phalanges. McGowan (1972) thought the Stewart Park *P. australis* fin was almost complete, although the thickness, shape and arrangement of the phalanges in the most posterior digit preserved indicated a broken trailing edge, and the length of the Richmond fin blade subsequently indicated that 2/3 of its length had also been broken away from the outer end of the Stewart Park fin. McGowan (1972) restored the limbs with a narrow attachment. I have come to think that the masses of lateral accessory bones indicate wide areas of cartilage each side of the fin, and a broad fin attachment, such as is sketched around the skeleton in the figure. Other workers have not yet had the opportunity to assess the two restorations, so both viewpoints are now 'current theories' and will remain so for some years at least.

The shapes of ichthyosaur tail fins tended to produce forward and downward movement; this downward tendency was countered by the slight upward set of the limbs, which were also used for turning, soaring or diving in underwater movement. The trim of the fins was crucial for steering at speed. Ichthyosaurs are therefore unlikely to have used the fins for paddling except when they were used to remain in one spot or 'hold station'. Fins also probably aided in regulating speed when feeding. The only marine animals that can gape while moving fast under water are those that have gill-openings to let the water out at the back (fish and some amphibians), so ichthyosaurs probably checked their speed as they struck prey.

It is not certain whether *P. australis* ever lived outside the Great Artesian Basin sea, for the ichthyosaur fossils so far known from Darwin and Western Australia are not diagnostic enough to be assigned to a genus. *P. australis* could have been the species which was spread around Australia's continental shelf. If it does occur on the shelf and not more widely, it will provide an argument in favour of the opinion that ichthyosaurs were largely confined to shelf depths of water. This suggestion is usually made together with the theory that they were poor swimmers, but perhaps retreating to the sea floor in bad weather was essential to them, and they could not survive oceanic gales without rest at the surface or on a shallow sea floor. Modern sea-going crocodiles are confined to inner continental shelves by this requirement, and only one species of sea-snake has become pelagic. Their depth limits for resting on the sea floor are yet unknown, for our own capability as underwater observers is limited. Ichthyosaurs were very capable marine reptiles, but oceanic crossings may have been rare.

Plesiosauridae
The Andamooka Sea Monster

CLASS: Reptilia
ORDER: Sauropterygia
SUBORDER: Plesiosauria
FAMILY: Plesiosauridae
GENUS: Unknown
SPECIES: Unknown
AGE: Early Cretaceous

Andamooka is a small scatter of buildings in a dusty, arid landscape just west of Lake Torrens in South Australia, some 640 kilometres north of Adelaide. From beneath the red plain that has been affected chemically in times past (during the Tertiary Period 65 – 2 million year ago) comes an unexpected prize, the precious opal that has made Australia famous to the connoisseur of high quality gems.

Opal is similar to quartz in make-up. It is a form of silica ($SiO_2\ nH_2O$), but it traps a variable amount of water (H_2O) in its structure. The amorphous silica in opal occurs as small spheres, organized in a regular pattern (Barnes and Townsend, 1982). Water is trapped in these empty spaces between the silica when cementation of the spheres occurs, and it is the regular arrangement of these spheres and spaces which diffracts light, thus breaking it into an array of spectral colours, giving rise to the 'fire' in opal. Colour itself depends on the size of the silica spheres; the greater the diameter, the further towards the red end of the spectrum. Breaks in the stacking patterns of the spheres, which occur during the growth of the opal from a colloidal solution, produces the abrupt changes in colour at different spots within an individual stone.

Very rarely opal replaces the organic material of animals that have died and been fossilized. Such specimens have not only a monetary value, but are of marked scientific interest as well. At Andamooka and Cooper Pedy, another opal town in South Australia, many different kinds of opalized fossils have been found: wood, gastropods (snails), belemnites (relatives of the living octopus and squid, but with a hard, internal skeleton), crinoids (sea lilies), as well as several different kinds of vertebrates. Among the vertebrates found at Andamooka are three different kinds of plesiosaurs and a dolphin-like ichthyosaur (also a marine reptile), as well as a few bones of a small carnivorous, theropod dinosaur. All of these groups were extinct by 65 million years ago.

Most of the vertebrates and invertebrates from Andamooka lived in a marine environment, which deposited the Early Cretaceous (110 million years ago) Marree Formation, the rock unit that bears the opal and opalized fossils. So, at the time these animals lived, Andamooka lay under a shallow, epicontinental (on-continent) sea that covered much of Australia, dividing it up into a series of islands. Because of the presence of the little dinosaur bones (*Kakaru*) (Molnar and Pledge, 1980) and fossil wood, land must have been nearby. The fragments of terrestrial animals were rafted out, perhaps from the mouth of a river emptying into the shallow epicontinental sea, and dropped on the sea floor along with the more complete remains of the marine forms.

The largest opalized vertebrate fossil ever to have been found at Andamooka is a real sea monster, discovered late in 1983 by miners John St. Alban and Stephan Bilka, who were bulldozing a new cut in search of saleable opal on Lunatic Hill. The fossil they discovered is a plesiosaur, a reptile highly adapted for a life in the ocean. Their fossil, nicknamed 'Plesie' by the locals, was very seal-like in habits, having its four limbs modified into paddles, no longer of much use on land. The Andamooka sea monster probably had a long neck with a small head, unlike a second group to which it was more distantly related, the pliosauroid plesiosaurs. The pliosaurs, such as the famous *Kronosaurus* known from the Cretaceous of Queensland, had rather short necks and very large heads.

Plesiosaurs used the broad paddles much as living sea turtles do and as rowing crews use oars. The paddles acted as hydrofoils, producing thrust on the backstroke and lift during the recovery stroke. These reptiles swallowed stones (gastroliths or 'stomach stones') perhaps in part to provide ballast that allowed them to sink despite the bouyancy brought about by air in their lungs. The stones may have been more important in helping to grind up food, in much the same way as 'crop stones' are used in birds, such as chickens. Apparently their backbone was not rigidly held in one position, but could be bowed, so that their lungs could be inflated were they to move out of the water. Some palaeontologists think that, unlike the contemporary ichthyosaurs, some plesiosaurs may have been able to come ashore at times, much as seals and sea turtles do, to give birth or lay eggs or even to sun themselves or escape aquatic predators.

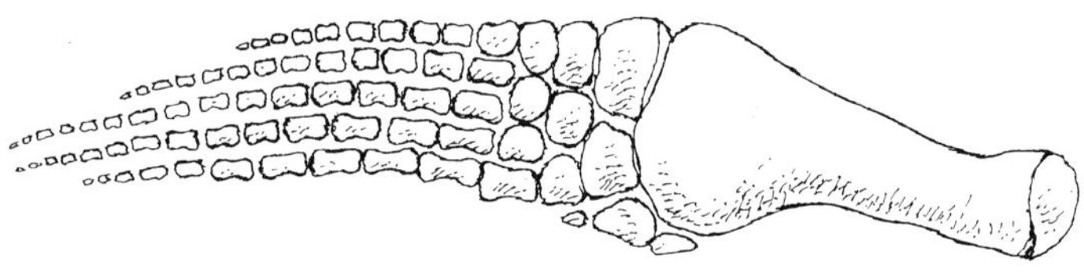

Paddle of a plesiosaur showing how a tetrapod limb has been altered into an efficient swimming organ. This paddle exhibits hyperphalangy, an increase in the number of bones of each digit, which results in a greater surface area of the paddle.

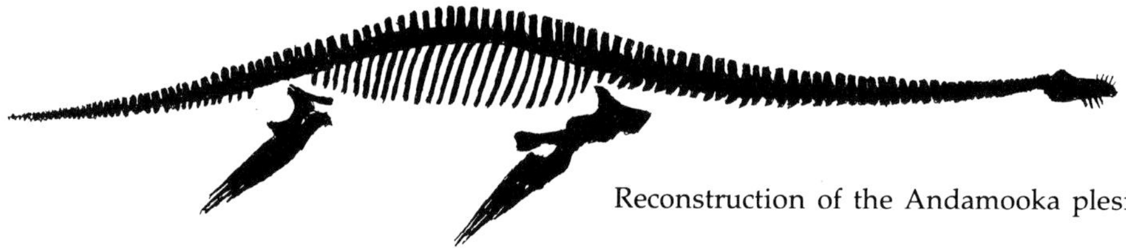

Reconstruction of the Andamooka plesiosaur.

Plesiosaurs, being mainly marine animals (although some invaded freshwater), are known from around the world. They have unique vertebrae, with a pair of foramina (or holes) on the bottom, unlike those in any other known reptiles. The bones of the paddles have undergone 'hyperphalangy', each of the five fingers or toes (digits) have many more than the usual number of bones in them when compared to other vertebrate animals, and unlike the toe and finger elements of ichthyosaurs, each

---

The Andamooka plesiosaur and its young in the cool, shallow sea that lapped onto the Australian continent in the Early Cretaceous. The female is gathering stones from the bottom, which are often found inside fossilized skeletons of these animals. The juvenile is watching a school of fish, which probably formed the main food of this species.

F. KNIGHT 84

bone has an hour-glass shape, with a major constriction in the middle. Once a person has seen the vertebrae or toe bones of plesiosaurs, these bones can easily be picked from those of other reptiles.

The discovery of isolated plesiosaur bones is not rare in Australia (Rich and Thompson, 1982). But good specimens, like that from Andamooka, are rare. Two different plesiosaurs, *Cimoliasaurus* and *Woolungasaurus*, are reasonably well known in Australia, coming from Queensland and New South Wales, and recently perhaps from South Australia. *Cimoliasaurus*, a cimoliasaur, had relatives in North America and New Zealand and was probably an animal with a short neck and elongate, slender, recurved teeth, excellent for catching fish. Some forms, such as the specimen, also opalized, from White Cliffs, N.S.W., were quite small, 3-4 metres in total length.

Woolungasaurus belonged to yet another group, the elasmosaurs, which had gracefully long necks, and to this form the Andamooka 'Plesie' may be related.

The Andamooka plesiosaur is not complete but is represented by all four paddles, much of the backbone from the base of the neck to the tail, part of the shoulder girdle and ribs. Perhaps there is more of it yet to be found, but thus far the skull and neck have eluded excavators. The total length of the skeleton would have been about 10 metres, which is within the size range of one form known only from bits and pieces elsewhere in Australia, *Woolungasaurus glendowerensis*. Only time, tedious preparation, and detailed comparison will later allow 'Plesie's' true identity to be determined — it may be *Woolungasaurus* or some entirely new form yet unreported from Australia.

It was fortunate that John St. Alban and Stephan Bilka were able to work together with a team of palaeontologists from several museums; the South Australian Museum (Adelaide), the Museum of Victoria (Melbourne), and the Institute of Vertebrate Palaeontology and Palaeoanthropology (Beijing, Peoples' Republic of China). The miners had the foresight to seek technical help, and this meant that the specimen could be removed from the rock in an associated context by someone who understood the anatomy of the fossil animal. Thus, all of the bony elements were collected that were preserved as fossils and exposed by the mining operation.

The main concern during the collecting of 'Plesie' was to damage it as little as possible during excavation and transport, to preserve the association of the bones. To do this, the Andamooka fossil was collected using the plaster-jacketing technique (see final chapter), which means that the bones were encased in plaster like that put on a broken leg. This stabilized the bones during transport and allowed later meticulous preparation back in the laboratory where conditions are far better — no flies, no heat and plenty of preservatives.

Before a plaster jacket could be applied it was first necessary to expose the fossil to determine its total extent. Then the bones were tightly covered with wet toilet tissue, that acted as a separator of the bone from the plaster, followed by hessian strips soaked briefly in liquid plaster. These strips were used to tightly encase the bones. Once the plaster set, a deeper trench was dug around the fossil, and then the whole block undercut. Finally it was rolled over and a plaster cap applied to completely encase the fossil. Only then was 'Plesie' ready to be transported to Adelaide.

This was the procedure used to 'catch' and remove the Andamooka monster. We know this monster was real, unlike the elusive 'Nessie' of Loch Ness in Scotland. 'Plesie' has now been donated to the South Australian Museum and is to be prepared and studied by palaeontologists who will determine its relationships to other plesiosaurs. Then it will be put on display for amateur and professional alike to appreciate for its sheer beauty as well as its scientific merit. This is one situation, where the cooperation between miners and palaeontologists of Australia and China has saved a unique and beautiful specimen.

Kronosaurus queenslandicus

Longman, 1930

The Giant Fish Lizard

CLASS: Reptilia
ORDER: Sauropterygia
SUBORDER: Plesiosauria
FAMILY: Pliosauridae
GENUS: *Kronosaurus*
SPECIES: *Kronosaurus queenslandicus*
AGE: Early Cretaceous

The rolling downs grasslands of central Queensland are not the ideal prospecting grounds for fossil hunters. Although this area, the Great Artesian Basin, covers a vast area of Queensland and even extends into the north of New South Wales and South Australia, there are very limited exposures of rock, and the very nature of the Flinders and Mitchell grass cover, except in drought, tends to mask what is there. The underlying rocks are tilted at a very low angle towards the centre of the Basin and erosion is very slow, increasing the depth of the weathered zone and often destroying any fossils which are released at the surface. Further, the number of streams cutting through the upper sedimentary rock layers is very limited.

It is, therefore, not surprising that the fossilized remains of animals preserved in these sediments are often fragmentary, damaged by weathering, or, when luck prevails, solidly encased in resistant concretions. These latter structures are inorganic in origin and result from water-borne chemicals — in this case calcium carbonate — being deposited around a nucleus, which is occasionally a fossil and, more rarely, a vertebrate fossil. Yet, these sediments have been the source of a large number of the more spectacular fossil reptiles found in Australia.

The Great Artesian Basin was a giant downwarp in the continent, which varied from being a freshwater inland sea to a vast, shallow embayment into which the ocean penetrated from the north. Muds and sands thousands of metres thick accumulated in the Great Artesian Basin as it gradually sank and, with them, the remains of land-living and marine animals and plants. Following compaction into sedimentary rocks and the area becoming land once again, marine and freshwater fossils began to appear as erosion slowly proceeded.

Most of the known fossils have come from the sediments deposited under marine conditions during the Early Cretaceous, dating from about 100 million years ago. These sediments are poorly exposed at the surface around the outer margins of the Great Artesian Basin and in the centre are masked by a cover of younger sediments.

The warm, inland sea of the Early Cretaceous teemed with marine life — a wide variety of shellfish including the coiled ammonites and goniatites, sea lilies, brittle starfish and other invertebrates, as well as sharks, primitive enamel-scaled fish, bony fish, turtles and other reptiles. Also present, but rarely preserved, are the remains of land-living reptiles, which were washed out to sea to become buried with the remains of the marine creatures.

Exploration and settlement of central Queensland and the spread of the grazing industry led to the first discoveries of fossils from the Lower Cretaceous marine rocks of the Great Artesian Basin. The first collections came to the notice of the scientific community by 1866. In subsequent years, finds continued to be made and more detailed knowledge of just what animals were present and their relationships began to accumulate.

Amongst material from the area around Hughenden presented to the Queensland Museum in 1899 by a Mr A. Crombie was a lower jaw fragment of an obviously very large, flesh-eating reptile. The shattered remains of six, deeply-socketed conical teeth were present in the jaw. These had a maximum diameter of 40 mm. This specimen was studied by Heber Longman, then the Director of the Museum, who in 1924 decided that its earlier identification as a part of an ichthyosaur, a dolphin-like marine reptile, was not correct. He suggested it was from a pliosaur, one of the short-necked plesiosaurs, which were contemporaries of the ichthyosaurs. Longman estimated that the larger teeth in the animal must have been up to 250 mm long. This specimen was the basis for the name *Kronosaurus queenslandicus*, the name derived from the Greek mythological figure, *Kronos*, who swallowed his first five children to prevent them deposing him, and an allusion to the predatory nature of the animal.

In a later study, Longman (1930) described parts of the post-cranial skeleton from isolated, incomplete bones forwarded to the Museum from a site or sites some three kilometres south of Hughenden by Messrs. H.A. Craig, W. Charles and N.E. Anderson. Amongst 15 fragments forwarded were the heads of two of the large bones of the flippers. From these, Longman was convinced he was dealing with a pliosaur and that *K. queenslandicus* apparently represented the largest marine reptile ever recorded.

During 1931-1932, an expedition from Harvard University, Cambridge, Mass., U.S.A. was sent to Australia. One of the party, W.E. Schevill explored the Early Cretaceous rocks of central Queensland in an area to the north of Richmond and succeeded in locating a further two specimens. One of these was regarded as a juvenile and included a partial snout and lower jaw. The other, from the grazing property Army Downs about 56 kilometres north of Richmond was found on the surface, buried in soil and was a much more complete specimen. This proved to be one of those rare occurrences where the animal had been preserved in an almost complete state. It was encased in a hard, calcareous concretion which partially protected it when it was exposed by weathering and erosion. Most of the top of the specimen was destroyed.

Dynamite was used to reduce the specimen's encasing rock, but a total of nearly four tonnes of specimen and matrix was finally crated and shipped to the United States. The rock proved to be very difficult to remove by the methods then available and it was not until three years later that White (1935) was able to present the first description of the skull of *K. queenslandicus*. White gives full acknowledgement and praise to Longman's 'remarkable acumen in allocating this species to the Pliosauridae'.

However, it was not until 27 years after its discovery that the full details of the preserved skeleton were revealed. Rich and others (Rich and Thompson,1982) relate an interesting story associated with the final preparation and display of this material. Work on the major part of the specimen was delayed because of lack of funds. After World War II, a member of a wealthy Boston family approached Harvard, offering

---

Kronosaurus was a large marine pliosaur from the Early Cretaceous of Queensland. With its large head and impressive array of teeth, it was probably the most voracious carnivore that lived in the shallow seas which lapped onto the Australian continent at this time.

F. KNIGHT 82

support towards the display of any 'sea monster' that might be available. Apparently, the family had been involved in the reporting of the great Boston sea serpent of the early nineteenth century and had afterwards 'been regarded with disbelief by the proper Bostonians for having been involved in such a dubious affair'. The Harvard staff were able to convince the potential donor that *Kronosaurus* was as good as a live sea serpent, and the funds were forthcoming for the final preparation and display of the specimen.

Romer and Lewis (1959) have described this material and concluded that *K. queenslandicus* was up to 14 metres long with a head approximately one quarter the length of the animal.

The neck was extremely short compared with pliosaurs from elsewhere, resulting in a relatively more elongate trunk region. These authors concluded that, while *Kronosaurus* was exceeded in length by some of the long-necked plesiosaurs, in bulk it certainly appears to be among the largest known. This skeleton, largely restored, is now one of the major exhibits in the Museum of Comparative Zoology at Harvard University (U.S.A.).

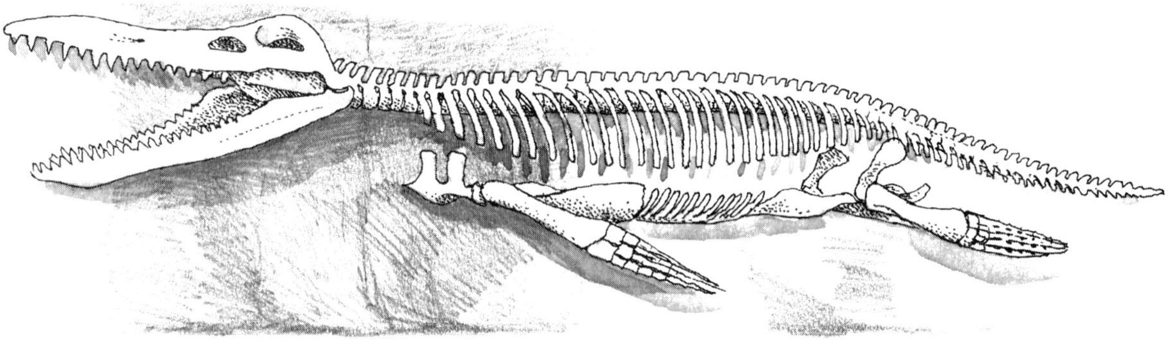

Skeleton of *Kronosaurus queenslandicus*, as displayed in the Museum of Comparative Zoology, Harvard, U.S.A. Like all pliosaurs it has a large head and a short neck; × 1/100

Jim Jensen preparing the skeleton of *Kronosaurus*, using drills to remove the limestone from around individual bones, at Harvard University.

Kronosaurus as it appeared in the field.

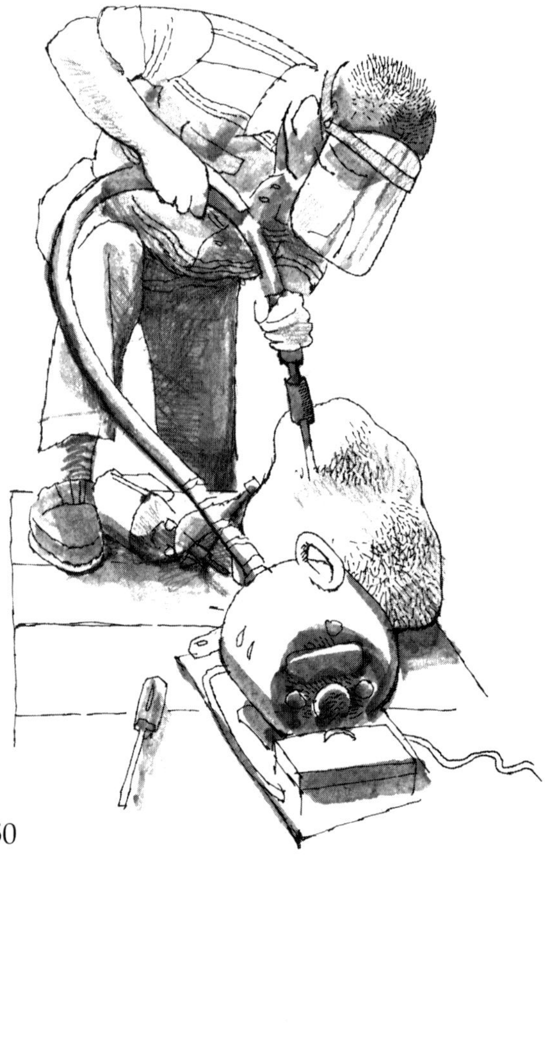

Fossils are now restricted exports under the Customs Regulations, and it is unlikely that any finds of this significance would be allowed to be exported permanently in the future.

Specimens of *Kronosaurus* have continued to accumulate in the Queensland Museum over the years, mostly as a result of donations from residents in central Queensland. In 1935, a portion of a skull was located by Mr J. Edgar Young on Telemon Station, halfway between Richmond and Hughenden, and Romer and Lewis (1959) refer to correspondence from Longman mentioning a series of eight caudal vertebrae. These latter specimens are now believed to be from a sauropod dinosaur.

The latest find occurred in July 1979, when guided by local graziers, I located and excavated a major portion of the skull and post-cranial skeleton of *K. queenslandicus* on Lucerne Station, to the northeast of Richmond. As with the specimen now in the United States, this material occurred both on the surface and, *in situ*, in soil. Some is fractured, while some is still solidly encased in concretionary limestone nodules. Preparation of this incomplete skeleton by a combination of mechanical and chemical means will take years.

Like pliosaurs from elsewhere in the world, *Kronosaurus* must have been a voracious feeder. Its bulk would have demanded the ingestion of large quantities of flesh daily and it must be inferred from the fossils of large primitive bony fish and other types of marine reptiles found in the same deposits, that these formed part, if not all, of its diet. Movement was probably generally slow, interspersed with very rapid bursts of speed when prey was being pursued. As the tail in *Kronosaurus* was relatively short and tapered, it apparently did not contribute greatly to movement through the water. The massive flippers were probably responsible for virtually all movement and most of its manoeuvering ability. Unlike its co-existing close relatives, the long-necked plesiosaurs, *Kronosaurus* could not augment its hunting ability by sharp, darting movements of its head.

The pliosaurs were mostly inhabitants of the shallow, marine waters of inland seas and the continental shelves. Being fully adapted to aquatic life, they probably had the capacity to have their eggs develop fully internally and to have their young born live and capable of fending for themselves. Such newly-born young must have been at least two metres long at birth, if this was the case. However, there is limited evidence presented by ideas on the musculature to suggest they may have been able to leave the water and, like turtles, could then have laid eggs in the more usual reptilian way. Unfortunately, evidence for either hypothesis will be virtually impossible to find.

Kronosaurus continues to represent one of the most impressive and interesting fossil vertebrates yet discovered in Australia, and the search for more and better preserved specimens is sure to continue. It appears that some of the specimens already collected have come from sediments of both Aptian and Albian age within the Early Cretaceous and a wider range of specimens will certainly show whether only a single species, *K. queenslandicus*, or several are represented.

Megalania prisca

Owen, 1859

The Giant Goanna

CLASS: Reptilia
ORDER: Squamata
FAMILY: Varanidae
GENUS: *Megalania*
SPECIES: *Megalania prisca*
AGE: Pleistocene

The 'top dog' or largest predator of the Australian Pleistocene terrestrial vertebrate community certainly was not a dog and probably was not even a mammal! Rather it may have been *Megalania prisca*, 'the great ripper', a gargantuan relative of the living goannas, members of the widespread genus *Varanus*.

One of the peculiarities of the Australian Pleistocene fauna is the absence of a lion-sized mammalian predator that could have preyed upon animals such as *Diprotodon* or *Zygomaturus*. *Megalania* may have filled part of this niche.

In the manner of capturing its prey, *Megalania prisca* probably was quite similar to the living Komodo Dragon. That animal typically ambushes its prey around a waterhole (Auffenberg, 1972). There it hides in the undergrowth and makes a short dash, hamstringing its victim which it can then kill and eat at its leisure. It is also a carrion eater, so presumably *M. prisca* would not have passed up the opportunity for a meal merely because it happened to be a little 'high'.

As discussed in the section on *Meiolania*, when Sir Richard Owen (1859a, 1860) first described *Megalania*, he inadvertently combined elements of the turtle *Meiolania* and a lizard into one skeleton. Eventually the confusion was straightened out (Woodward, 1888), but even today after the animal has been known for more than a century, there is no reasonably complete skeleton of this ferocious predator available. In the late nineteenth century a specimen was apparently found that, even though not complete, was by far the closest to such that has ever been discovered. Unfortunately, parts of this specimen appear to have been sold to the Australian Museum in Sydney; the Queensland Museum in Brisbane; and possibly to Baron von Nopsca of Hungary, a well-known palaeontologist, secret agent, and after World War I, almost King of Albania. Herman Lau collected the partial skeleton preserved today in the Australian Museum, and R.W. Frost, the Queensland Museum specimen. What is uncertain is whether or not they collected different parts of one individual or bits from two separate skeletons. No parts are duplicated in the two collections; for instance, there are not two left femora.

Because of the incomplete knowledge of *Megalania prisca*, in order to make a

---

After lying in wait in the undergrowth near a billabong, a *Megalania prisca* has successfully ambushed a *Diprotodon* sp. that came down to the water for a drink. This is the same tactic used by its largest living relative, the Komodo Dragon which will attack a cow in the same manner. *M. prisca* was probably the only predator large enough to prey upon the biggest marsupials such as *Diprotodon* and *Zygomaturus*.

F.KNIGHT '82

reconstruction of its entire skeleton (Rich and Hall, 1979), it was necessary to rely heavily on information about the skeletons of its closest living relatives, the goannas. Taking 47 measurements on various parts of the skeleton of several different living species of *Varanus*, from the smallest available to the largest; the impressive Komodo Dragon of Indonesia, it was possible to determine how pairs of measurements varied with one another. Plotting the data on log-log graph paper, the points always fell close to a straight line. With this information, it was possible to work out a network of relationships between the known dimensions of the *Megalania* material available and the unknown ones and thus by extrapolating the line on the graph, to predict what the dimensions of the, as yet unknown, bones of *Megalania* would be.

It was not possible to simply multiply all dimensions of a Komodo Dragon skeleton by the same amount to arrive at a reasonable hypothesis as to what that of *Megalania prisca* would have looked like. As animals become larger, their proportions necessarily change in response to the differing demands placed on their bodies. For example, an elephant's limbs are much stouter relative to the body length of the animal than are those of a mouse. This is because if an object is held constant in shape but length, width and height are all doubled, its mass is increased eight times. If any columnar supports for this larger object are to be equally as strong as those of the smaller one, their cross-sectional area must be increased eight times as well. This is not done by merely doubling their diameters, which would only increase their cross-section by 4, but by increasing the diameter by 2.83, which results in a column with 8 (2.83^2) times the cross-sectional area.

With this information, the next step was to sculpt the missing bones in wax at the appropriate size using *Varanus* bones as models, changing the proportions as indicated by the changing relationships observed in the *Varanus* species of various sizes. Once the shaping of the wax bone model was finished, it was moulded and a cast was poured in polyurethane or epoxy resin, a more durable medium than wax, which could be coloured to closely resemble the real bones of *Megalania*.

On the outline drawing of the skeleton, the bones or parts of the bones of

Comparison of the relative trunk and tail length of a small and large species of *Varanus* illustrating the relative reduction in the size of the tail with increased total length. The larger *V. komodensis* (the Komodo Dragon lizard) (above) has a total length of 3.0 metres; the smaller *V. eremius* (below) is 0.4 metres long.

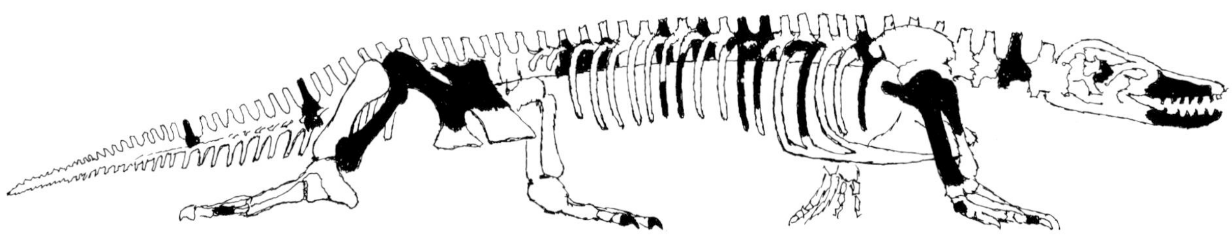

An outline drawing of the reconstruction of the skeleton of *Megalania prisca*. The parts of the skeleton represented by fossil material are shown in solid black, the parts in outline are only inferred by comparison with modern specimens of *Varanus*.

Megalania prisca that are known are drawn in solid black. Those in outline are inferred from measurements of *Varanus* specimens. The skeleton that was produced as a result of this work was 5.5 metres long. Although massive, Hecht (1975) estimated that the largest *Megalania prisca* for which there is some fossil evidence was 7 metres long and weighed between 600 and 620 kilograms, nearly eight times the mass of the living Komodo Dragon.

Although a living *Varanus* was used as a model to fill in the unknown parts of the skeleton of *Megalania prisca*, this is not to say that the only difference between the two genera was size alone. The teeth of *M. prisca*, for example, are more widely spaced and recurved with the front edge smooth and the rear serrated, unlike the straight, unserrated teeth of *Varanus* species. In addition, the skull may have been less flexible than that of *Varanus* species, suggesting that the mode of feeding might have been somewhat different.

Besides the one (or possibly two) skeletons known from the Darling Downs of Queensland, records of *Megalania prisca* are based almost exclusively on the discovery of vertebrae either singly or in small groups. This meagre record indicates that *M. prisca* had a distribution pattern similar to what may have been an important prey species for it, the much more abundant *Diprotodon*. Remains of both occur primarily in the inland areas of southeastern Australia and neither is known from Tasmania.

The family Varanidae to which *Megalania prisca* belongs may have entered Australia relatively recently. It may have come into the continent from the north, as Australia drifted close to Asia in the mid-Tertiary. This is suggested by the fact that although nineteen of the thirty-two species of living *Varanus* occur in Australia, only two of the eleven sub-genera known in the world do. The great proliferation of species of *Varanus* in Australia therefore,seems to have been a relatively recent phenomenon. Most of the sub-genera of varanids are in Eurasia, suggesting that the earlier differentiation of *Varanus* had occurred in that region. The earliest varanids appear in the Late Cretaceous of North America and Eurasia (Holarctica) implying that the diversification of *Varanus* did not get underway before then.

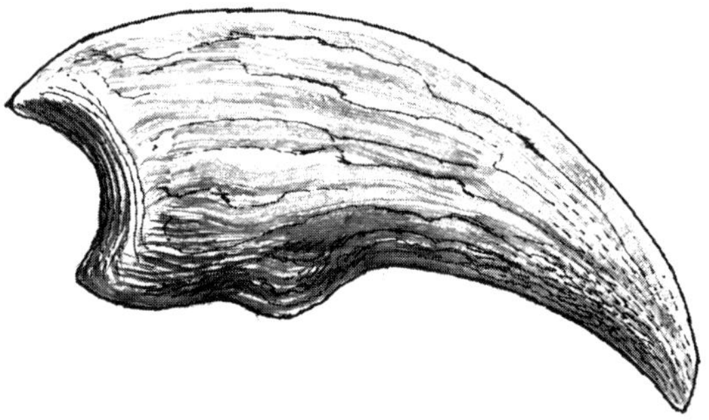

Claw (terminal phalanx) of *Megalania prisca* from Pleistocene deposits along Cooper Creek, South Australia, x 1.

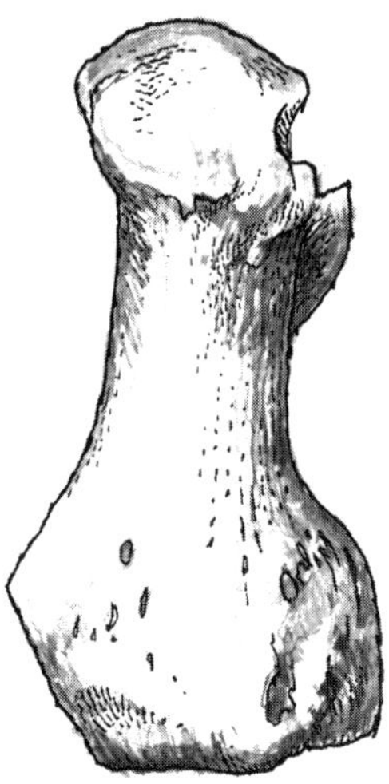

Fifth metatarsal of *Megalania prisca* from Pleistocene rocks at Kings Creek, southeastern Queensland, x 7/10.

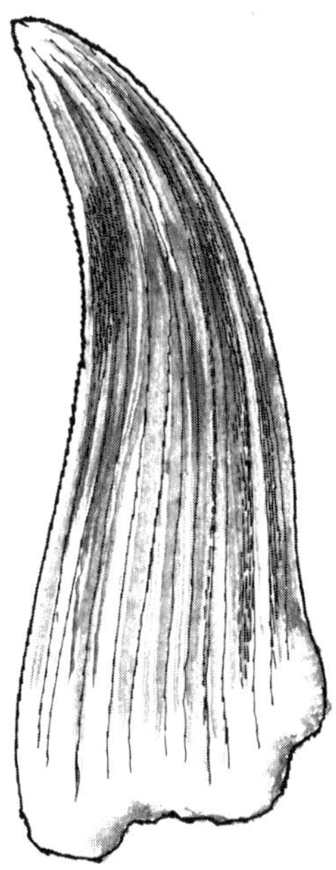

Isolated, unworn tooth of *Megalania prisca*, x 3.6.

Wonambi naracoortensis

Smith, 1976

The Giant Australian Python

CLASS: Reptilia
ORDER: Squamata
FAMILY: Boidae
GENUS: *Wonambi*
SPECIES: *Wonambi naracoortensis*
AGE: Pleistocene

Snake skulls and jaws, being made up of many individual bones, are seldom preserved as fossils, although fragments of toothbearing bone may remain. The backbone consists of a large number of vertebrae — generally more than 100 and over 400 in some pythons. Fortunately vertebrae are robust and are often found well-preserved in fossil deposits. One pair of ribs is attached to each vertebra between the neck and the cloaca. The strength of the ribs and hence the likelihood of their being preserved varies between species. Of necessity, the palaeontologist studying fossil snakes relies heavily on characteristics of the vertebrae for identification and reconstruction.

In keeping with the highly specialized locomotion of snakes, the vertebrae show adaptations distinguishing them from other backboned animals. Each vertebra interacts with the one in front at two contact areas in addition to the paired lateral facets (zygapophyses). One contact area is a 'ball-and-socket' coupling, the socket (cotyle) facing forwards and receiving the ball (condyle) on the back of the vertebra in front of it. The other contact area is a forwardly projecting tenon that fits snugly into a mortise on the back of the vertebra in front. This latter articulation, known as zygosphene (the tenon) and zygantrum (the mortise) is a feature of all snakes, and is otherwise found only in a few lizard species.

Excavation and sieving of the Pleistocene deposit in Fossil Chamber of Victoria Cave, Naracoorte, South Australia in 1969 and subsequent years by Rod Wells and his colleagues yielded many hundreds of snake vertebrae. Most were indistinguishable from those of the living brown snake, a member of the family Elapidae. However, several vertebrae were found, very much larger than the others, that bore features associating them with the family Boidae (the pythons and boas). The most readily seen characteristic diagnostic of the Boidae is the absence of a deep keel (hypapophysis) beneath the vertebrae of the posterior two-thirds of the pre-cloacal vertebral column. Closer study of these vertebrae showed that they were unlike those of any living boid, and they were described as a new genus and species *Wonambi naracoortensis* Smith, 1976.

At the time of its description, *W. naracoortensis* was known from eight vertebrae and a fragment of left upper jaw bone (maxilla) with three teeth. Since then a further seven vertebrae, four ribs (all incomplete), and an incomplete right maxilla

---

Wonambi stalking a young Toolache Wallaby at a waterhole.

F. KNIGHT
84

have been recovered. Were it not for their scattered locations in Fossil Chamber, all these bones could conceivably have been derived from a single individual! Hence *W. naracoortensis* has proved to be an extremely rare species, especially considering that many tonnes of sediment have been sieved and sorted and that some extinct mammals, for example *Thylacoleo carnifex,* are represented by tens of specimens. In Henschke Fossil Cave, another rich Pleistocene deposit (Pledge, 1981) not far from Victoria Cave, *Wonambi* has not been found at all.

However, *Wonambi* has been identified at two other locations (Smith, unpublished). In a faunal list for Mammoth Cave in the extreme southwest of Western Australia, Merrilees (1968) listed 'Snake, probably *Python sp.*'. Examination of two vertebrae from Mammoth Cave showed that they were of *Wonambi* and almost certainly the same species as from Victoria Cave. Similarly two vertebrae of a 'large snake' from Koala Cave, near Yanchep, Western Australia (Archer, 1972) are indistinguishable from *W. naracoortensis.* On the other hand, boid vertebrae from Pleistocene deposits at Lake Tandou and Willandra Lakes, New South Wales, are not *Wonambi* (Smith, unpublished) nor is a large vertebra from a Lower Pliocene fauna from northeastern Queensland (Archer and Wade, 1976).

While the evidence to date is that *Wonambi* played a minor role in Australian palaeoecology, it may take a more significant part in zoogeography, for the characteristics that distinguish *Wonambi* vertebrae from other Australian boid species are precisely the characteristics of species of the fossil subfamily Madtsoiinae (Hoffstetter, 1961b). This subfamily is known from three or four species — *Madtsoia bai* from the Palaeocene and Eocene of Patagonia (Simpson, 1933; Hoffstetter, 1959), *M. madagascariensis* from the Late Cretaceous of Madagascar (Hoffstetter, 1961a), *Madtsoia* sp. from the Late Cretaceous of Niger, West Africa (Rage, 1974) and *Gigantophis garstini* from the Eocene of Egypt and Libya (Andrews, 1906; Hoffstetter, 1961b). A close relationship between *Wonambi, Madtsoia* and *Gigantophis* probably will not be confirmed until skulls of all three genera are known. Meanwhile the evidence of the vertebrae provides compelling reasons for including the genera within one subfamily despite the disparity of millions of years in the ages of the fossils. Such relationship would indicate that Australia was first colonized by boids before the drift of Australia from Antarctica.

Madtsoia, Gigantophis and *Wonambi* share not only the features used by phylogenists to associate them (such as the presence of a small opening for a blood vessel on each side of the 'socket' of the ball-and-socket coupling; and the upturned facets of the lateral articulation), but also their overall shape is similar and remarkably different from vertebrae of modern boids. The fossil vertebrae are cumbersome. The neural canal is relatively small and the facets for articulation of the ribs are very large and set wide apart on the sides of the vertebrae. The greatest width of the vertebra is that measured across the rib articulation facets, whereas in all living Australian boids the lateral intervertebral articulations are set wider than the rib facets. The ribs themselves are massive in *Wonambi,* as in *Madtsoia* and *Gigantophis,* and it is likely the body was much thicker relative to its length than in modern boids. On the assumption that *Wonambi* had about the same number of vertebrae as the Carpet Python, the total length of the *Wonambi* from Victoria Cave would have been about 5 metres (Smith, 1976). *Gigantophis garstini* was about 9 metres long (Andrews, 1906), while *Madtsoia bai* was 7-8 metres (Hoffstetter, 1959). The largest living boids have been measured at more than 9 metres long.

While the estimated length of *Wonambi* does not approach the known length of some modern boid snakes, it raises interesting questions about the animal's habits. Except in Australia, medium to large boids are found only in or near the tropics.

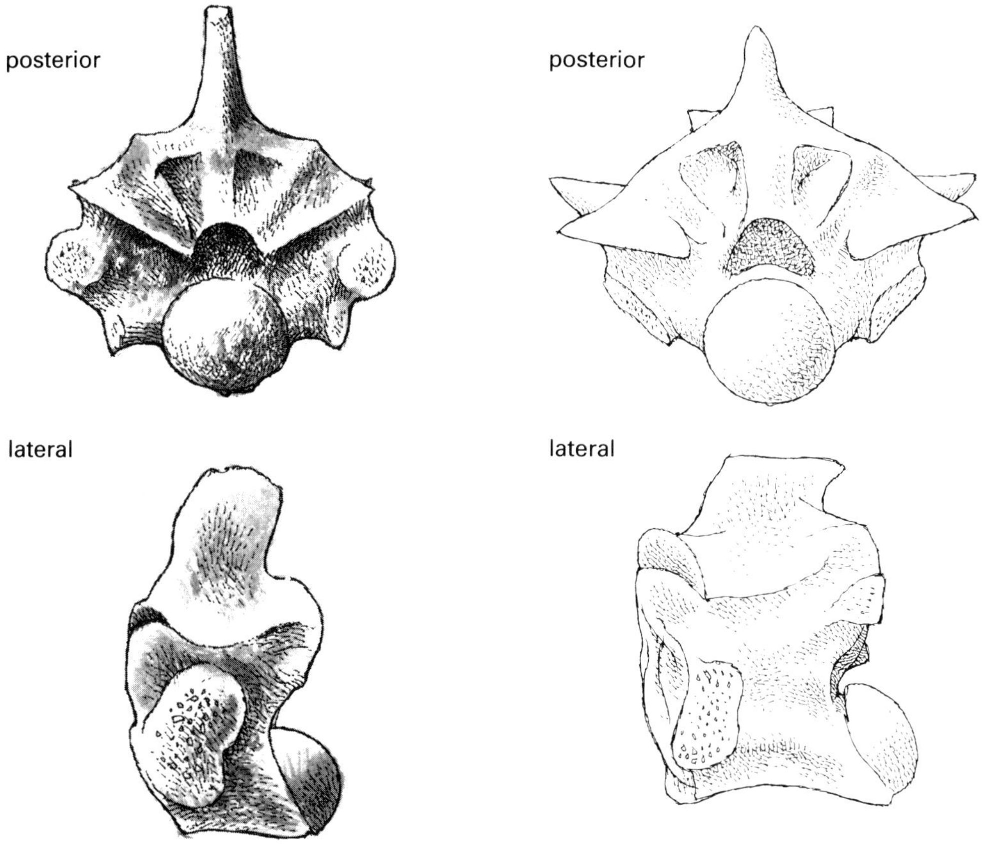

A mid-dorsal vertebra of *Wonambi naracoortensis* (on left) compared with one from a similar region of *Python reticulatus*. x 1.

None of the Australian species found in temperate latitudes is of a large size and only one, the Carpet Python, exceeds 3 metres in maximum length. This is entirely to be expected, because snakes rely on external heat absorbed through the skin to maintain their body temperature at a level sufficiently high for locomotion, food capture, digestion and reproduction. The larger the snake the lower the ratio of skin area to body volume and hence the longer the exposure to warm air and warm ground required to raise the deep body temperature. For *Wonambi* to have existed in the temperate climate of southern Australia it must have been able to be fully active at temperatures much lower than living boids require. There is no evidence from the associated fauna in Victoria Cave nor from Pleistocene faunas from southern Western Australia that the Pleistocene climate was markedly warmer than now.

Its relatively thick body suggests that *Wonambi* may have been at least partly aquatic. Most of the large boids live close to water and the largest, the Anaconda, is always found near permanent water.

There are still many gaps in our knowledge of *Wonambi*. We do not know if it laid eggs or produced live young; whether it had a wider distribution than the present records indicate; nor the geological time span of the species. Nevertheless the few fragments found show that *Wonambi* is significant in the radiation and adaptation of the boids.

Quinkana fortirostrum

Molnar, 1981

The Quinkan

CLASS: Reptilia
ORDER: Crocodylomorpha
FAMILY: Crocodylidae(?)
GENUS: *Quinkana*
SPECIES: *Quinkana fortirostrum*
AGE: Pleistocene

Near Chillagoe, north Queensland, is a series of unique and striking, small, but very steep limestone 'hills', locally and appropriately known as towers. Many are penetrated by caves, but at least one cave, Tea Tree Cave, lies beneath, rather than within, one of these towers. During the exploration of this cave in 1970, Lyndsey Hawkins of the Sydney Speleological Society, found a crocodile skull lying upside down on the floor of the cave. At some time, not too long ago (geologically speaking), Tea Tree Cave became filled with rock fragments, a breccia, among which were the bones of various animals, some now extinct. Later this breccia was partly eroded away and presumably released the crocodile skull, which came to rest on the floor of the cave. Lyndsey Hawkins collected the skull, and on her return to Sydney donated it to the Australian Museum.

The skull was found to have two most unusual features: although broad, the snout was not flattened but as deep as it was broad; and the alveoli, sockets for the teeth, were elongate, as if the teeth were not round in section like modern crocodile teeth, but instead blade-like. Crocodiles with blade-like teeth have been long known, chiefly from South America, but also from North America, Europe, Asia and Africa and have been called ziphodont crocodiles.

Ziphodont crocodiles have a snout at least as deep as it is broad, and sometimes nearly twice as deep as broad; and have laterally compressed, serrate and blade-like teeth (Langston, 1975). Modern crocodilians (and most extinct ones as well) have flattened snouts broader than deep (as in the saltwater crocodile, *Crocodylus porosus*) or narrow snouts about as broad as deep (as in the freshwater crocodile, *C. johnsoni*), but never snouts that are both broad and deep. Again modern crocodilians may have stout teeth of round cross-section (like the saltwater crocodile) or thin, gracefully recurved teeth, also of round cross-section (like the freshwater crocodile), but do not, as adults, have flattened blade-like teeth. Clearly the skull from the Tea Tree Cave resembled that of a ziphodont, but no teeth remained in it when found. In 1976, a Queensland Museum field party, under the direction of Mike Archer, reported finding a partial crocodile skull in another cave, this time in the Texas Caves of southeastern Queensland (Hecht and Archer, 1977). This skull, when compared with that from Tea

---

Quinkana fortirostrum, probably a terrestrial crocodilian from the Pleistocene of eastern Queensland. It would have been one of the largest carnivores of this epoch.

F. KNIGHT

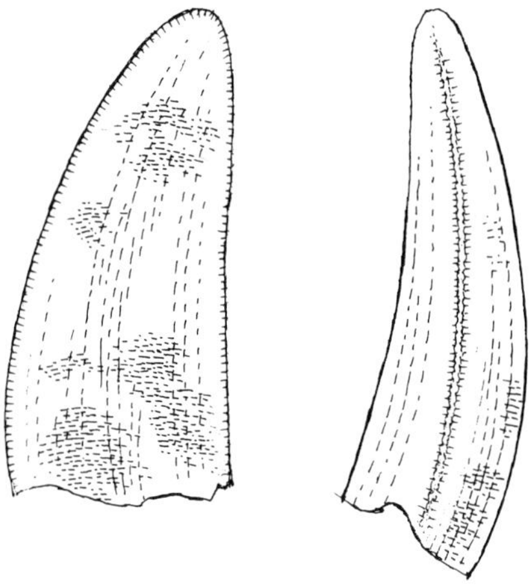

The tooth of a ziphodont crocodile, viewed from the side (left) and from in front (right).

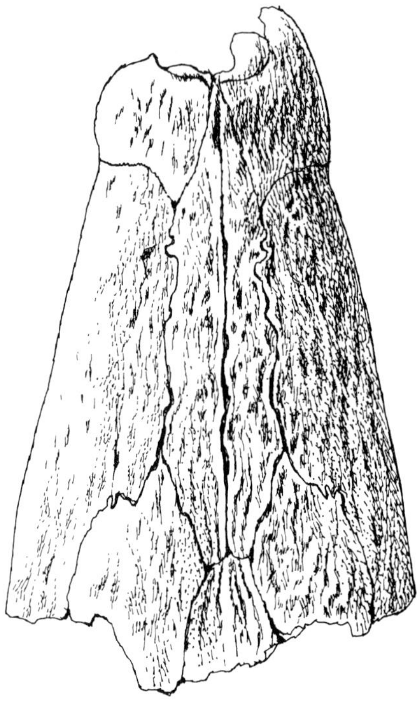

The snout of *Quinkana*, viewed from above; × 0.4

Tree Cave, proved to be quite similar, enough to refer both to the same genus of crocodiles. These were later named *Quinkana*, after the quinkans, mythical humanoids of the Dreamtime. The Texas Caves specimen retained some teeth still in their sockets, and these teeth were indeed flattened, blade-like, serrate teeth. Thus, these specimens represented the first described ziphodont crocodilians from Australia.

Neither of the two skulls was particularly complete, but then relatively few of those known from overseas were complete either. Indeed, ziphodont crocodiles are often represented only by their teeth. Ziphodont crocodile remains had been discovered first in France and later in Wyoming (U.S.A.) during the nineteenth century. But while they were recognized as unusual in both cases by the workers who studied them, both, the Baron Cuvier and O.C. Marsh respectively, had so much other work to keep them occupied that neither paid further attention to ziphodonts (Langston, 1975).

Near the end of the nineteenth century a brilliant, dedicated Argentinian, Florentino Ameghino, claimed that in Argentina, where all of the modern mammals originated (or so he had thought, and not without some reason), relatively advanced mammals had co-existed with dinosaurs. However, during the 1930s G.G. Simpson, a North American vertebrate palaeontologist exploring for fossils in Patagonia, discovered that there was indeed evidence to support Ameghino's contentions. Ameghino had claimed that dinosaurs lived together with relatively advanced (Tertiary) mammals because the teeth of carnivorous dinosaurs were associated with the remains of the mammals. Simpson, also found such a tooth associated with a mammal skeleton. However, Simpson pointed out that neither the tooth he had found, nor those found by Ameghino, looked precisely like those of carnivorous dinosaurs — generally, yes ... but in detail, no (Simpson, 1932). At the time Simpson did not know from what animal these teeth derived, but five years later, in 1937, he described an unusual crocodile skull from the Eocene rocks of Argentina. This form he called *Sebecus* (after the Egyptian crocodile deity), and it possessed just such teeth as he and Ameghino had found (Colbert, 1946). So Ameghino's observations were validated, but his interpretation was not.

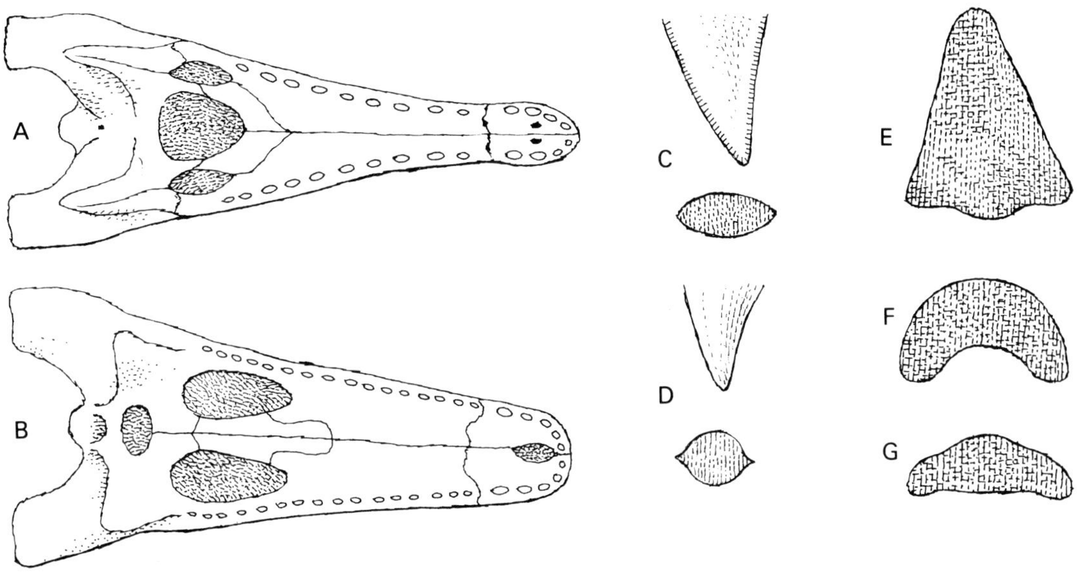

Comparison of ziphodont and non-ziphodont crocodiles. The palates of *Sebecus* (**A**) and *Pristichampsus* (**B**), showing that the internal nares (nostrils) are further back relative to the bones of the palate in the latter (an eusuchian) than in the former (a sebecosuchian). Ziphodont teeth (**C**) are serrate and flatter than normal crocodilian teeth (**D**). Sebecosuchian snouts (**E**) are deep in cross-section, while pristichampsine snouts (**F**) are less deep, and normal crocodilian snouts (**G**) are flat.

Had anyone at this time gone through the papers of Cuvier or Marsh they would have found that *Sebecus*-like crocodiles were already known, or more accurately had been described, but ignored. However, neither worker had made a point of their peculiar crocodilians, and so they remained unrecognized for some years more. Immediately recognized on the other hand, was a crocodile skeleton (Kuhn, 1938) from the marvellous coal deposits of the Geiseltal, in Germany, found in the early 1930s. This skeleton belonged to *Pristichampsus rollinati*, the crocodile originally described by Cuvier (although this was not realized for some years, and the Germans gave it a new name, *Weigeltisuchus*). It demonstrated that not only was the ziphodont's skull unusual, but that the tail was rounded in section, not flattened like the typical crocodile tail, and that the toes had flattened hooves rather than claws. These features all suggested that ziphodont crocodiles were less aquatic than their living relatives (although even the living crocodiles often roam farther from water than most people think — Cott, 1961).

Later discoveries showed that ziphodonts had lived in Argentina, Peru, Brazil, Colombia, U.S.A., Portugal, France, Germany and China. Isolated teeth had also been found in Kenya, Pakistan, India, Burma, China and New Guinea. They had even been found in Queensland, but were not recognized as belonging to ziphodont crocodiles until the *Quinkana* skulls were studied.

This abundance of material led at first to some confusion, for it was quickly realized that there were more than a single kind of ziphodont crocodilian. There are two basic types of palatal structure found among crocodiles (although the earliest crocodiles known had neither of these two types), which correspond to two basic groups of crocodiles. (There are more than two major groups of crocodiles, but the two referred to here are the most frequently encountered). These two types of palate differ in the position of the choanae, that is the internal nostrils, relative to the bones of the palate (Langston, 1973). All living crocodiles have an eusuchian palate and are classified as eusuchian crocodilians: in the eusuchian palate the internal nostrils are surrounded entirely by the pterygoid bones. Most of the Mesozoic crocodiles belonged to the mesosuchians, from which eusuchians are derived. In the mesosuchian palate the internal nostrils are surrounded either entirely by the palatine bones, or in part by the palatines and in part by the pterygoids.

The confusion about ziphodonts arose because some of these crocodiles, such as *Baurusuchus* and *Sebecus*, had clearly mesosuchian palates, while others, *Pristichampsus*, had clearly eusuchian palates. So, it is now generally accepted that there are at least two independent groups of ziphodont crocodiles. Indeed there may be more than two; Eric Buffetaut (1979), suggests that the hsisosuchians from Eurasia were a third group of Mesozoic ziphodonts.

An obvious question raised concerns the relationships of *Quinkana*; to what group of crocodilians is it most closely related, sebecosuchian or eusuchian? The diagnostic features concern palatal structure, and neither of the two incomplete skulls of *Quinkana* retain the appropriate portions of the palate. However, considering only the sebecosuchians, they may be distinguished from the (eusuchian) pristichampsines by certain features of the top of the snout. The nasal bones, which roof the snout, form a high crest in sebecosuchians, and are plainly visible when viewed from the side. Those of pristichampsines, however, form a flat, horizontal roof, visible only from above. Those of *Quinkana* also form a flat, horizontal roof, and hence suggest that *Quinkana* is related to *Pristichampsus*. Other features of the known material of *Quinkana* corroborate this suggestion, for example, length of the tooth-row, number of teeth in the maxilla, shape of the nasal bones (Molnar, 1981).

Pristichampsus, a ziphodont crocodile from western Europe, probably related to *Quinkana*, about 2m in length.

Quinkana extends the distribution of eusuchian ziphodont crocodilians into the southern hemisphere. At some time there was either a continuous distribution of such crocodilians from Asia, where the oldest pristichampsines have been found, or else some population of these crocodiles slowly expanded its range to encompass Australia, while at the same time its range retracted out of Asia — the population itself thus migrating out of Asia to Australia. Since pristichampsines are not certainly known from either South America or Africa, we may presume that they were also absent from Antarctica, leaving Asia as the only potential land of origin. The only relevant evidence is the isolated ziphodont teeth from the valley of the Watut River in Papua New Guinea. These teeth could be either sebecosuchian or eusuchian, so this only shows that such an hypothesis is plausible, which is just about as much as was known already.

The situation is complicated by the discovery of what appears to be a fragment of a sebecosuchian crocodilian skull at Lake Palankarinna, South Australia. This specimen, yet to be properly described, is Pliocene in age, and hence older than

Quinkana. Since sebecosuchians were most abundant in South America, and were already present there while South America was still indirectly linked to Australia via Antarctica, this crocodile's ancestors probably inhabited Antarctica. Thus, the history of ziphodont crocodiles in Australia will in part be clarified by work in Antarctica and in Southeast Asia.

Quinkana survived in Australia long after its kin and similar forms (the sebecosuchians) became extinct elsewhere, probably because of the isolation of Australia and the apparently peculiar ecology of the large land vertebrates. There were very few large mammalian carnivores at any time in Australia, both today and in the past. To be sure, there were thylacines (or Tasmanian tigers) and *Thylacoleo* (the marsupial lion) and its relatives. But there were probably no more than two large mammalian carnivores at any given place and time. This contrasts with the rest of the world. For example, in North America there are coyote, wolf, cougar and lynx, and in Africa jackal, leopard, cheetah, lion and hyaena — not to mention the sabre-tooth cats, which are (geologically speaking) recently extinct on both continents. It may be that *Quinkana* (and the giant goanna, *Megalania*) were reptilian predators upon the herbivorous mammals of Pleistocene Australia.

Quinkana would not have been very large, maybe three metres long overall. It ranged throughout eastern Queensland, at least from the area of Laura on Cape York Peninsula to the New South Wales border in the south, but so far has not been found west of Croydon. It is not certain that ziphodont crocodiles were land dwellers as opposed to amphibious forms like the living crocodiles, but there is some evidence consistent with this view (the hooves and rounded tail) and none opposing it. There is little doubt that these forms were carnivores, probably preying on large vertebrates, although very likely eating smaller forms as well.

Rhoetosaurus brownei

Longman, 1926

The Giant Queensland Dinosaur

CLASS: Reptilia
ORDER: Saurischia
SUBORDER: Sauropoda
FAMILY: ?Cetiosauridae
GENUS: *Rhoetosaurus*
SPECIES: *Rhoetosaurus brownei*
AGE: Middle Jurassic

Among the most impressive of Australia's extinct backboned animals is *Rhoetosaurus*, a giant plant-eating dinosaur that lived in southeastern Queensland in the early part of the Jurassic period, about 180 million years ago. *Rhoetosaurus* was a sauropod dinosaur — a member of the reptile group (Sauropoda) that includes the largest land animals of all time. Other, and perhaps more familiar sauropods were the American dinosaurs *Diplodocus* and *Brontosaurus* (more correctly called *Apatosaurus*).

The first inkling that a great sauropod dinosaur had once lived in Australia came at the beginning of 1924, when some large fossil bones came to light in a creek on Durham Downs Station, near Roma in southeastern Queensland. Some pieces of those bones found their way to H.A. Longman, then Director of the Queensland Museum, who recognized them as fragments from the skeleton of a sauropod dinosaur. In the following weeks of 1924 more than 190 kilograms of fossil bone was collected from the Durham Downs site, much of it unfortunately in the form of scattered fragments. The fossil bones were then transported by road to Roma and were sent on by rail to Brisbane.

Longman published the first scientific account of the specimens in 1926. His raw material was, by any standards, rather scanty; it consisted of sections of the tail skeleton together with a few pieces of bone from the hip and the hind limb. Nevertheless it was clear that all these fragments had originally come from the skeleton of a single dinosaur. The sheer size of the bones was astounding; they were far bigger than those of any animal, living or extinct, that had ever been found in Australia. For that reason Longman named the dinosaur *Rhoetosaurus* — derived from *Rhoetos*, a giant in Greek mythology, and *saurus*, meaning 'lizard'. In the meantime further collecting was under way at Durham Downs. This time the collectors were rewarded with the discovery of some 950 kilograms of fossil bones, including a single neck vertebra, parts of the back and fragments of a femur. These additional finds were described by Longman in 1927a, and 1927b. By that time the skeleton of *Rhoetosaurus* must have seemed like an enormous jigsaw puzzle...with most of the pieces still missing.

The bones of *Rhoetosaurus* are immense. The femur, for example, is 1.5 metres long. From the dimensions of the bones known in 1927 Longman estimated that the living *Rhoetosaurus* had a total length of about 12 metres, but he later admitted that this was probably a conservative estimate.

Outline diagram of a sauropod skeleton. Bones identified for *Rhoetosaurus* are shown in black (or slightly paler shading where they are concealed). Locations of the vertebrae and rib fragments are only approximate.

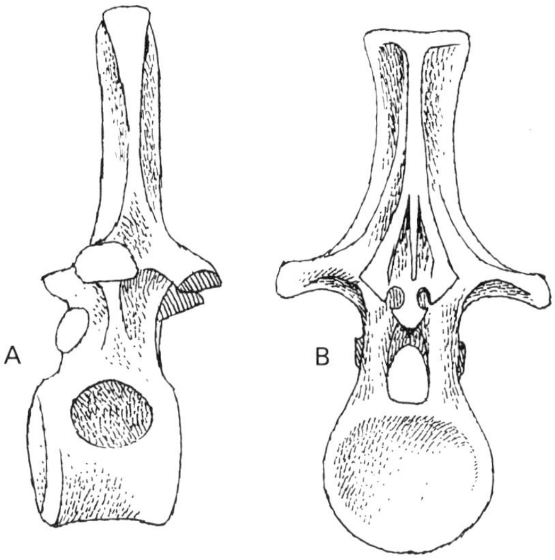

Vertebra from the back region of a sauropod dinosaur to show characteristic weight-saving and weight-bearing adaptations. The vertebra is shown from the left side (**A**) and from the rear (**B**). The neural arch is reduced to a complicated system of buttresses combining maximum strength with minimum weight. On each side of the centrum a large cavity (pleurocoel) also serves to reduce weight. Each vertebra is linked to the next not only by zygapophyses (as in other land vertebrates) but also by a hyposphene — a wedge of bone that fits into a slot in the front face of the following vertebra. For even greater strength the convex front face of the centrum fits into the concave rear face of the centrum in front. All these distinctive sauropod features are seen in the vertebrae of *Rhoetosaurus*.

Mr Marshall, preparator at the Queensland Museum, stands beside a plaster cast of the *Rhoetosaurus* thigh bone. The year is 1927. From a photo provided by courtesy of the Queensland Museum.

Subsequently there came a decline of scientific interest in *Rhoetosaurus*, and for nearly 50 years there was no further research into the structure, relationships and probable life-habits of this giant Australian dinosaur. In his original descriptions Longman hinted that parts of the *Rhoetosaurus* skeleton might still lie buried at Durham Downs, and in 1975 this possibility was followed up by Mary Wade, Curator of Geology at the Queensland Museum. Mary succeeded in relocating the exact site, where she discovered additional fragments of the skeleton. There is no doubt that these fragments are from the same *Rhoetosaurus* skeleton described by Longman, for some of them fit exactly on to the bones collected 50 years before. In the following year a full-scale excavation of the site produced even more parts of the skeleton, including a near-complete hind foot.

Thus, over more than a half century, and through the efforts of many field-workers, our knowledge of the *Rhoetosaurus* skeleton has been improved piece by piece. Today the skeleton comprises a neck vertebra, parts of the back, pieces of ribs, the sacrum, a sizeable section of the tail, a portion of the hips and an almost complete hind limb. Many other fragments of the skeleton were collected, but all are so badly weathered and broken that it may be impossible to identify them with certainty. This assortment of bones and fragments is very far indeed from a perfect skeleton, yet many of the world's sauropod dinosaurs are represented by even scantier fossil evidence. The final search of the Durham Downs site was very thorough, and it now seems unlikely that any more bones will come to light in the future. The missing parts of the *Rhoetosaurus* skeleton were probably destroyed by weathering long before palaeontologists arrived on the scene. Many of the bones were found embedded in tough sandstone, and the formidable task of freeing them from the rock has been undertaken at the Queensland Museum. The bones are rather fragile, in spite of their size, and they need to be reinforced with thin solutions of transparent plastic. Then the surrounding sandstone is carefully chipped away with a pneumatic engraving tool. When the bones are finally revealed it will be possible to give a new and fuller description of *Rhoetosaurus* and to compare it to the sauropods discovered on other continents.

Despite its incompleteness, the skeleton of *Rhoetosaurus* is of major scientific interest. It is among the most ancient of sauropod skeletons yet unearthed, and it may well shed some light on the mystery that has always surrounded the origin and the evolutionary history of the great sauropod dinosaurs. Even the barest outlines of sauropod history remain unknown, for in all too many cases it has proved impossible to determine the relationship of any one sauropod to another. This is simply because most sauropods are known from fragmentary fossils: one species may be represented by tail bones, another by a shoulder-blade, yet another by a skull ..... and so on. In this situation it is difficult to decide which bones belong to which species of sauropod, let alone to compare one species to another. For example, most reconstructions of *'Brontosaurus'* (*Apatosaurus*) include a skull that probably came from a quite different sauropod dinosaur (*Camarasaurus*) according to Berman and McIntosh (1978). Moreover no one has yet devised a satisfactory classification of the sauropods; some palaeontologists have recognized two family groupings, whereas others have listed more than a dozen. Even so it is possible to identify several distinctive 'types' of sauropod dinosaur. First there are the relatively long and slender sauropods, such as *Diplodocus*, with their distinctive 'whip-lash' tails. Then there are the shorter but much heavier sauropods such as *Brachiosaurus*, unusual in being taller at the shoulder than at the hip. Between the two extremes of body design seen in *Diplodocus* and *Brachiosaurus* is an entire range of sauropod 'types' that were neither exceptionally long and slender nor exceptionally short and tall. In its overall body design *Rhoetosaurus* was probably similar to some of these intermediate 'types', such as *Cetiosaurus* and *Camarasaurus*.

A giant plant-eating dinosaur from terrestrial Jurassic rocks of southeastern Queensland. *Rhoetosaurus* was a sauropod, a member of the dinosaur group that includes the largest of all land animals.

F.KNIGHT '83

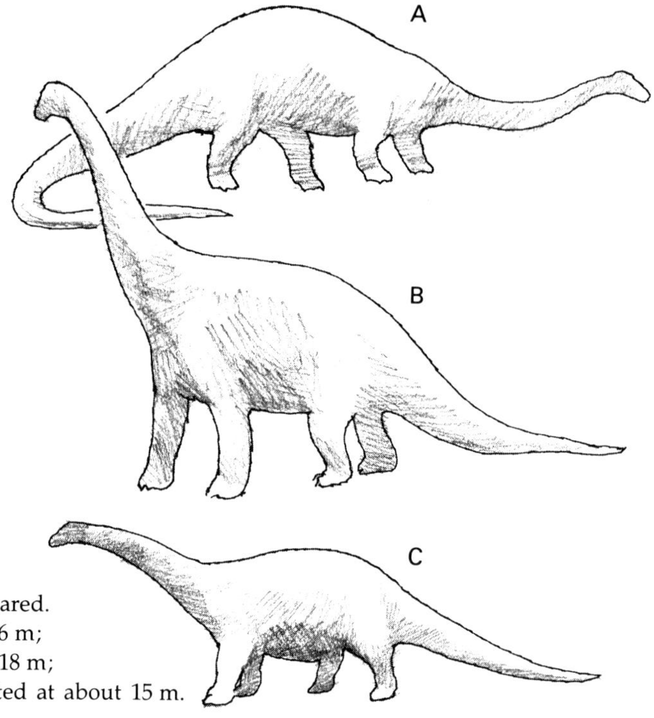

Some Jurassic sauropods compared.
(**A**) *Diplodocus*, length about 26 m;
(**B**) *Brachiosaurus*, length about 18 m;
(**C**) *Rhoetosaurus*, length estimated at about 15 m.

How big was *Rhoetosaurus*? What can we tell of its appearance and life-habits? The fossil evidence cannot provide exact answers, but it does allow some realistic estimates. Its total length, from snout to tail-tip, was probably in the region of 15 metres; its height at the hip, or shoulder, would have been about 4 metres. The neck was probably quite long, for the single vertebra from the neck is about 2½ times as long as a vertebra from the back region. The bones were, of course, very big and very strong, but many of them were riddled with cavities serving to reduce the total weight of the skeleton in the live animal. (The bones were mineralized during fossilization and are now very heavy indeed; one tail vertebra weighs nearly 14 kilograms).

All sauropods seem to have been inoffensive plant-eaters, with rather small skulls and relatively weak teeth, and there is no reason to suspect that *Rhoetosaurus* was an exception. The long neck was presumably flexible enough to have allowed the animal to feed at ground level or to have stretched up into tree-sized vegetation, rather like a modern-day giraffe. The stout and pillar-like legs had broad spreading feet with large blunt claws on the inner toes. *Rhoetosaurus* probably moved as slowly and as deliberately as a living elephant, using its clawed feet to gain secure purchase on soft or water-logged ground; it is literally true that an animal as big and as heavy as *Rhoetosaurus* could not have risked putting a foot wrong! The living *Rhoetosaurus* may have weighed as much as 20 tonnes — the equivalent of four full-grown elephants combined — but even this figure is far short of the sauropod weight record: the African sauropod *Brachiosaurus* may have weighed between 60 and 80 tonnes, and two American sauropods, '*Supersaurus*' and '*Ultrasaurus*', may have been even bigger. Some palaeontologists have reasoned that the sauropods were too ponderous to have supported themselves on dry land, and that they were obliged to buoy up their massive bodies by dwelling in the shallow waters of lakes and swamps. To what extent these giant dinosaurs might have ventured on to land is still a matter for speculation; the fossil evidence is inconclusive, and no living animal is big enough to provide any useful comparisons. Sauropods may, at least, have left the water to lay their eggs. No one has yet found any dinosaur eggs in Australia, but they are known from every other continent except Antarctica. If sauropods were egg-layers they must have been extremely long-lived. This is because the largest dinosaur eggs, presumably those of sauropods, are about 30 cm long and could have contained hatchlings no bigger than domestic cats. It must have taken a very long time for cat-sized hatchlings to grow to the enormous sizes attained by their parents, and it is conceivable that the normal life-span of a sauropod dinosaur was well over a century.

The sheer size of fully-grown sauropods might have rendered them immune from attack by predatory dinosaurs, but small and defenceless juveniles might have been easy prey. Consequently it is not surprising to find some evidence of herding behaviour in sauropods; a series of fossil tracks, discovered in Texas, has been attributed to a herd of sauropods on the move. The tracks seem to show that smaller animals occupied the centre of the herd, with their flanks protected by bigger animals.

Rhoetosaurus probably resembled other of the world's sauropods in its general appearance and life-habits, but its skeleton is too incomplete to add important new information to current understanding of sauropod biology. The real significance of *Rhoetosaurus* is in providing factual evidence about the anatomy of an early sauropod — evidence that may have a direct bearing on the origin and evolutionary history of the great sauropod dinosaurs.

Minmi paravertebra

Molnar, 1980

The Minmi, an Armoured Dinosaur

CLASS: Reptilia
ORDER: Ornithischia
FAMILY: (uncertain)
GENUS: *Minmi*
SPECIES: *Minmi paravertebra*
AGE: Early Cretaceous

Armoured dinosaurs have been reported twice in Australia: once due to the misidentification of bones of a sauropod dinosaur; and once due to the misidentification of fossil shells. The latter, although amusing, is not as far-fetched as it might at first seem, as the shells themselves were not preserved, only the impressions of their inside surfaces. These were somewhat incomplete and distorted and could be mistaken for pieces of dinosaurian armour. They were identified as not being bone since they lacked the characteristic microstructure of bone.

When first found in 1964, the specimen that later was identified as Australia's first ankylosaur was encased in hard, calcareous concretions and could be identified only as an armoured reptile. The specimen was not removed from the concretions for study (by dissolving away the concretions with acetic acid) until the late 1970s. It was quite incomplete, only eleven vertebrae from the back just in front of the pelvis, some ribs and associated bones, the right foot and some of the armour were preserved. Because so little of the animal was present and because the armour was from the belly, a region not usually represented (or at least not described) from other armoured dinosaurs, a detailed comparison had to be carried out with all of the known groups of armoured dinosaurs, not to mention other armoured reptiles, such as tortoises and crocodilians. The nature of the belly armour eliminated tortoises as a possible identification. The form of the vertebrae eliminated crocodilians, and other armoured dinosaurs. In ankylosaurs, the last vertebrae just in front of the pelvis (corresponding to the lumbar vertebrae of humans) are fused together to form a stiff rod. This fusion, not unlike the arthritis that sometimes occurs here in humans, presumably helped strengthen the vertebral column to better support the weight in ankylosaurs; beasts which could be described as very portly. These fused vertebrae are termed the presacral rod, and such a rod is known only in ankylosaurs. This rod was clearly preserved and confirmed that it was from an ankylosaur (Molnar, 1980). It was named *Minmi* after nearby Minmi Crossing where it was found.

The oldest ankylosaurs are known from the Late Jurassic of Portugal, England and Utah (U.S.A.) . So far they are poorly known, but appear to have been broad, portly

---

Minmi paravertebra, a small, armoured dinosaur known from the Early Cretaceous of southeastern Queensland. These forms were herbivorous and the armour provided protection from contemporaneous carnivorous dinosaurs, such as *Allosaurus*.

F. KNIGHT '82

beasts, long-tailed and covered with low, oval dermal plates (Galton, 1980). During the Early Cretaceous they had spread across North America and Eurasia, where they formed a prominent part of the fauna by the Late Cretaceous. They penetrated south into north Africa, but may have been rare — only a few bones of ankylosaurs have been found there. They did reach both Madagascar and India, although the specimens from Madagascar have yet to be studied and those from India, again are only a handful of individual bones. They have been reported from armour plates from Argentina, but in all likelihood these are actually from titanosaurs, and not from ankylosaurs at all. So they appear to have been common in the north but rare in the south. However, the fact that they had spread to Madagascar (at that time lying along what is now the coast of Kenya), India and Queensland, shows that they did inhabit at least the eastern part of the old southern land of Gondwanaland, although they may have been absent from the western part.

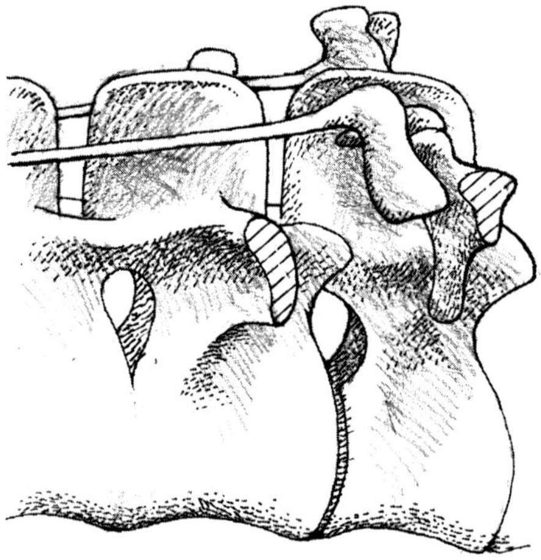

The dorsal vertebrae of *Minmi* viewed from the side, showing the paravertebra, with its long trailing rod, near the top of the vertebra.

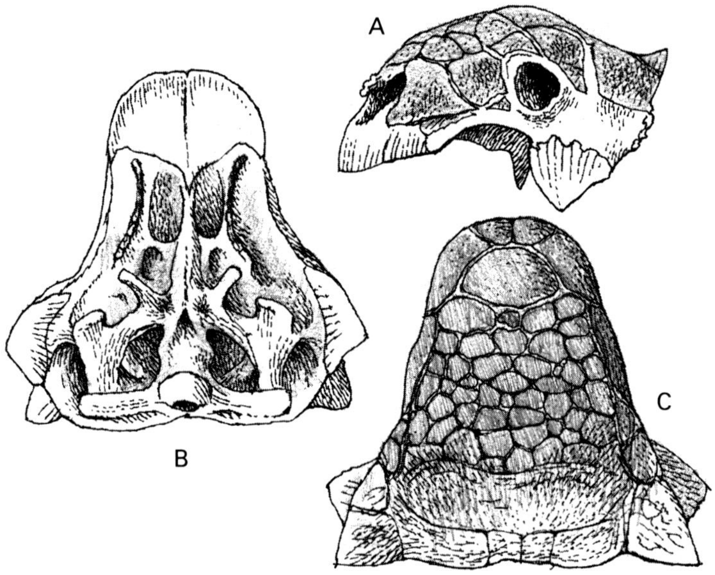

The skull of an advanced ankylosaur. Shown here is the skull of *Euoplocephalus* viewed from the side (**A**) and below (**B**), and the skull of *Ankylosaurus* viewed from above (**C**).

Ankylosaurs tend to have broad, flat skulls, well-covered with armour plates so that the openings (fenestrae) of the back of the skull are completely covered. Some ankylosaurs (such as *Ankylosaurus*) have low, broad, triangular horns at the back of the skull and in the cheek region, while others (such as *Pinacosaurus*) have two supplementary openings near each nostril.

The limb bones of ankylosaurs are stout, massive and short. The upper bones of the pelvis, the ilia, are greatly expanded forwards and laterally — perhaps to help support the armour in this region, or perhaps just the weight of the viscera. The ribs are T-shaped in cross-section probably to help support the armour. In the later forms, particularly *Ankylosaurus* and *Lametasaurus*, the tail vertebrae have complex, overlapping articulations that stiffen the tail and support a massive block of bone that tips the tail. This club was doubtless swung about by any disturbed ankylosaur and would effectively fracture the limbs or ribs of any creature approaching incautiously close. Ankylosaur skulls, while appearing quite solid on the outside, have a remarkable and mystifyingly complex series of sinuses and convoluted nasal passages on the inside (Coombs, 1978). The function of such chambers is unknown, although sharpening the

Various armoured dinosaurs as they probably looked in life.
(**A**): *Scelidosaurus,* a scelidosaur from England;
(**B**): *Lexovisaurus,* a stegosaur from France;
(**C**): *Tuojiangosaurus,* a stegosaur from China;
(**D**): *Struthiosaurus,* an ankylosaur from Transylvania;
(**E**): *Ankylosaurus,* an ankylosaur from the U.S.A.,
(**F**): *Saltasaurus,* an armoured sauropod (titanosaur) from Argentina.

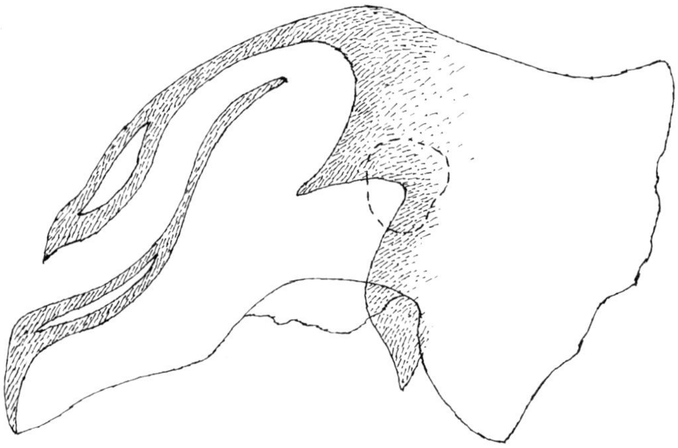

The skull of *Euoplocephalus* in cross-section, showing the convoluted nasal passage.

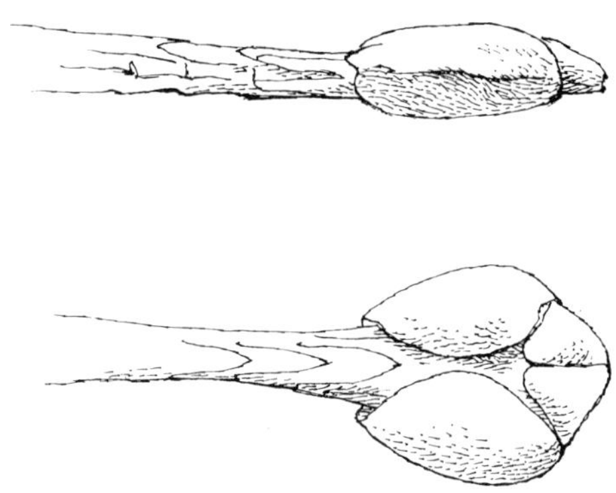

Tail club of *Ankylosaurus*, an advanced ankylosaur.

sense of smell by increasing the available area of olfactory epithelium comes immediately to mind.

Minmi has one peculiar feature not found in any other known ankylosaur, or for that matter among any other known reptile. Alongside the vertical neural spines of the vertebrae of the back were flat plates of bone about 2 cm long by 4 cm wide. Lying horizontally, these plates had little flanges bent downwards along the edge adjacent to the spine and a long cylindrical rod extending back and slightly down from the posterior margin. Since no other known vertebrates have such bones they were given the new name of paravertebrae, which (in Latin) means that they were situated alongside of the vertebrae. Their function cannot be determined. The long rod possessed by each is very similar to the ossified tendons found in other dinosaurs, and suggests a muscle attachment. If these plates were somehow also attached to the armour scutes of the back, and this is not certainly identified, then the paravertebrae would serve as movable bases and also allow *Minmi* to move its armour.

Belly armour is not described in detail for any other ankylosaur, although it was present in at least one other, the Mongolian *Saichania*. Apparently the whole underside of *Minmi* was coated with small, rounded scutes a little over ½ cm in diameter.

Because so little is really known of *Minmi*, the reconstruction given here is based very largely on animals assumed to be related such as *Silvisaurus* and *Nodosaurus*, which are known from more complete skeletons. We have used the head of *Silvisaurus* and the body armour of *Nodosaurus*. Since the cranial horns and tail club seem restricted to Late Cretaceous ankylosaurs and since *Minmi* dates from the Early Cretaceous, *Minmi* probably lacked these refinements.

The first specimen of *Minmi* died near a river and was washed into the Australian inland sea before being buried. The specimen was found lying upside down, and the body cavity had apparently been filled with clay before the limestone concretion, in which the specimen was found, had formed. Unfortunately, the form of the body cavity was greatly distorted before the specimen became preserved and hence tells us nothing of interest. Oddly enough the second known Australian ankylosaur, found near Boulia and yet to be studied, was also washed out to sea before burial. Even though so little is preserved of *Minmi*, it is the most complete ankylosaur yet described from the southern hemisphere (including India), which indicates how little is known of the African and Indian forms. It was a small form, no more than 3 metres long, but shows that these beasts once lived in Australia.

Dicynodontia
The Queensland Dicynodont

CLASS: Reptilia
ORDER: Therapsida
SUBORDER: Dicynodontia
FAMILY: Kannemeyeriidae
GENUS: Indeterminate
SPECIES: Indeterminate
AGE: Early Triassic

For many years there was one curious deficiency in the list of fossil reptiles known from Australia; there was absolutely no trace of therapsids, the so-called mammal-like reptiles. The absence of therapsids was very puzzling, because at the peak of their history, in the Permian and Triassic periods, these reptiles had been the most common of land animals. Mammal-like reptiles had been discovered on every other continent, even in Antarctica — yet they had never been identified in Australia. Very recently this problem was solved, when the first fragmentary remains of a mammal-like reptile were found in Queensland (Thulborn, 1983).

These fragments came from the Arcadia Formation, a thick sequence of sandstones and mudstones, at a site close by the Carnarvon National Park, in southeast Queensland. These rocks accumulated as sediments in freshwater pools, lakes and meandering streams during the early part of the Triassic Period, roughly 230 million years ago. In most places these rather soft sedimentary rocks have now weathered into a thick layer of soil which supports dense vegetation; but in the few areas where the Arcadia Formation is exposed by constant erosion, it appears as a brilliantly coloured succession of sandstone and mudstone layers — red, purple, green, chocolate and orange. Over the last ten years palaeontologists from Queensland and Victoria have collected hundreds of fossils from these colourful sediments, thereby building up a detailed picture of the animals that lived in and around the ancient watercourses of the Early Triassic. Most of these fossils are no more than weathered scraps of bone, which are sometimes very difficult to identify, but from time to time more informative specimens, skulls or parts of skeletons, have been discovered before they were too seriously damaged by weathering.

By far the most common animals in the Arcadia fauna were amphibians. From the abundance of bones and teeth, it seems that the freshwater environments of the Arcadia Formation were once literally teeming with amphibians, large and small. These were not modern-looking amphibians, like frogs and salamanders, but were bulky sprawling creatures known as labyrinthodonts on account of their distinctive tooth structure, as described by Anne Warren in an earlier chapter.

The Arcadia environment seems to have been something of an amphibian paradise; so far labyrinthodonts belonging to no fewer than seven families have been identified. No other rock formation in the world has produced such a diversity of labyrinthodonts. Together with these amphibians were several types of fish. These included lungfish, some of them similar to the living Queensland lungfish, and savage-looking predators resembling pike. Reptiles were not particularly common, but seem nonetheless to have been very diverse. Among the smaller and outwardly lizard-like reptiles were some very primitive types; procolophonids, which retained a pattern of skull structure like that of the very first reptiles, and some very advanced types; prolacertids, which may be close to the ancestry of modern lizards. In addition, there were large crocodile-like predators; these were not true crocodiles, but thecodontians, primitive socket-toothed reptiles, allied to the ancestors of dinosaurs and pterosaurs.

In many cases the bones and teeth of the smaller animals, be they fish, amphibians or reptiles, are found in coprolites — fossilized droppings — of the bigger animals. These coprolites can give a fascinating insight into who was eating whom more than 200 million years ago!

Some of the animals in the Arcadia rocks, especially the reptiles and some of the fish, have close relatives in other parts of the world. These particular animals give some idea of the age of the Arcadia Formation. More specifically, the reptiles indicate a correlation with the Early Triassic *Lystrosaurus* Zone of southern Africa, and with its equivalents in South America, Antarctica, India and China. In all these geographic regions there are Early Triassic faunas, which include some or all of the distinctive types of reptiles discovered in the Arcadia Formation. From the presence of these widely distributed reptiles it seems fairly safe to conclude that the Arcadia fauna is also of Early Triassic age. This conclusion is reinforced by evidence from the labyrinthodonts and from the study of fossil spores (palynology).

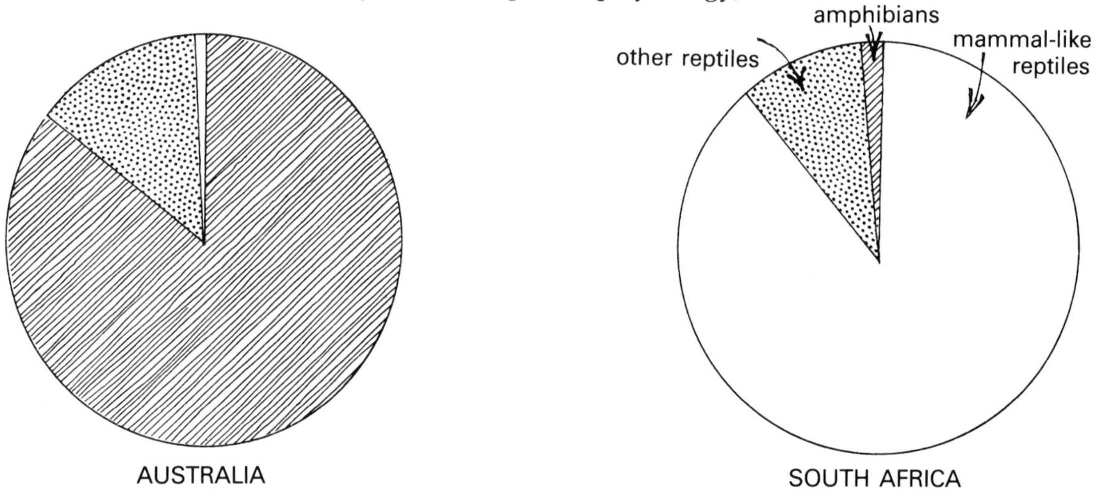

The reptile and amphibian fauna of Australia (left) compared with that of South Africa (right) during the Early Triassic. The African fauna is dominated by reptiles (especially mammal-like reptiles), whereas the Australian fauna is composed mainly of amphibians. Faunas of similar age in India, China and Antarctica are generally similar to the South African fauna. The abundance of amphibians in the Australian fauna is an ecological peculiarity yet to be explained.

Nevertheless, there remains one awkward problem. In other parts of the world the fauna of the *Lystrosaurus* Zone is dominated by therapsids, the mammal-like reptiles, while labyrinthodonts are rare or completely unknown. In South Africa, for example, about 85% of the fossil vertebrates found in the *Lystrosaurus* Zone are examples of the single therapsid genus *Lystrosaurus*. In fact, this animal is so common and so distinctive that it serves as a zone-fossil — a convenient indicator to the age and identity of the sediments. The same genus, *Lystrosaurus*, has also been found, and sometimes quite commonly, in India, China and Antarctica. By contrast, the fauna of the Arcadia Formation is dominated by labyrinthodonts, and until very recently there seemed to be no trace at all of therapsids. In other words this Australian fauna seemed to be completely anomalous; animals that were rare in other continents (the labyrinthodonts) were extremely abundant, and animals that were common elsewhere (the therapsids) were entirely absent. Several explanations were possible, although none of them seemed particularly convincing. Perhaps the Arcadia Formation was not Early Triassic in age; this seemed unlikely because all the reptiles that had been found were characteristic Early Triassic types. Or could it be that Australia was separated from the other continents by some ecological barrier during the Early Triassic? It was difficult to imagine any barrier that would allow free passage to fish, amphibians and most reptiles, but would simultaneously bar the entry of therapsids. Or could it be that therapsids *were* present in the Arcadia fauna, but that palaeontologists had failed to recognize them? This also seemed unlikely; the bones of therapsids such as *Lystrosaurus* are very characteristic in shape and would not be easy to overlook, even if they had been weathered into fragments.

Then, at last, came the discovery of an unmistakable therapsid fossil. This was no more than a single bone — the quadrate, which in reptiles is located at the lower back corner of the skull, where it forms the upper part of the jaw joint. It was, indeed, a stroke of luck finding a quadrate, rather than any other bone. This is because the quadrate has an unusual and highly distinctive shape in the one particular group of therapsids that includes *Lystrosaurus* and its allies. In this group of therapsids, the dicynodonts (Suborder Dicynodontia), the quadrate is elaborated into a pair of large knuckle-like condyles, so that the jaw joint is almost pulley-shaped. This peculiar jaw joint, which allowed the lower jaw to slide backwards and forwards during feeding, is a feature unique to the dicynodonts. So the discovery of this unusual quadrate bone was conclusive evidence that a dicynodont, something like *Lystrosaurus*, did exist in Australia. The name dicynodont (literally meaning 'two canine teeth') refers to the fact that many of these therapsids had only two teeth, a large downwardly projecting tusk on each side of the snout. Recently a part of one of these characteristic tusks was discovered among the fragmentary fossils collected from the Arcadia Formation.

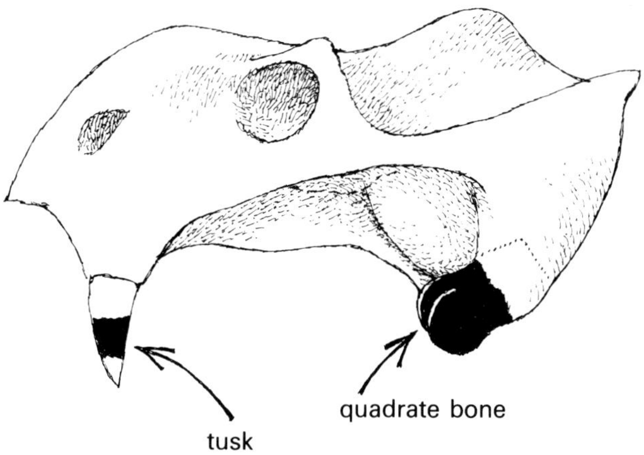

Outline diagram of a generalized dicynodont skull. Note the beak-like snout and the prominent tusk. Elements so far identified for the Queensland dicynodont are shown in black (quadrate bone and part of a tusk).

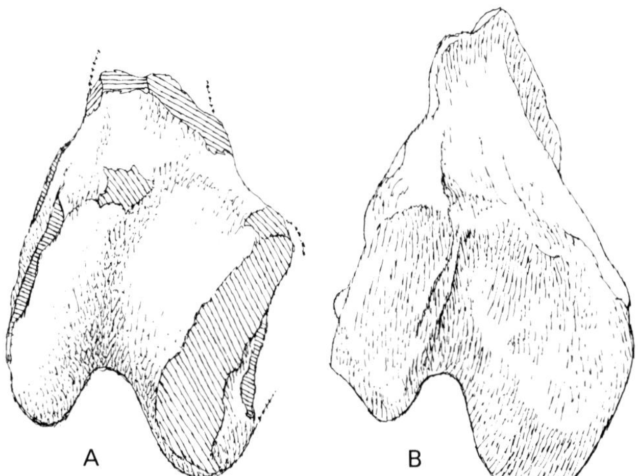

Left quadrate bone of the Queensland dicynodont (A) compared with that of an African dicynodont (B) *Kannemeyeria*. Both are shown in front view. These distinctive pulley-like bones formed the upper part of the jaw joint, and allowed the lower jaw to slide backwards and forwards during feeding.

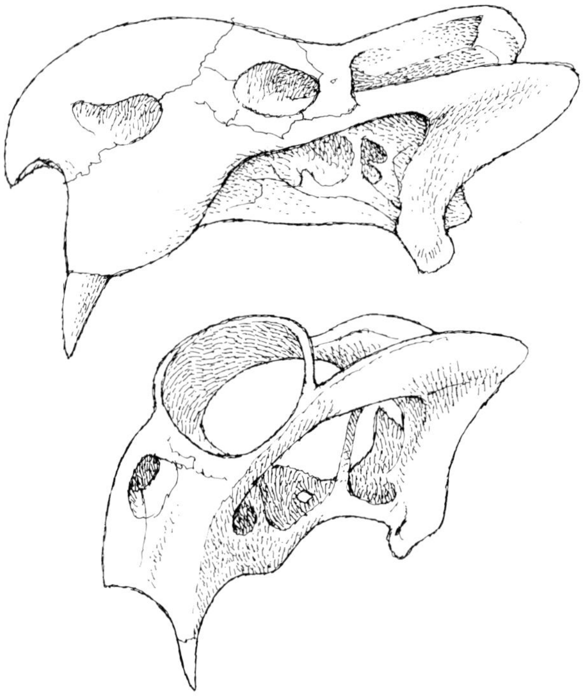

Skull of the dicynodont *Kannemeyeria* (above) compared with that of the dicynodont *Lystrosaurus* (below). The *Kannemeyeria* skull is quite large (about the size of a cow's skull) and has a pronounced beak-like snout; the *Lystrosaurus* skull is smaller (about the size of a sheep's skull), with a very short snout, and comes from slightly older sediments. The Queensland dicynodont probably resembled one or other of these African reptiles, but it is difficult to decide which, because the quadrate bone is rather similar in both.

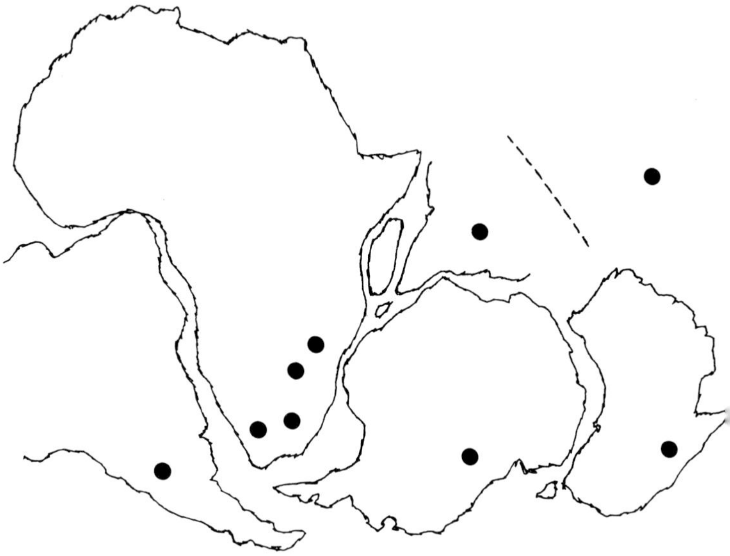

Outline reassembly of the southern (Gondwana) continents during the Early Triassic. Spots indicate localities where dicynodonts of the Family Kannemeyeriidae have been discovered (including *Kannemeyeria, Lystrosaurus,* and the Queensland dicynodont). The position of China (marked off by dashed line) is uncertain.

Dicynodonts were not flesh-eaters, despite their tusks. At the front of the skull the upper and lower jaws came together in a sharp-edged beak, rather like that of a turtle, so that the animal could slice and shred its resistant plant food. The function of the tusks is not at all clear; they may have been digging tools, weapons or display structures. One interesting theory is that tusk-bearing dicynodonts were males whereas those forms without tusks were females.

In overall appearance a dicynodont would have been a robust and rather clumsy-looking quadruped, something like a miniature hippopotamus. Some mammal-like reptiles may have possessed hair, or fur, but it is not known if this occurred in dicynodonts. Average specimens of *Lystrosaurus* were about the size of sheep. In fact, this is a useful analogy, because herds of dicynodonts probably filled much the same ecological role as do herds of sheep and goats today.

What exactly is the identity of the Australian dicynodont? Unfortunately the animal cannot be identified with certainty on the basis of isolated fragments, particularly since the quadrate bone, although distinctive enough, seems to have been rather uniform in structure through a wide range of dicynodont genera. The Australian

---

A mammal-like reptile from Early Triassic freshwater deposits in southeastern Queensland. Although dicynodonts had tusks, they were not carnivorous, but instead may have used these structures for digging up food, as defence weapons, or in display.

F. KNIGHT 82

fossils could possibly represent a large example of *Lystrosaurus*, which would tie in very neatly with the presumed *Lystrosaurus* Zone age of the Arcadia Formation. Alternatively the fossils could be from *Kannemeyeria*, a large dicynodont that occurs in abundance in the rocks immediately above those of the *Lystrosaurus* Zone. *Kannemeyeria* seems to have been about as widespread as *Lystrosaurus*, having been reported from South America, Africa and India. At present the Australian mammal-like reptile can be identified no more precisely than 'dicynodont'. Only when more complete fossils are found will it be possible to decide if this animal is *Lystrosaurus* or *Kannemeyeria*, or some other dicynodont.

Nevertheless the identification 'dicynodont', inadequate as it might seem, is of great significance; it provides important clues to the remote history of Australian vertebrate animals, and it points the way forward for new research into that history. Firstly, the dicynodont provides evidence for the former existence of the southern supercontinent Gondwana. It confirms that Australia was once part of that supercontinent, and that it once supported an ancient vertebrate fauna similar in many respects to that found in Antarctica, Africa, South America, India and China. Secondly, the discovery of the dicynodont fills an obvious gap in the Australian fossil record. But it simultaneously draws attention to another problem: the apparent rarity of mammal-like reptiles on this continent. Was the rarity of mammal-like reptiles really an ecological peculiarity of Australia during the Early Triassic? Or is it simply that the sediments of the Arcadia Formation accumulated in an unusual environment that was rarely visited by mammal-like reptiles? Only further searching, both in the Arcadia Formation and elsewhere, can provide answers.

Lastly, and most importantly, is the fact that in all other parts of the Gondwana supercontinent dicynodonts are found in association with a second group of therapsids — the cynodonts ('dog-toothed' reptiles). The cynodonts are of the greatest interest, because they were the most mammal-like of all the reptiles, and probably included the ancestors of mammals (Kemp, 1982). Since dicynodonts occur in Australia, one might expect to find cynodonts as well. In other words, the discovery of a dicynodont raises the hope that very ancient mammals and perhaps even mammal ancestors, may one day be discovered in Australia.

Anthropornis nordenskjoeldi

Wiman, 1905

Nordenskjoeld's Giant Penguin

CLASS: Aves
ORDER: Sphenisciformes
FAMILY: Spheniscidae
GENUS: *Anthropornis*
SPECIES: *Anthropornis nordenskjoeldi*
AGE: Late Eocene to ? Early Oligocene

Living penguins are highly specialized marine birds of the southern hemisphere (Stonehouse, 1968; Simpson, 1976). The majority inhabit islands and parts of continents between about 30° and 68°S, with the largest number of species and individuals associated with sub-Antarctic islands, 48° to 62°S. Sea temperatures in this region are generally cool to cold, while the Antarctic Ocean has rich concentrations of marine plankton or 'krill' that provide a critical link in the food chain of larger marine animals, including penguins (C.S.I.R.O., 1979). The more northerly distribution of penguins on the west coast of South America and the Galapagos Islands is related to the cool, northward flowing Humbolt Current.

Certain well-ossified bones of penguins form frequent fossil remains known to date from the Early Tertiary onwards. Without exception, the fossil finds are near places where penguins live today (Simpson, 1946, 1976), and it is apparent that ancient penguins had major ecological requirements comparable to those of modern forms.

The fossil remains are chiefly of extinct species not closely related to living ones. Some represent small forms, but many are of robust birds, approaching or exceeding the size of the largest modern species, the Emperor Penguin *Aptenodytes forsteri*, which stands 90 cm tall and inhabits Antarctic shores. Comparatively huge bones eroded from Tertiary rocks on Seymour Island, Antarctic Peninsula (Wiman, 1905), and the finding of massive chest and wing bones of *Pachydyptes ponderosus* in the Late Eocene (approx. 40 million years ago) of southern New Zealand, catalysed speculation that early penguins reached giant size, perhaps 'as tall as a man' (Oliver, 1930). The great fossil penguin specialist G.G. Simpson originally countered such reports to an extent (Simpson, 1946), but later estimated standing heights for these birds of between 134 and 169 cm (Simpson, 1975, 1976).

In 1932 L.W. Parkin, recently Director of the Australian Mineral Foundation, Adelaide, discovered several fragmentary bones in the Late Eocene, Blanche Point Formation (Jenkins *et al.*, 1982), at Blanche Point, 37 kilometres SSW of Adelaide, but this material went unrecognized for some time. The first identification of a fossil penguin from Australia was made by Finlayson (1938) from a wing bone (humerus) found by Mr. W. Burdett in the same formation at nearby Witton Bluff, Port Noarlunga. Mr. Burdett was so pleased by his discovery that he made a bequest to the South Australian Museum for an award which has several times been granted to further additional research on fossil penguins, surely a most unusual prize! M.F.

Outcrops of Late Eocene, Pliocene and Pleistocene rocks at Blanche Point, South Australia. Fossil penguin remains occur within the Tuketja Member of the Blanche Point Formation just above the shelf at the base of the cliff, and in the Gull Rock Member of the same formation at a level about half the height of the lower cliff.

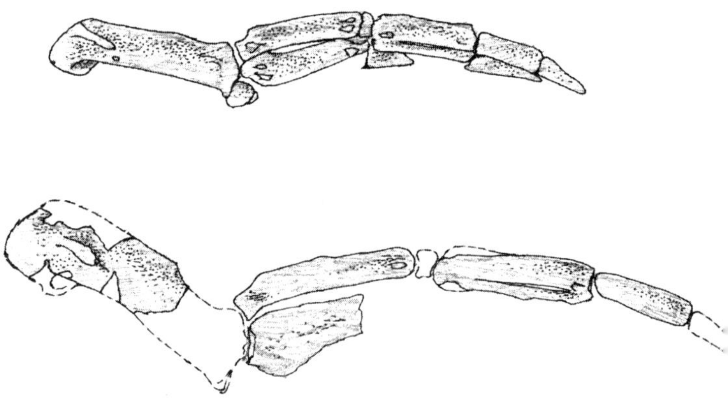

Comparison between the bones of the wing of *Anthropornis nordenskjoeldi* (bottom) and the wing of the largest living species of penguin, *Aptenodytes forsteri* (top); about 35 cm in length.

Glaessner (1955) found and described a penguin leg bone, also from Witton Bluff. On clambering down a rather dangerous cliff face to a 'fishermans ledge' on Blanche Point during 1968, schoolboy collectors Bret Robinson and Harry Eames discovered the partial skeleton of yet another penguin; bones of the flipper resembled the fragments collected by Parkin. Clearly the form was another 'giant', and it was described as *Pachydyptes simpsoni* by Jenkins in 1974. Weathering of the cliff later revealed more bones of this same skeleton. The collection was nearly interrupted when a young lady student peering over the cliff edge at the proceedings accidentally dropped a hammer, which narrowly missed my head! A leg bone collected on this occasion proved closely comparable to Glaessner's Witton Bluff find.

Finlayson's bone is of an Antarctic form also known from Seymour Island. A smaller third form since found at Blanche Point (one bone), has also been found in the Late Eocene rocks of Browns Creek, Otway Peninsula, Victoria.

Fossil remains of penguins consist mostly of isolated bones remaining in sediments after cadavers have been dismembered by scavengers and washed about on the sea floor. The fossil bones are assigned to particular species on the basis of peculiarities of shape and size. Associations of several parts of one bird are rare. The numerous finds from the probably Late Eocene to Early Oligocene sediments of Seymour Island consist mostly of such isolated bones (Marples, 1953; Simpson, 1971b; Elliot *et al.*, 1975; Zinsmeister, 1982). The 15 or so largest bones have been variously described as *Anthropornis nordenskjoeldi* and *Orthopteryx gigas*. It is simplest to assume that only one exceptionally large kind of penguin inhabited this area, and that these particular bones all belong to *A. nordenskjoeldi*.

On reconstruction of frost-shattered material and careful comparison with the Blanche Point bones, it became apparent that *P. simpsoni* is the same species as *A. nordenskjoeldi* and should be referred to by this latter name, which has historical priority. [*Anthropornis* is from the Greek *Anthropos* for 'man' and *ornis*, 'bird', in reference to the size approaching that of humans; *nordenskjoeldi*, is for the Swedish

Giant, primitive penguins, *Anthropornis nordenskjoeldi*, congregating on the eastern shore of the ancient Gulf St. Vincent during the Late Eocene. The recurved shape of the flipper and long neck are traits probably inherited from the flying ancestor of penguins.

F. KNIGHT 81

geologist and polar explorer Professor Nils Otto Gustaf Nordenskjöld (1869-1928), leader of the expedition which recovered the bones from Seymour Island (Simpson, 1971b)]. The preserved parts enable reconstruction of the wings and legs, but not the body outline. A single neck vertebra is relatively elongate, suggesting a lengthy neck.

A large penguin occurring in the mid Oligocene (approx. 30 million years ago) of the South Island of New Zealand is represented by several partial skeletons (Marples, 1952; Oliver, 1930; Simpson, 1971a), giving sufficient material to make a reasonable reconstruction. A single bone of the same species bearing grooves apparently made by the bite of a shark, is known from rocks of similar age near Mount Gambier, South Australia (Glaessner, 1955; Simpson, 1957). This species discloses that Oligocene penguins looked rather similar to modern ones. However, an extraordinary attribute of this Australasian form is that it had a long tapering beak, nearly twice the length of the skull.

The high body weight of penguins and the streamlining of their body form necessitate a near vertical stance on land, and hence determines the geometry of the legs. A vertical stance seems already to have been fully developed in the Oligocene species, and what is known of the Eocene forms indicates a comparable posture. The reconstructed form is amongst the largest of known fossil penguins, standing about 1.2 metres high in a 'normal' terrestrial stance.

A. nordenskjoeldi was evidently fractionally larger, and was only slightly exceeded in size by *Pachydyptes ponderosus* and possibly other Oligocene New Zealand forms known from only a few bones and fragments. The heights of these birds may have approximated Simpson's smaller estimate of 134 cm, but were certainly shorter than his maximum of 169 cm, a reflection of the fact that the relative proportions of the bones of ancient penguins differ from those of their modern counterparts. The body weight of *A. nordenskjoeldi* was possibly of the order of 90 kilograms.

On the basis of its large body size, hydrodynamic constraints dictate that *A. nordenskjoeldi* was a rather slow swimmer with speeds perhaps of the order of 7-8 kilometres per hour. Hence the particular advantage of the long neck in the capture of motile prey. Its geographic range was quite extensive, from Seymour Island and presumably through the southern South Atlantic and Southern Indian Oceans to southern Australia. On Seymour Island, it was part of a varied penguin population of about five genera and eight species. It is notable that the densest assemblage of living penguins (four genera and seven species) occurs nearby in the Scotia Arc, and the Antarctic Peninsula, where Seymour Island lies, and is sustained by concentrations of Antarctic krill and small squid.

Reconstructing continents to their Early Tertiary positions, the known finds of Late Eocene penguins can be interpreted as located at about 45° to about 72°S. The more northerly finds are apparently restricted to narrow stratigraphic intervals. Seas of the Late Eocene were warmer than present, and it is possible that the visitations of penguins to southern Australia and New Zealand may have coincided with cooler fluctuations in climate (Jenkins, 1974).

Simpson (1946, 1976) championed the view that penguins evolved from flying ancestors, probably a member of the seabird group, the Procellariiformes, which includes the petrels, shearwaters and albatrosses. He observed that the earliest known fossil penguins were already essentially 'penguin-like' and hence argued that their ancestry must have been much earlier, probably within the Cretaceous. The oldest known penguins, such as *A. nordenskjoeldi*, certainly retain evidence of their probable flying ancestors in certain skeletal features; thus the coracoid bones, the main prop between the attachment of the wing and the breast-bone or sternum, are relatively short and broadly flared, characteristics present in volant seabirds. As well, the elbow

Approximate positioning of southern hemisphere lands in the Late Eocene and localities where fossil penguins of this age have been found. *Anthropornis nordenskjoeldi* and Finlayson's penguin (a new genus and species) both occur on Seymour Island, Antarctica, and in southern Australia, and were possibly distributed around the coast of Greater Antarctica and in the South Indian Ocean. *Palaeeudyptes antarcticus* evidently ranged from New Zealand via the southernmost Pacific Ocean to Seymour Island. An hypothetical pattern of ocean circulation is shown, with cool (solid arrow heads) and warm currents indicated. Inner circle is 45° of latitude.

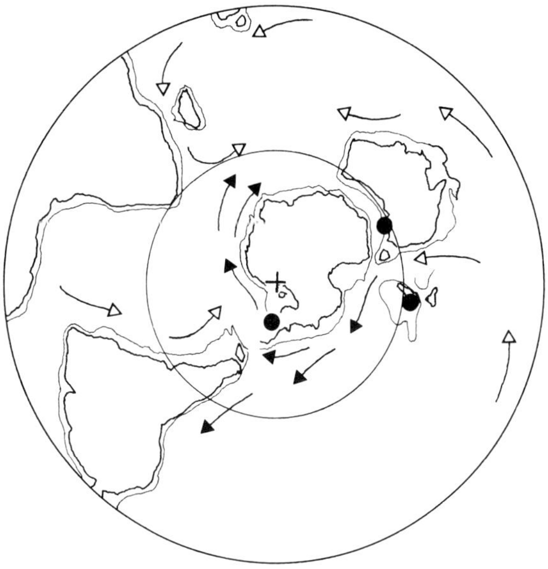

of the wings is still angled, a trait of flying birds, not straightened as in later penguins. The mode of underwater 'flight' of penguins is clearly one of their major adaptations to an aqueous realm (Clark and Bemis, 1979), and the remains of *A. nordenskjoeldi* reveal that the flipper was already flattened and heavy-boned for this specialized function.

Fossil penguins show several evolutionary grades or steps trending towards the modern forms. I consider it possible that the initial adaptation of penguins to an aqueous realm following loss of flight may have been a relatively rapid event. This transition was likely accompanied by intense selection, with resulting accelerated changes in physiology and form related to the need to conserve body heat in a much more conductive medium, and the requirements of a comparatively heavier body and greater functional streamlining necessary for deep-diving. A pronounced change in the oceanographic regime of far southern seas occurred at about the Early-Middle Eocene boundary (McGowran, 1978), and it is tempting to suppose that this event triggered the evolution of penguins, perhaps through the development of a proto-krill in near Antarctic seas. Such an optimum in food resources would have minimized the necessary ranging capacity of birds to feed their chicks, and allowed transition to slower aqueous travel. An Eocene origin of penguins is also consistent with the lack of known earlier fossils.

The terminal Eocene and Early Oligocene were characterized by sharp climatic deterioration linked to the progressive separation of Australia and Antarctica and the strengthening of the Circum Antarctic Current. The effect of these changes on the oceanic food chain is unknown, but the extinction of *A. nordenskjoeldi* and several other large Early Tertiary penguin species broadly coincide with these events.

The reconstruction here shows a parade of *A. nordenskjoeldi* on the eastern shore of ancient Gulf St.Vincent, with the ancestral Mt. Lofty Ranges behind. The plumage is based on the adaptive colouration of modern penguins. Shells of *Mikadotrochus* (wide coil), *Cimomia* (orange patterning) and *Spiroculpus* (narrow cone), all known to be living in Australian coastal waters during the Late Eocene, are shown washed up on the beach. The branch in the foreground is of *Banksieaephyllum*, one of various kinds of plants likely to have grown in near-shore areas at the time *Anthropornis* was alive.

Genyornis newtoni

Stirling and Zietz, 1896

A Mihirung

CLASS: Aves
ORDER: Uncertain
FAMILY: Dromornithidae
GENUS: *Genyornis*
SPECIES: *Genyornis newtoni*
AGE: Pleistocene

The day is warm, not hot, and without a breeze. Several men are walking slowly about on the flat claypan. A few have long, slender metal rods in their hands and between brushing flies away, they stop and push these rods firmly into the lake surface. They continue for some time before someone calls the others to his side and after discussion, several leave and walk back toward a Land Rover parked at the claypan's edge. The two men left on the lake bed continue to probe with the rods that seem to stop after being pushed into the soft clays about half a metre. Obviously, something lies beneath the lake surface that is resistant enough to stop the probes. Soon several men return, only to set about excavating a large pit; before long the bones of a large, long dead animal begin to appear. The bones are what the probes have struck and the paydirt of vertebrate palaeontologists who, like 'Mad dogs and Englishmen', are often found out in the midday sun in search of their elusive prey!

The cameo is real, not imaginary, and actually took place on the dry surface of Lake Callabonna in the early 1950s. The cast was a group of South Australians and Californians led by the late Ruben A. Stirton, an indefatigable palaeontologist from the University of California at Berkeley, who brought about a renaissance of field work, the main aim of which was to find the bones of fossil marsupials older than a few million years. Stirton and his colleagues were following up earlier discoveries at Lake Callabonna that started on 10 January 1892, when an Aboriginal informed F.B. Ragless of Callabonna Station that whole skeletons of very large animals were present in the bed of the lake (Hale, 1956; Stirling, 1896; Stirling and Zietz, 1899, 1900). Ragless informed E.C. Stirling and A.H. Zietz, then Honorary Director and Assistant Director of the South Australian Museum in Adelaide and they in turn sent, H. Hurst to Lake Callabonna to report on the situation there. He travelled to Leigh Creek and then to Callabonna at the end of November 1892, returning to Adelaide in December with the news that fossils were much more abundant than originally supposed. The Museum was enthusiastic and donated £250, despite the stringent economic times. Hurst, with three assistants, left Adelaide again in January 1893 to obtain a representative collection. When he returned to Adelaide by 'buck-board buggy' in June, he brought back (in a series of three trips) remains of many *Diprotodon*, giant extinct

---

A *Genyornis* picking up pebbles which were swallowed to aid in the mechnical breakdown of food. The bird in the foreground is retrieving stones left behind from a dead specimen, whose bones lie bleaching on the surface of Lake Callabonna some 30,000 years ago, in preparation for fossilization.

F. KNIGHT 80

marsupials. Besides the big mammals, however, as the government geologist, H.Y.L. Brown (1894) noted on his visit to the lake, there were 'the legbones of a gigantic bird as large as the moa' — a bird which Stirling (1896) named *Genyornis* (from the Greek words *genys*, meaning 'jaw', and *ornis*, meaning 'bird'). Further work was carried out by Zietz, with the result that a large collection of *Genyornis* bones was made and transported by camel back to the South Australian Museum for study and storage.

First announced in 1893 by Alfred Newton in the British scientific journal *Nature*, further studies by Stirling and Zietz (1896, 1900) led to a detailed understanding of *Genyornis*. Lake Callabonna had produced several nearly complete skeletons, and when these were compared to those of other large ground birds such as the Australian emu, the extinct New Zealand moas and the extinct Malagassy elephantbirds (Wetmore, 1967), they were clearly quite distinct from them. Although still known only from fragments, the skull of *Genyornis* appeared to be very different from all of the other big, ground-dwelling birds, having a deep and massive lower jaw and a compressed quadrate (the hinge bone that connects the lower and upper jaws in birds) (Rich and Balouet, 1984). The wings were very reduced — quite obviously *Genyornis* did not fly — and the sternum or breast-bone had no median ridge (keel) on it, which in flying birds forms a large surface of attachment for the powerful flight muscles. The shape of the hind limb bones, the femur (thigh bone), the tibiotarsus (drumstick) and the tarsometatarsus, were distinct from those of other birds, and the inside toe (digit, II) was more reduced than that of the living emu and the moas. Unique too were the 'claws' or ungual phalanges of the feet, which resembled the shape of small horses' hooves. Although many authors in the past have called *Genyornis* a 'giant emu', the bones of this bird clearly demonstrate that it is not an emu. For instance, the femur of *Genyornis*, has a very compressed, not a deep, distal end and the trochleae (articular surfaces for the toes) of the tarsometatarsus of the dromornithids are deep and tapering, not shallow and parallel-sided. It is so distinct from other birds that it belongs in its own separate family (Dromornithidae) and perhaps its own separate order (Dromornithiformes), whose relationships to other families and orders of birds are not at all understood (Rich and Balouet, 1984). This group of birds has been given the 'common' name mihirung in order to point out its difference from emus.

Besides just bones, the 'unctuous blue clay containing abundant quantities of saline minerals and concretions of carbonate of lime' (Stirling and Zietz, 1900) contained other information about *Genyornis*. Many of the skeletons were found with the legs in a vertical position suggesting that they had become bogged on an unpredictable lake surface and thus died. Often the birds were found in small groups, as if several had become bogged together, perhaps inferring that they may have lived in groups like the emus, rather than as solitary individuals like cassowaries. Associated with the skeletons and often resting on the top surface of the breast-bone (sternum) were masses of gravel, similar to stomach stones (gastroliths) found in chickens and many other kinds of birds, both herbivorous and carnivorous. These stones ranged from coarse sand to pebbles up to 2 cm in diameter and averaged about 0.4 kilograms per mass, consisting of sandstone, jasper, claystone, black and clear quartz, chalcedony and even hardened clay — a great variety indeed. Although no plant material has ever been found associated with the skeletons of *Genyornis* that could give a clue as to what these birds were eating, plant fragments and pollen in the fossil-bearing clays indicate that at the time *Genyornis* was alive, the vegetation at Lake Callabonna was not remarkably different from that there today. Perhaps the climate was slightly wetter, more like the area south and east of Lake Frome today, but it was still an arid land, certainly not the lush forest that probably covered this area 20 million years earlier.

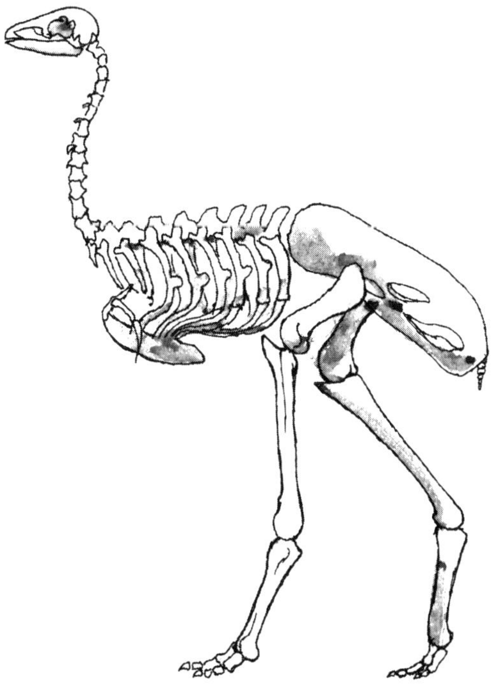

Skeleton of *Genyornis newtoni* from Pleistocene sediments at Lake Callabonna in South Australia. Notice the small wing and the deep lower jaw.

Mr. Ranken descending into Breccia Cave in 1838 in the Wellington district of New South Wales. His rope is attached to a bone, the first dromornithid ever found.

Although the fossils at Lake Callabonna provide more information about *Genyornis* than any other locality elsewhere in Australia, other sites have their value. Perhaps the first 'observation' of *Genyornis* was by Aborigines, for European man never saw it alive. Among others, the Tjapwurong tribe of western Victoria had a name for 'Giant Emus' — *mihirung paringmal* — and oral tradition implied that these birds existed in the Dreamtime when volcanoes were active in the Western District — that could have been as recent as 6000 years ago, or even less. For a fact, something very like *Genyornis*, if not *Genyornis* itself, was alive in the Lancefield area of Victoria 26,000 years ago, as radiocarbon dates on charcoal at this site confirm (Gillespie *et al.*, 1978). Man and mihirungs definitely overlapped in time, and the traditional tale of the Aborigines could be correct. Cave paintings in the Cape York region discovered by Percy Trezise with Quinkan figures (Rich, 1979, 1981) and carved tracks of giant birds at Mootwingee and Pimba (Tindale, 1951) all are suggestive of contact between mihirung and man.

The first European to collect bones of mihirungs was probably Major Thomas L. Mitchell (1838) when he was exploring the Wellington Valley Caves in New South Wales. As Mitchell noted in his diary, this discovery was not without humour as is discussed earlier. Unfortunately the bone found by Mitchell and Ranken, the original mihirung, had subsequently been lost, and the only record of it is the drawing of it on Plate 51 in Mitchell's book, provided by the well-known British comparative anatomist of the day, Sir Richard Owen, well remembered for his work on the New Zealand moas. More material of mihirungs from the Wellington Caves has recently been discovered by Tim Flannery and Lyndal Dawson of the University of New South Wales, so their presence there is definite.

Over the next few decades bones continued to be found by government geologists (H.Y.L. Brown, R. Etheridge, R. Tate), priests (J.E.T. Woods) and amateurs, and such fossils gradually filtered into museum and university collections. In 1869 W.B. Clarke, the rector of Willoughby, New South Wales, and a well-known early geologist in Australia, and G. Krefft, the Curator of the Australian Museum, reported on a strange thigh bone that was recovered from a well sunk 55 metres at Peak Downs, Queensland. Sir Richard Owen (1872, 1874a) called this *Dromornis australis*, the Latin scientific name (meaning 'southern running bird') for the first described mihirung. Later, in 1888, based in part on this specimen, the family Dromornithidae was set up. Still later, the Callabonna discoveries led to a much better understanding of these feathered leviathans.

After the turn of the century, although fossils of mihirungs continued to turn up sporadically, little anatomical information was added to that gained from the Callabonna discoveries. Because of these twentieth century discoveries, the geographic range of *Genyornis* was expanded, but no more *Dromornis* were found. In the early 1950s a series of major joint expeditions, manned primarily by Australians and Americans, initiated by R.A. Stirton and R.H. Tedford of the University of California at Berkeley and coordinated by Paul Lawson of the South Australian Museum, began a fruitful exploration of the Australian interior — the 'Dead Heart' of J.W. Gregory (1906), who explored it for similar reasons with a team of Melbourne University students. Stirton and his associates had phenomenal luck combined with persistence, and investigated many sites in South Australia, Queensland, and the Northern Territory that ranged in age from sometime in the Miocene (15 million years ago) to very recent. Besides the collections of marsupials which was their prime aim, these expeditions brought back fossils of lungfish, crocodiles, turtles, flamingoes and several species of mihirungs, including more of Owen's *Dromornis* from 10 million year old river sediments near Alice Springs, and a new species called *Dromornis stirtoni* that was much bigger than the Peak Downs *Dromornis*. In fact, it is probably the largest bird that ever lived, rivalling even the elephantbirds of Madagascar in bulk.

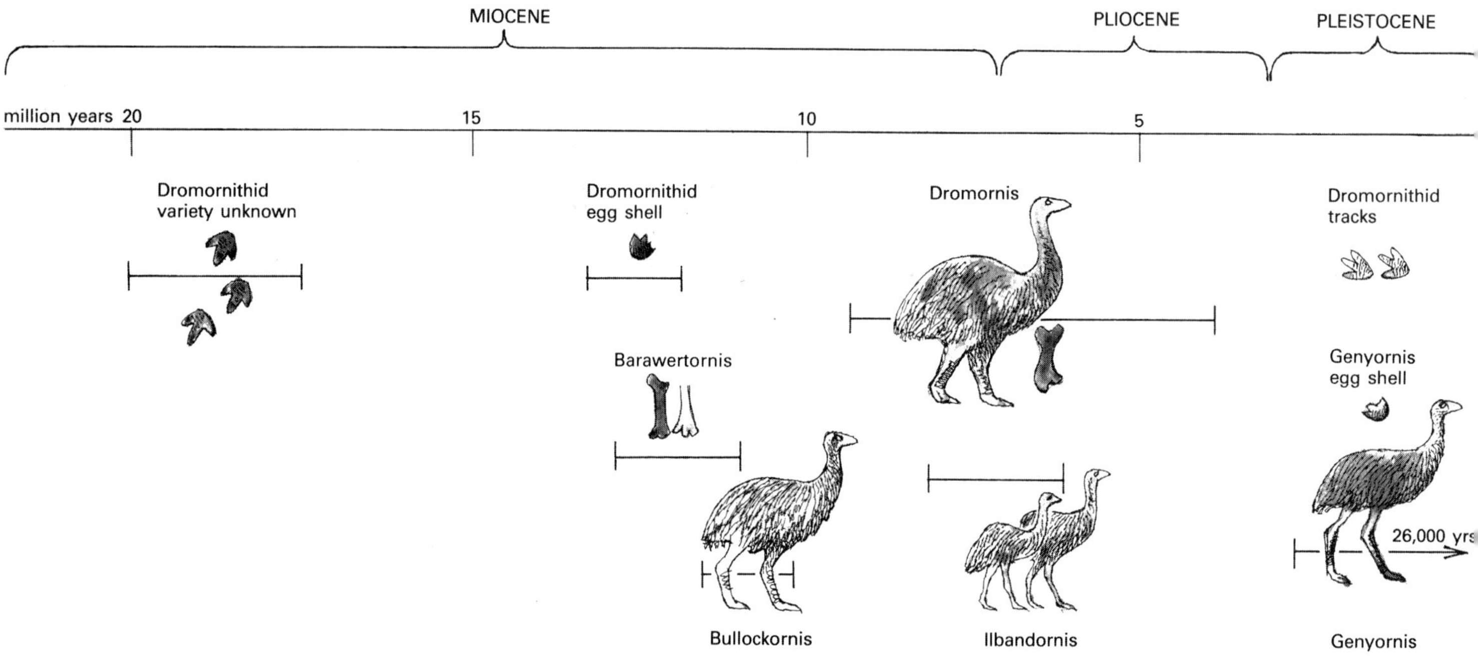

Distribution in Australia of the different dromornithids through time.

At least 6 genera and 8 species of mihirungs are known to have lived during the past fifteen million years in Australia. Some like the geologically youngest, *Genyornis*, and an older form called *Ilbandornis* (*Ilbanda*, Aboriginal for 'ground '; *ornis*, Greek for 'bird') were only slightly taller than the living emu; some were streamlined and gracile, while others were massive and elephantine, such as *Dromornis stirtoni*. In short, the different species of mihirungs had a variety of locomotary styles, perhaps some adapted to savannah or forest conditions (*Dromornis*), while others moved into the grasslands and more desert-like country (*Genyornis* and some species of *Ilbandornis*). When they first appear in the fossil record about 15 million years ago they already show a high diversity of forms, suggesting a long history of this group that is not yet known. During their 15 million years of known existence the main evolutionary trends of mihirungs were to reduce the size of the internal toe, increase the size of the outside toe and increase the side to side compression of the proximal end of the tibiotarsus, probably all correlated with increased cursoriality.

Most of the knowledge about *Genyornis* and other mihirungs comes from fossil bones, but besides the bones and stomach stones already discussed, there are a few other traces. In lake sediments, mined for use in the production of paper in northeastern Tasmania (Rich and Green, 1974) and in the coastal sandstones of southern Victoria (Rich and Gill, 1976), tracks have been found that were probably made by mihirungs. The tracks of mid-Tertiary age (15-20 million years old) in Tasmania are about the same size as those of the living emu but much broader and with larger angles between the toes. A fossil emu (*Dromaius ocypus*) known from rocks slightly younger than those bearing the Tasmanian tracks show that the contemporaneous emu was distinctly smaller than the living emu (*Dromaius novaehollandiae*) and thus markedly smaller and less robust than the bird making the fossil tracks.

Eggshell fragments have also been found in Pleistocene dune sands of South Australia that belong to birds decidedly larger than the modern day emu (Williams, 1981). The shell fragments indicate that the eggs, presumably of *Genyornis*, were about the same shape as those of the emu, about 120 mm wide and 150 mm long, and about 1.3 kilograms in weight. Because the shells are common in sand dunes and rare in river deposits, it has been suggested that perhaps *Genyornis* nested in sand dunes. Two remarkable egg fragments, from sediments thought to be about 15 million years old near Marree in South Australia, are incredibly massive, measuring 4.5 mm in thickness, more than 3 times that of the living emu with an estimated width of about 130 mm (Williams, *pers. comm.*). Although its surface texture somewhat resembles that of an emu egg, it is distinct from it, and very distinct from that of other large ground birds such as moas, ostriches and elephantbirds. The egg size would fit nicely with a bird such as *Dromornis stirtoni*, being about 32-42 cm long, 24-29 cm wide, with a volume of 12 litres. (Williams, and Rich, unpublished).

Clues given by bones, tracks, eggshells and stomach stones have allowed reconstructions, such as those on display at the Museum of Victoria in Melbourne and the South Australian Museum in Adelaide. Fertile imagination is required on the point of feather colour, but much can be inferred about the lifestyle of fossil animals from such clues. Mihirungs appear to have been birds with varied habits, from cursorial to graviportal styles of movement; they lacked highly recurved claws on the foot, typical of carnivores, and often form a large part of the fossil assemblages (a large part of the biomass), very typical of herbivores and omnivores and not of carnivores, except under very peculiar circumstances. Mihirungs probably inhabited both forested and savannah-grassland environments, but were extinct before the arrival of European man in Australia. In part their extinction may be tied to their unsuccessful competition with the kangaroos, which moved into the ever-expanding grasslands during the past 5 million years or more, as the mihirungs steadily declined in diversity through the last 10 million years, despite their partial invasion of the grasslands. Aboriginal man may have contributed to the demise of the last mihirung, *Genyornis*, as the two overlapped in time, but this is not a certainty.

Despite what is known of the mihirungs, there is still much to be learned. Their relationships to other birds are still unclear (Rich and van Tets, 1982), and just how they came to be in Australia is yet a mystery. It is a pity that none, not even one species, survived until the present day, for then the palaeontological puzzle posed by these gargantuan birds would have been somewhat lessened.

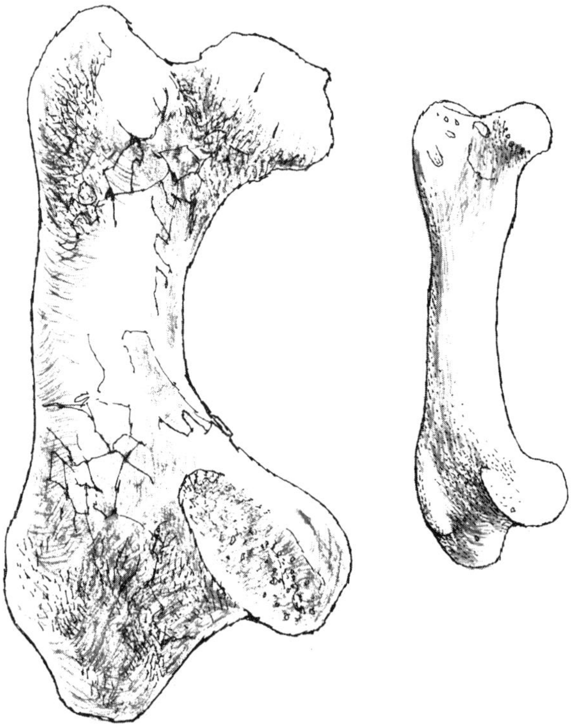

Front view of the thigh bone (femur) of *Dromornis stirtoni* (left) from the Miocene of the Northern Territory compared with that of the living emu. *D. stirtoni* was probably the heaviest bird that ever lived.

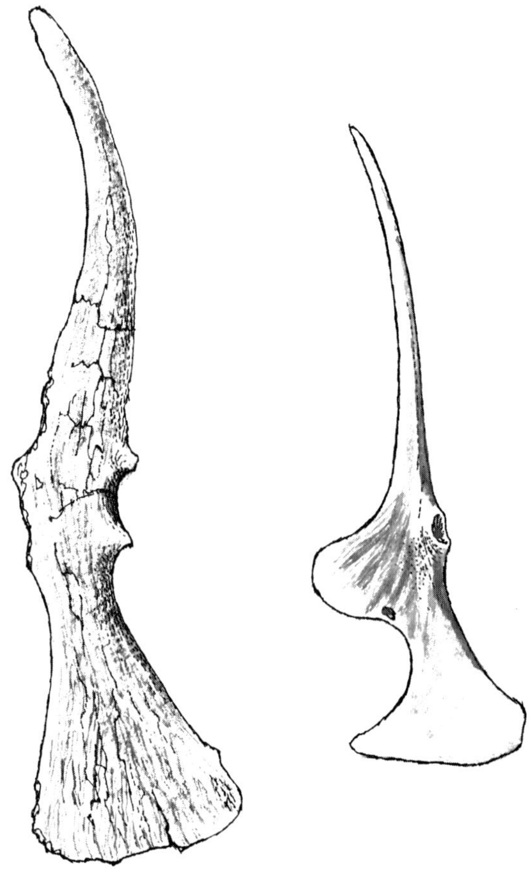

The scapulocoracoid, a shoulder bone, of *Dromornis stirtoni* (left) and the living emu. In most birds, except some large non-flying types, the scapula and coracoid are not fused.

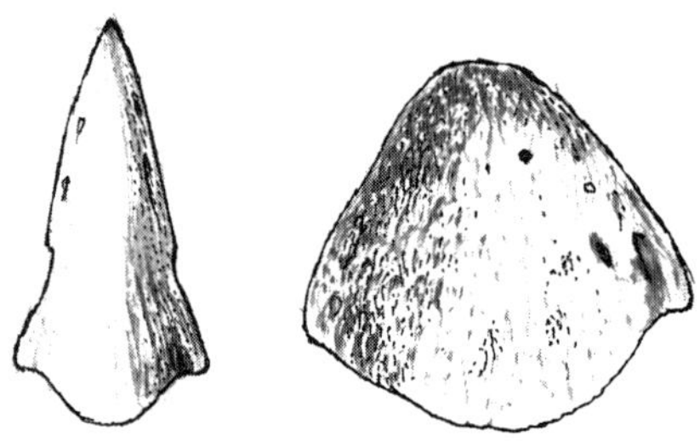

'Claw' of a dromornithid (right) and the living emu (left), viewed from above. Dromornithids have hoof-like ungual phalanges.

Progura gallinacea

De Vis, 1888

The Australian Giant Megapode

CLASS: Aves
ORDER: Galliformes
FAMILY: Megapodiidae
GENUS: *Progura*
SPECIES: *Progura gallinacea*
AGE: Late Pliocene to Late Pleistocene at least up to 34,000 years B.P.

When Aboriginal man reached southeastern Australia he found there a large turkey-sized fowl, with large breast and leg muscles, that should have been easy prey and good eating. This fowl is now extinct, a relative of the several genera and species of chicken-sized megapodes (big feet), or incubator birds, that still occur in Australia, New Guinea and some islands off southeastern Asia and in the South Pacific Ocean.

The extant megapodes use solar, geothermal or fermentive (rotting vegetation) heat in mounds or burrows to incubate their eggs (Frith, 1962), and presumably the giant megapode in southeastern Australia had a similar method of incubation. Aboriginal man may have prized not only its flesh but also its eggs as food. Unfortunately the egg shells of the giant megapode have not yet been found, or have not yet been recognized as such.

Anatomically the megapodes resemble the curassows (Cracidae) of Central and South America, and the guineafowl (Numididae) of southern Africa, Madagascar and Arabia in some characteristics, more than the turkeys (Meleagrididae) of North and Central America, the grouse (Tetraonidae) of North America and Eurasia and the peacocks, pheasants, chickens and quail (Phasianidae) of mainly the northern hemisphere. In particular the megapodes, curassows and guineafowl lack a bony internal spur, called an intercarpal process, on the carpometacarpus (fused wrist and hand bones). The intercarpal process does occur in the other gallinaceous (chicken-like) families, as well as in songbirds (Passeriformes) and a few other orders of landbirds. The lack of an intercarpal process may be an advanced character that, with their present mainly southern hemisphere distribution, suggests a common Gondwanic origin for the megapodes, curassows and guineafowl (van Tets, 1974). However, Olson (1980) postulates northern origins for these three families and suggests that the Megapodiidae were forced out of the mainland of southeastern Asia by the Phasianidae. Archaeological evidence suggests that prehistoric and unrecorded mariners transported and introduced chickens (*Gallus*) and scrubfowl (*Megapodius*) to many islands in the Indopacific area on either side of Wallace's Line. DNA hybridization tests by Charles Sibley and Jon Ahlquist may resolve the phyologenetic affinities of the megapodes and by the timing of their separation from other galliform groups indicate whether they are of Gondwanic or Southeast Asian origin. A better fossil record of birds in Southeast Asia and Australia for the early Cainozoic would be of great help too.

The realization that there had been giant megapodes came slowly. De Vis (1888a, 1888c, 1889, 1891) classified different bones of *Progura gallinacea* and even matching halves of the same bone as pigeons, storks, bustards and megapodes. De Vis (1888a) started with some tarsometatarsi (fused foot and ankle bones) which he regarded as being of a very large, chicken-like pigeon ancestral to *Goura*, the giant crowned pigeons of New Guinea, hence the name *Progura gallinacea*. The holotypes of *Chosornis praeteritus* (De Vis, 1889) and *Palaeopelargus nobilis* (De Vis, 1891) are matching fragments of the same bone, a carpometacarpus (fused hand and wrist bone). De Vis considered *Chosornis praeteritus* ('ancient moundbird') to be a megapode and *Palaeopelargus nobilis* ('noble ancient stork') to be a stork. De Vis (1888c) considered a scapula (shoulderblade) to be of an undetermined genus and species of bustard. All the De Vis megapode material is from the Darling Downs and Chinchilla in southeastern Queensland and has been referred to *Progura gallinacea* by van Tets (1974).

The first material of *P. gallinacea* to be collected was obtained from a cave in the Wellington Valley, New South Wales, and was presented in 1870 to the British Museum (Natural History) by the Trustees of the Australian Museum. It was a coracoid (shoulder bone) that Lydekker (1891) referred to the extant brush-turkey, *Alectura lathami*. A tarsometatarsus of a *Progura*, smaller than the De Vis material, was collected on 24 May, 1945 by E.T. O'Roarke at the Gore Limestone Quarries in southeastern Queensland. Longman (1945) also referred it to the brush-turkey.

My interest in fossil megapodes started in 1972 when Meredith Smith asked me to help identify a small collection of bird bones that had been found by Rod Wells and other members of the Cave Exploration Group of South Australia at Victoria Cave near Naracoorte, South Australia (van Tets and Smith, 1974). The largest and the first bone I examined, a coracoid, appeared to be of a chicken-like bird larger than any of the native galliforms living in Australia. Subsequent enquiries led to a re-examination of the coracoid from the Wellington Valley in the British Museum (Natural History), and another similar coracoid that Jeannette Hope of the Australian National University had found at the Wombeyan Caves, New South Wales. She had passed it on to Pat Rich for identification, who had independently concluded that it was a megapode.

On the basis of size, the three coracoids appeared to be referrable to De Vis' *Chosornis*, but as a manuscript was gradually being drafted and redrafted, more material became available by instalments from caves in the Naracoorte area through Neville Pledge and Meredith Smith, and from the Wellington and Wombeyan areas through Jeannette Hope. During a visit to the Queensland Museum in 1973, Pat Rich and I checked all the avian fossil material we could find in the De Vis collection, and in the process found the misassigned material mentioned above.

The additional material suggested that there had been two species of giant megapodes in southeastern Australia, a large long-legged form, *Progura gallinacea*, and a somewhat smaller short-legged form, *P. naracoortensis* (van Tets, 1974). By using the coracoid as an indicator of size, the weight of *P. gallinacea* was estimated to be about 5 to 7 kilograms and of *P. naracoortensis* about 4 to 5 kilograms.

Since 1974 more material has continued to come in of both forms, extending and filling in their geographic distribution and the known range of elements of their skeletons. They appear to have been restricted to the high country of inland southeastern Australia from southeastern South Australia to southeastern Queensland, including eastern New South Wales. There is no evidence that giant megapodes occurred in Tasmania and southwestern Australia during the Late Quaternary.

What were thought to have been two species, now appear to represent large males and small females of a single species, *P. gallinacea* and '*P. naracoortensis*' was

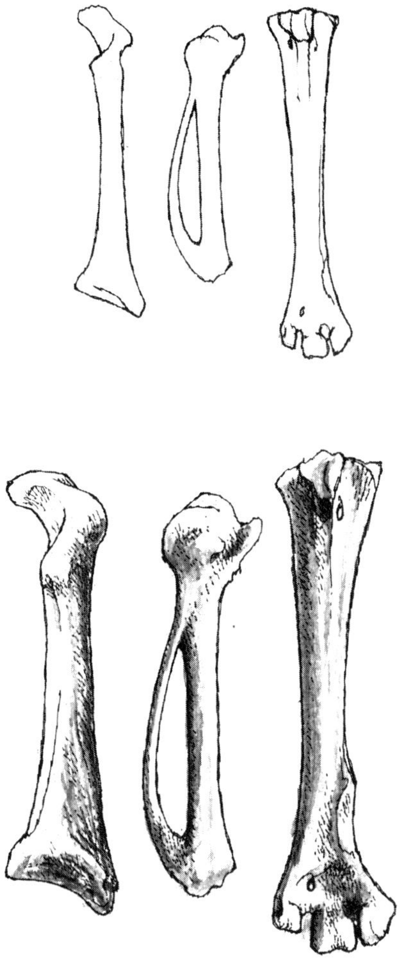

Tarsometatarsus (right), coracoid (left), and carpometacarpus (middle) of *Leipoa* (above) and *Progura* (below).

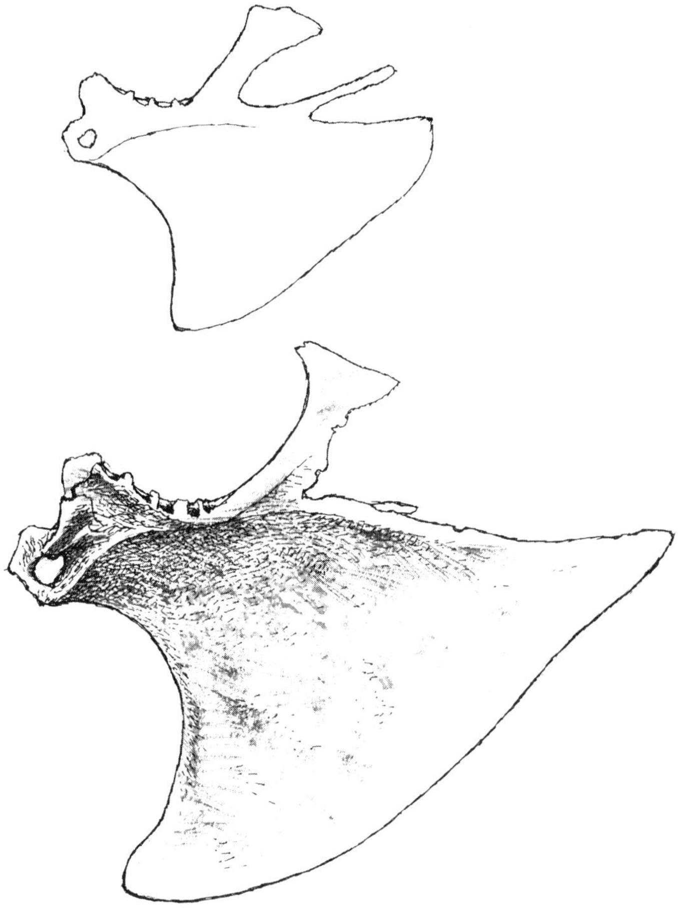

Side view of sternum of *Leipoa* (top) and *Progura* (bottom),

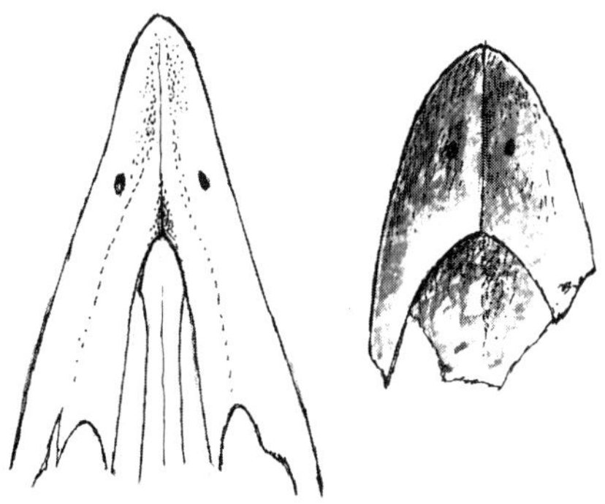

Ventral view of maxilla (upper bill tip) of *Leipoa* (left) and *Progura* (right),

Relative sizes and shapes of *Leipoa* (top) and *Progura* (bottom).

probably the female.

Most of the skeleton is now known, although only fragments are available of the bill, cranium and pelvis. The toe bones have probably been collected but only a few have been recognized because they are difficult to identify. The proportions of the long bones of *Progura* are similar to those of the Mallee fowl or Lowan, *Leipoa ocellata*, a living megapode of the dry scrublands of southern Australia. Presumably they shared this habitat. Bones of young and fully grown individuals of both species have been found together in the caves near Naracoorte. Besides being much larger than *Leipoa*, *Progura* had a relatively broader bill, head and body. The keel on the sternum (breast-bone) of *Progura* is very deep. Presumably its powers of flight were similar to those of other large gallinaceous birds, such as turkeys and peafowl. It would have been able to fly up into trees, onto rock ledges, and down slopes to avoid ground predators. The reconstruction of *Progura* on the colour plate is based on the Mallee fowl with an emphasis on greater size and robustness.

Leg bones of a new, extinct genus and species of giant megapode have been found as artifacts in a Polynesian midden on a tiny island in Fiji. It had very short and broad tarsometatarsi and appears to have been a larger bird than *P. gallinacea*.

At the Isle of Pines, New Caledonia, bones have been found of an extinct emu-sized bird, *Sylviornis neocaledoniae*, that on the basis of some poorly preserved bones was initially considered to be a ratite (Poplin, 1980). But the discovery and study of much more and much better-preserved material have shown it to be a megapode that dwarfs even the giant megapodes of Australia and Fiji (Poplin *et al.*, 1983). A carbon date of 3470 ± 210 years BP (before present) suggests that *Sylviornis* may also have been known to the prehistoric people of Australasia.

Two *Progura* at a nesting mound for incubating eggs. As with living megapodes, these extinct forms probably used the heat produced by rotting vegetation to hatch the eggs.

F KNIGHT 82

Phoeniconotius eyrensis

Miller, 1963

The Robust Eyrean Flamingo

CLASS: Aves.
ORDER: Phoenicopteriformes.
FAMILY: Phoenicopteridae.
GENUS: *Phoeniconotius*
SPECIES: *Phoeniconotius eyrensis*
AGE: Miocene.

Flamingoes are strikingly beautiful birds that today grace the landscapes of the East African rift valleys and high Andean lakes as well as the tropical Caribbean and subtropical Mediterranean. They are unknown in Australia and Southeast Asia however, and until the middle of this century had never been expected in the Antipodes.

R.A. Stirton, the indefatigable Californian palaeontologist, who set his mind to finding pre-Pleistocene mammals in Australia, changed that misconception. In the 1950s he and several colleagues, including a well-known ornithologist, Professor Alden Miller, found fossil flamingoes on the floor of a dried salt pan in northern South Australia. Their finds not only extended the geographic range of this avian family but also brought to mind how this arid area had changed since the flamingoes lived there; from once having a permanent, predictable water supply spread throughout the year from one year to the next, to that of today, with unpredictable times of drought and flood (Hope, 1982). The flamingoes, unlike a number of other Australian birds, were unable to cope with the recent climatic flippancy and lowered humidity, and so became extinct, sometime during the Pleistocene, perhaps only in the last 400,000 years.

Although Stirton and Miller (Miller, 1963) were the first to recognize flamingoes as fossils in Australia, bones of these birds had actually been collected long before but had remained undetected, being described as ibises, storks and waders by C.W. De Vis (1905), then a Curator at the Queensland Museum. De Vis was working in isolation from other ornithologists and had only a small osteological collection that did not include flamingoes, so he assigned the bones to the closest things he had. Flamingoes, of course, did not occur as living forms in Australia, and he did not expect them as fossils. One wonders if his contemporaries in other parts of the world would themselves have recognized the bones as flamingoes, since these birds were such an unexpected element in the fauna of Australia. Perhaps, perhaps not. We will never know!

Some of the fossils De Vis described had been collected by J.W. Gregory. The bones came with little information about where they occurred in the rock sequences. The place names, Wurdulamankula and Unduwampa, with their various spellings, on the specimen labels, were provided by an Aboriginal guide, who was the last survivor of the tribe that once lived along the lower Cooper Creek (Gregory, 1906). Stirton and his co-workers found more bones along the lower Cooper as well as on lake pans to the north of the Cooper a half a century later. But neither Gregory nor Stirton were able to precisely date the occurrences. For the most part the fossils appeared to be of Late Tertiary or Pleistocene age, but because there are no volcanic rocks in the area and little charcoal for radio-carbon dating (which itself only allows dating back to 40,000-50,000 years), no one really knows exactly when the flamingoes became extinct.

It was not just one kind of flamingo that inhabited Australia either; there were several different types at any one time in the past. The oldest flamingoes are two species in separate genera, *Phoenicopterus novaehollandiae* and *Phoeniconotius eyrensis*, both from Miocene lake sediments in a saltpan north of Marree. *Phoenicopterus novaehollandiae* (meaning 'purple wing from New Holland'), although a species unique to Australia, was thought by Miller to be closely related to the living Greater Flamingo (*Phoenicopterus ruber*) that has a nearly cosmopolitan distribution today. *Phoeniconotius eyrensis* (meaning 'southern purple from Lake Eyre'), belonged to a genus that was restricted completely to Australia. It was a large, robust flamingo that apparently had a well-developed hind toe that is either tiny or entirely missing in modern flamingoes. Unfortunately, only small fragments of the tarsometatarsus, the lower-most leg bone, and a part of the foot are preserved, so it is difficult to guess precisely at the proportions. It is reconstructed here using the limb proportions of living flamingoes, but these have been scaled up to match the robustness indicated by the fossil hind limb. Only new specimens of other parts of the skeleton can help to confirm or deny this reconstruction.

In size, both Miocene fossil flamingoes are within the range of the Greater Flamingo which has a tremendous variation. *Phoeniconotius eyrensis* is, however, at the large end of the range, whereas *Phoenicopterus novaehollandiae* is at the bottom of the range.

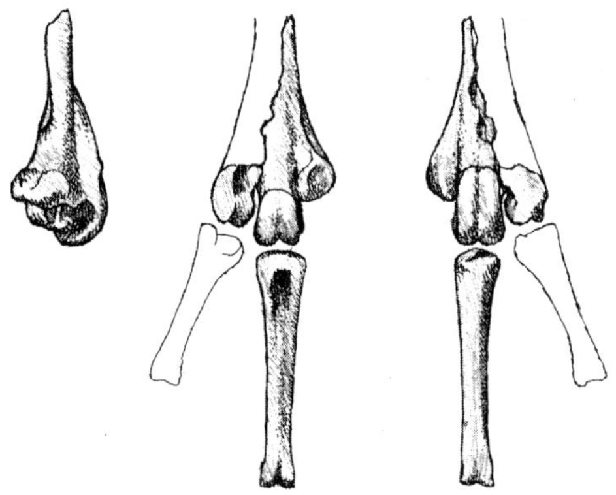

Type specimen of *Phoeniconotius eyrensis*, a fragment of the tarsometatarsus and toe bones, from Miocene lake deposits in northern South Australia.

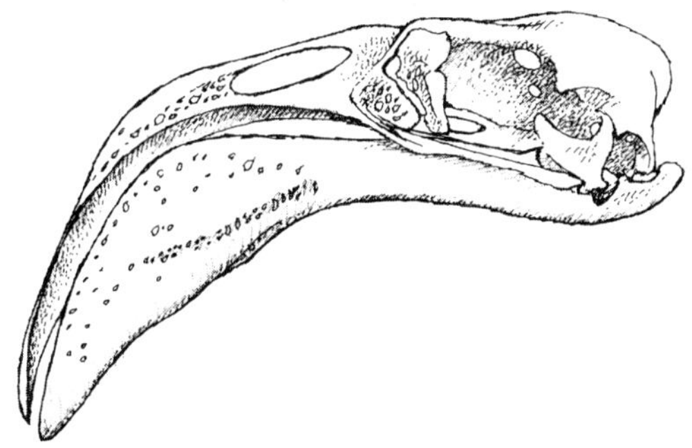

Skull of the living Greater Flamingo, *Phoenicopterus ruber*. The down-turned bill allows these birds to filter-feed, much like living baleen whales.

Phoenicopterus novaehollandiae, the more gracile of the Miocene flamingoes, like *Phoeniconotius*, also had a well-developed hallux or 1st toe. This is indicated by a distinct scar or depression on the posterior face of the distal leg bone, the tarsometatarsus. The middle and outer toe lay closer together and the inside toe was probably more slender than in any of the living forms. Unfortunately, little more than distal leg and foot bones are reported in published papers, although more material has recently been recovered from Lake Palankarinna in South Australia.

Flamingo bones are also known from Late Miocene or Early Pliocene river channel sediments near Alice Springs, Northern Territory. The only other reasonable selection of flamingoes, however, comes from the Late Pliocene-Pleistocene river and lake deposits in the Lake Eyre Basin of South Australia. There were possibly four different species of flamingo present during that time. The Pliocene-Pleistocene species (Rich, van Tets, Rich, and McEvey, in prep.) range in size from those smaller than the smallest living to larger than the largest living forms, with one species perhaps in the size range of the living East African Lesser Flamingo (*Phoeniconias minor*) and another similar in size to the Greater Flamingo (*Phoenicopterus ruber*). Unfortunately, all of these different species are known only from a few fragments of limb bones, and none are known from bill fragments. For living flamingoes, it is the bill in which significant differences between species occur, clearly related to their feeding styles. In those flamingoes that live together today direct competition is avoided by different species utilizing different foods in their environment of highly saline, brackish water lakes (Jenkin, 1957). The overall bony structure of the bill reflects whether they were surface water filter-feeders or mud-grubbing vacuum cleaners — and so far the precise food habits of the Australian flamingoes are simply unknown. Since all flamingoes alive today, however, utilize some form of 'filter-feeding', in some ways like whales, prions and ducks, one can presume that the Australian forms fed in this general way. Water is taken into the mouth, which is then partially closed. By using the tongue, the birds force that same water back out of the mouth across a series of lamellae in the upper and lower bill. These lamellae act like the baleen filter in a whale and trap small food items, which can then be licked off by the tongue and eaten.

Fossil flamingoes in Australia are known only from central Australia mainly in the area around Lake Eyre and Lake Frome and not from coastal sites. When the flamingoes lived in the Centre, at least early in their history, the country had forests of *Nothofagus* (Southern Beech) and associated plants, which provided shelter for a wide variety of arboreal marsupials, such as *Ektopodon*, cuscus-like animals, koalas, and possums. Permanent lakes teemed with bony fish, lungfish (such as *Neoceratodus eyrensis*), crocodiles, turtles, river dolphins, and a wealth of other water-bird life. Crocodilians, perhaps some of the carnivorous marsupials and raptorial birds, were perhaps the only predators capable of taking flamingoes, probably preying mainly on chicks and eggs.

During the Pliocene and Pleistocene the surrounding country dried out; flamingoes managed to cope with this dessication until there were no more permanent lakes. Rainforest vegetation gave way to more open country, savannah-like floras that may at times have retained some pockets of rainforest. Then in the Pleistocene more open grasslands and finally an arid zone vegetation developed (Truswell and Harris, 1982). Pliocene and Early Pleistocene conditions in central Australia may have been more like those of the East African rift valley today, where the Greater and Lesser Flamingoes co-exist in a moderately dry environment but with a dependable permanent water source in the shallow, salt-rich lakes, which provide suitable breeding and feeding sites for flamingoes, and not much else.

Pink-coloured, stately birds in the midst of shallow central Australian lakes must have painted quite a spectacular mid to late Cainozoic scene, but mixed with these, another group, perhaps closely related, added to the diversity of flamingo-like birds. Known from a single locality of Miocene age north of Marree, the palaelodids (Rich and van Tets, 1982; Rich and Baird, in prep.), have only recently been recognized even

---

Many species of flamingoes or flamingo-like birds graced the landscape of Lake Palankarinna in northern South Australia 15 million years ago. In the foreground, *Phoeniconotius eyrensis* tends a young. The more gracile *Phoenicopterus novaehollandiae* feeds in the mid-distance, while the white flamingo-like palaelodids stand in the background. The nearest relatives of the palaelodids lived in Europe and North America at this time.

though they were collected nearly three years ago. Identification was hampered by the need for lengthy, delicate preparation to remove the highly crushed, fractured bones from the lake clays. In the field, palaelodid bones so closely resemble those of phoenicopterid flamingoes that they were thought to be in this family. Both have elongate, slender tibiotarsi and tarsometatarsi. There are differences, however, in these leg bones, especially in the all important tarsometatarsus, a fusion of ankle (tarsals) and foot (metatarsals) bones, in the shapes of articular ends. In flamingoes the proximal end, that nearest the body, has two prominent ridges running down the back of the shaft. In palaelodids, the area defined by these ridges has a distinct canal in it, which is lacking in flamingoes.

The palaelodids had previously been known only from Europe and North America, and what is so exciting about finding them in Australia is that they offer the chance to absolutely date the central Australian rocks. By comparing the fossil palaelodids with the closest relatives in Europe where rock sequences are well dated, using radiometric techniques as well as direct tie-ins with marine sequences, an age estimate of the Australian rocks can be made. A similar correlation is possible using the Nooraleeba bats (*Hipposideros*) reported on by Sue Hand in this book, from a locality in northwestern Queensland. Palaelodids are well-known in France in rocks deposited during the mid-Tertiary and extend into the Pliocene of North America. It appears that at least during Miocene times these birds were moving back and forth between Europe and Australia. At present we are not sure if this was a movement originally north from Australia, or south from the northern hemisphere. Only by finding an earlier record in Australia comparable to that from Europe or North America will we be able to make a cogent guess.

One thing is certain: the living Australian avifauna is sadly lacking some of its most lovely former inhabitants. 'Tis a pity that the fitful Late Pleistocene climate took such a toll.

Lower leg bone (tarsometatarsus) of an Australian palaelodid. Top and front views. Scale equals 1 cm.

Reconstructed skeleton (after Lambrecht, 1933) of the flamingo-like palaelodids, known from Miocene-Pliocene of North America and Europe, and recently from the Miocene of Australia.

Centropus colossus

Baird, 1985

The Giant Coucal

CLASS:	Aves
ORDER:	Cuculiformes
FAMILY:	Cuculidae
GENUS:	Centropus
SPECIES:	*Centropus* n.sp.
AGE:	Late Pleistocene

Donning wetsuits, tanks, flippers and masks, two undaunted divers, B. Brawley and G. McKensie, enter the depths by slowly backing down the muddy slope of a drowned cave. Once in the water, torches alight, using only their hands for propulsion, they follow the slope down some thirty metres to their destination. From above, it looks like a subaqueous graveyard, with all the skeletons exhumed. Rows of vertebrae, partially articulated limbs and skulls, are all clear to view. As Brawley and McKensie descend, it becomes apparent that the torches will be totally useless for this job, because the cave soil is held in suspension, and small amounts of movement produce a water so murky that they can barely see their hands.

The collection begins by their groping through previously designated areas and placing the soft, friable bones in a cotton-lined basket. Once full, the basket is slowly moved to the surface where the bones are dried, immersed in a plastic reinforcer, and divided into mammalian and non-mammalian groups. Then each element is carefully wrapped, to reduce damage during transport, and labeled with the date, collector and site of collection. This is the end of the field work aspect. From here the material either goes to the museum or to a specialist in each particular group, mammals, birds, reptiles, or amphibians.

The mammals, for example, were worked on by Neville Pledge, Curator of Palaeontology at the South Australian Museum. In a 1980 paper he recorded the presence of *Thylacinus* (Tasmanian Tiger), *Simosthenurus* (Short-faced Kangaroo), *Macropus* cf. *giganteus* (a form closely related to the Grey Kangaroo), as well as a number of smaller forms from this Green Waterhole.

The bird fauna (Baird, in prep.), however, was much more diverse and contained everything from the large, flightless Tasmanian Native-hen(*Gallinula mortierii*) and the Long-billed Corella (*Cacatua tenuirostris*), to extinct species. Of particular interest was a bird of large proportions, belonging to the cuckoo family.

Cuckoos are one of the most widely recognized bird families in the world, particularly for their role as the time keepers in 'cuckoo' clocks. The model for these noisy wooden beasts is the European Cuckoo (*Cuculus canorus*), a wide ranging species with a characteristic call. They are also represented in the cartoon world with the ever troublesome 'Roadrunner', which although very much stylized, does have a basis in nature. Two species of this cursorial genus (*Geococcyx*) live in North and Central America, inhabiting arid and semi-arid lands. They run about catching lizards, insects, and the like demonstrating an omnivorous diet much like the many other members of the Cuculidae. Unfortunately, although these cuckoos can be recognized instantly by much of the western world's population, little is commonly known about the rest of this diverse family. Until recently the fossil record of the cuckoos has been restricted to a handful of Tertiary specimens from North America and a very small number of Late Pleistocene representatives of living species from Europe, North America, and Australia — not a good record for a group which is considered to be close to the origin of the landbirds (Olson, in litt.).

Luckily, two unique specimens have just recently been found. The first and most interesting specimen is considered to be near the base of the cuckoo family (Baird, Rich, and Walker, in prep.). This bone (tarsometatarsal) was found in Brazil, South America and is of Palaeocene age. The foot structure in this specimen is unique in that it is intermediate between having three toes forward and being permanently yoke-toed. This means it would not be classified as a cuckoo today, since all of the species in the Cuculidae share a characteristic yoke-toed (zygodactylous) foot. In this type of foot two toes point forward and two backward. This type of toe arrangement can also be seen in the parrots, hornbills, and woodpeckers, but was derived separately in each group (Olson, 1983).

Besides their diagnostic foot bones, cuckoos also have a characteristic upper arm bone (the humerus) with small pneumatic foramina and a head which is tilted backwards. Some cuckoos also have a pectoral crest on the humerus that is twisted backwards with respect to the humeral head. These last two characters are also associated with reduced powers of flight (Olson, 1973; Olson and Wetmore, 1976; and Olson and Steadman, 1979). Although no extant cuckoo is flightless, several are approaching such a condition, including the Coral-billed Cuckoo of Southeast Asia (*Carpococcyx renauldii*), the Greater Roadrunner of North America (*Geococcyx californicus*), and all species of Coucals (*Centropus*). This can be seen in their wing structure and in their flight patterns, (e.g. laboured flight and or frequent gliding). The same pattern can be seen in many other Australian bird groups, like the Bristlebirds (*Dasyornis*), Grass-wrens (*Amytornis*), Scrub-birds (*Atrichornis*), and Lyrebirds (*Menura*) (Rich, McEvey, and Baird, in press).

The second specimen lies at the other end of the cuckoo time spectrum and is thought to be a recently extinct species of *Centropus*. This new species of *Centropus* is known from one beautifully preserved humerus, that measures one third again larger than the extant Australian representative of the genus, *Centropus phasianinus*. Besides its massive size, this new species differs from other *Centropus* in having a more reduced power of flight. This is inferred from the smaller area of muscle scars on the dorsal side of the humeral pectoral crest.

The reconstruction is, therefore, of a bird spending most of its time on the ground looking for insects, amphibians and small lizards (Thompson, 1964) perhaps with short laboured flights into the air only in times of emergency, in order to escape predators, for example. In the colour plate, the artist has shown the bird in front of the sinkhole from which it was recovered, with the sinkhole located in low scrub. Of course, the artist's rendition is speculative, but it is not unreasonable to assume that the plumage would be similar to that of present-day coucals.

Coucals have two different plumages, one for the breeding season and one for the non-breeding season. Of the breeding season plumages there is a basic pattern, which carries through to most of the species, and that is to have the body and tail a single,

---

Two individuals of the Giant Coucal at the mouth of Green Waterhole demonstrating hypothetical breeding (top) and non-breeding (bottom) plumages. Long-billed Corellas, the most abundant species in the deposit, can be seen in the upper right.

F. KNIGHT '83

uniform colour, with the back and wings another. These colours range from white to black for the body and usually brown, rust, or mottled tans for the back and wings. During the non-breeding season, the plumage tends to be drab, mottled brown. It is a matter of artistic licence that both of these plumages are shown in the same illustration.

The generic name of the Giant Coucal is made up of two Latin roots, *centro*, meaning 'spine', and *pus*, 'foot'. All species of the genus *Centropus* are characterized by having an elongate nail on the hind toe (hallux), hence 'spine-foot'. The new species name, will reflect the large size of this recently discovered species.

With increased study of the Late Pleistocene avifauna, the number of bird species known from Australia continues to expand. Some examples of these are the Giant Coucal, the Giant Megapode, a Mihirung, and a Giant Eagle, among others. The importance of this lies in the fact that the present day avifauna is depauperate, with smaller numbers of species and genera when compared with those of other continents (Keast, 1981). Was this always so, or is it just a result of Australia's most recent fluctuations of aridity (Bowler, 1982)? Clues to the reasons for this low faunal diversity lie in the fossil record and will be found in the years to come.

Comparison of the living Pheasant Coucal (*Centropus phasianinus*) and the extinct Giant Coucal (below).

Quipollornis koniberi

Rich and McEvey, 1977

The Bird of the Dusk

CLASS: Aves
ORDER: Caprimulgiformes
FAMILY: Aegothelidae
GENUS: *Quipollornis*
SPECIES: *Quipollornis koniberi*
AGE: Miocene

In 1961 at Chalk Mountain, Bugaldie, west of Coonabarabran, New South Wales, a small fossil specimen of a bird was uncovered by a mine-worker, Mr. G. Mills, in diatomaceous earth, that is, earth composed of the literally countless remains of single-celled algae known as diatoms. These deposits had long been known to contain fossil leaves of the genus *Cinnamomum*, suggesting that rainforest had existed not far away at that time, together with fossil fish that were similar to the Murray cod (*Maccullochella macquariensis*). The newly found fossil animal was presented to the Australian Museum, Sydney, by J.W. Burton, a teacher at the Bugaldie Public School, where H.O. Fletcher, the Curator of Fossils, identified the specimen as 'a fossil bird of possible Pliocene age' (from 2-5 million years old) (Jones, *pers. comm.*).

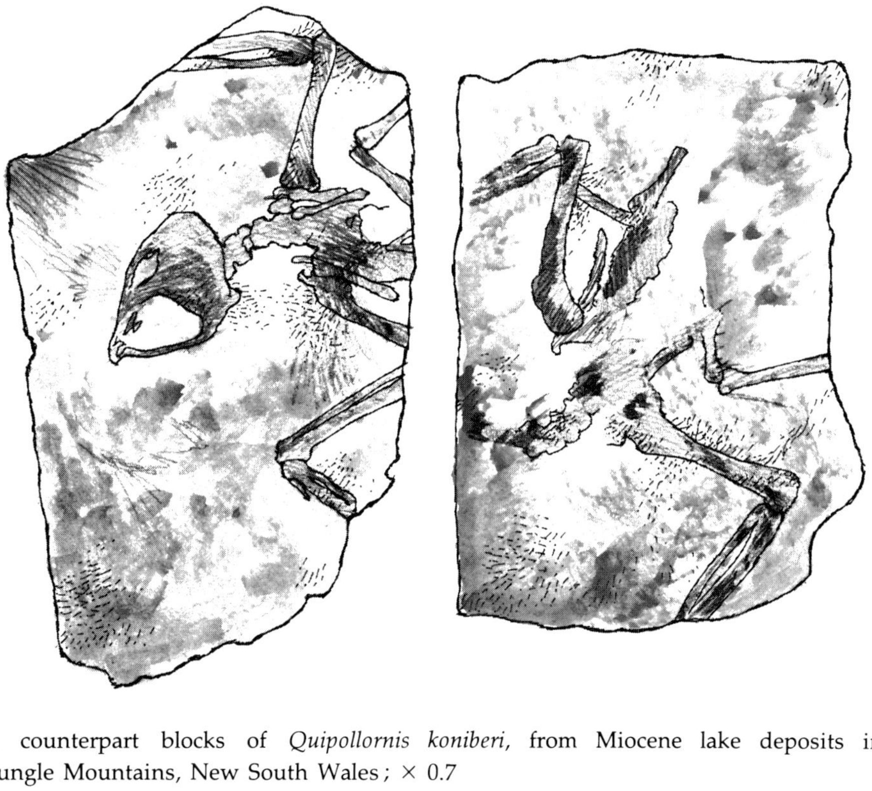

Part and counterpart blocks of *Quipollornis koniberi*, from Miocene lake deposits in the Warrumbungle Mountains, New South Wales; × 0.7

There the matter rested. For more than ten years the specimen lay without attention. Then research on fossil birds began to increase rapidly in Australia. The fact that our country is large and could contain extensive fossil beds as yet undiscovered, increasingly encouraged the search for such finds. Also the Australian avifauna is, in many of its groups, so different from that found in other zoogeographic regions of the world that Australian fossil birds whenever found are of particular interest to research workers overseas who are engaged in this field. Exemplifying the growth of such activity Pat Rich, an American who is now a lecturer in geology and zoology at Monash University, Clayton, Victoria, had by the 1970s already published a number of papers on Australian fossil bird forms. At the same time, as Curator of Birds at the Museum of Victoria, I was building up a bird-skeleton collection at the Museum for osteological research.

The late R.A. Stirton, a palaeontologist formerly of the University of California, U.S.A., drew Rich's attention to the bird fossil from Chalk Mountain, then housed at the Australian Museum in Sydney. Thus, it eventuated that in the mid-1970s this fossil specimen was re-examined and has proven to be of considerable interest and importance.

First, it is very rare to have a fossil bird so well-preserved, both with evidence of feathers and with the skull; an element of the skeleton of paramount importance in any osteological research that is concerned with determining the relationships of animals. Secondly, it is important that the deposit in which the specimen was found has now been fairly well dated, using the potassium argon technique, as being from 13.5 – 17 million years old. This extends the age of the fossil back beyond the Pliocene, suggested by H. Fletcher, to the Mid- to Early Miocene.

Re-examination of the specimen quickly showed it to possess certain significant characteristics, such as the broad skull with slender-boned lower and upper jaws and the rounded braincase; the highly curved internal edge of the coracoid, (bone extending from the humerus (wing-bone) down to sternum or breast-bone) and the short, broad, neck vertebrae. In combination, and quite apart from the fossil's obvious likeness to the group, these are characters that distinguish the Owlet-nightjar, family Aegothelidae; it appeared that an undescribed and extinct species of Owlet-nightjar had been found.

The family Aegothelidae is confined to the Australasian Region where some eight species now live, including one Australian species (*Aegotheles cristatus*). A gigantic fossil specimen has been described from New Zealand by R.J. Scarlett (1968) and named *Megaegotheles novaezealandiae*. It is very similar to *Aegotheles cristatus* but has much larger leg and wing bones and its wings were decidedly shorter relative to the length of the legs suggesting that it was a very terrestrial bird, probably not flying much, if at all (Rich and Scarlett, 1977). The Bugaldie fossil being discussed is about the same size as the Australian Owlet-nightjar (*Aegotheles cristatus*).

Comparison of the fossil specimen with the two known genera of Owlet-nightjars, *Aegotheles* and *Megaegotheles*, revealed distinct differences in several features of the skull and bones of the shoulder. The humerus, or upper wing bone, in the fossil form is about twice the length of the femur. In *Aegotheles* and *Megaegotheles* the two bones are much nearer the same length. Thus, the Bugaldie fossil had a long wing and short leg by comparison with the proportions of leg and wing found in *Aegotheles* and *Megaegotheles*. It is quite clear then that the fossil form, although

---

The little Owlet-nightjar, *Quipollornis koniberi*, as it might have appeared hunting at dusk. This form, unlike its living Australian relative, was more adapted to aerial hawking for insects than a life closer to the ground.

F. KNIGHT 82

considered to be of the family Aegothelidae, was on the basis of the differences listed, quite separate from the genera *Aegotheles* and *Megaegotheles*. It would also appear likely that *Quipollornis* is the most primitive form known in the Aegothelidae, an inference befitting its older age.

However, there are other and related families in the order of birds known as Caprimulgiformes; the order containing the Oilbirds, Frogmouths, Potoos, Owlet-nightjars and Nightjars, and comparative examination of the fossil skeleton and the skeletons of species in those other families had to be made, to be sure that identification as an aegothelid was correct. Such families were Steatornithidae containing only the Oilbird, genus *Steatornis*; that strange species from the Neotropical (South and Central American) Region that possesses an ability in echo-location, or the means of navigating in the dark by sending out a stream of short pulses of sound that rebound from any object in front of the bird's flight, thus informing it of obstacles in the way. Next came the Podargidae, the Frogmouths of the Indian and Australasian Regions that characteristically perch lengthwise on dead branches and, because of their stillness, their stretched-out posture, and their pencilled grey-brown marking, perfectly assume the appearance of a piece of dead branch. After that came the family Nyctibiidae, the Potoos of, chiefly, the Neotropical Region that adopt a vertically stiffened posture rather like that of the Frogmouths. Finally there were the Caprimulgidae, the Nightjars or Goat-suckers that are worldwide in distribution.

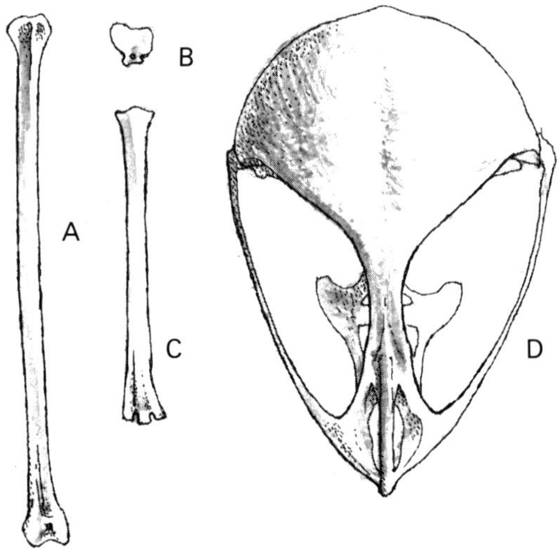

Different bones of the living Owlet-nightjar (*Aegotheles cristatus*). (**a**) drumstick or tibiotarsis, front view; (**b-c**) lower leg and foot bone, tarsometatarsus, top and front view; (**d**) top view of skull. Mouth gape of these birds is wide; good for catching insects × 1.7.

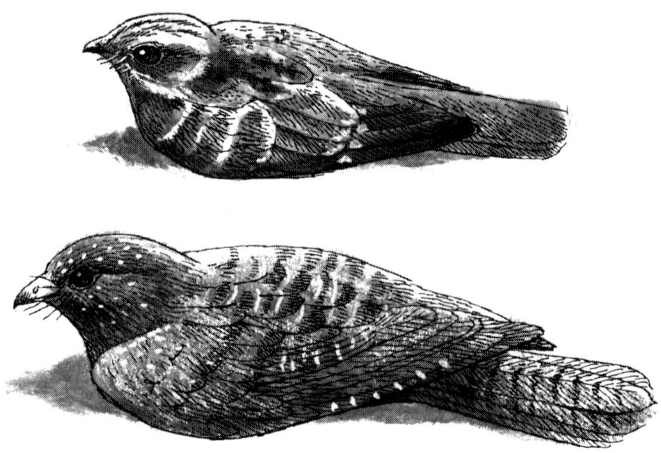

Two living birds that are related to the Owlet-nightjars; the Nightjars (*Caprimulgus*) (above) with a worldwide distribution, × 1/4; and the Oilbirds (*Steatornis*) from South America ; × 1/7.

All of these birds are active mainly at night or twilight, and while some have very distinctive and interesting habits, their association with the dusk and the shades of night have long invested them with fanciful and mysterious attributes. Thus, the Nightjars, so-called possibly from their jarring notes uttered at night, have been credited in folk-lore with the ability to suck milk from goats (*capri-mulgus* meaning in Latin, 'goat-milker') and therefore have been given the name Goat-suckers.

The fossil specimen could only be compared in dorsal view (that looking down on the crown of the head and the back of the bird) because of the way it was preserved. Ventral or underside elements could not be seen. Also, although at first glance the skeleton appeared to be extremely well-preserved, as is often the case with fossils, closer examination showed that the only parts distinct enough for close comparison were the broad outlines of the skull, the cervical or neck vertebrae and a part of the shoulder girdle and forelimb. The other bones while not contradicting the placing of the fossil in the family Aegothelidae did nothing to support this conclusion.

Close study of the fossil and comparison with skeletons of species in all the families of the Caprimulgiformes mentioned above revealed many differences including the following examples:

Skull: In the fossil the occipital (curved rear) region is more rounded than the rather flattened rear region found in the Nightjar (*Caprimulgus*), while the well-developed paired openings (temporal fossae) found on each side of the skull in the Frogmouth (*Podargus*) and in the Oilbird (*Steatornis*) appear to be lacking. The upper jaw bones are slender and not fused together as in the Frogmouth (*Podargus*) or expanded as in the Nightjar (*Caprimulgus*).

Cervical or neck vertebrae: These are all very broad in relation to their length and this feature is not found in the Frogmouth (*Podargus*), Nightjar (*Caprimulgus*), Oilbird (*Steatornis*) or the Potoo (*Nyctibius*). However, in the Owlet-nightjar (*Aegotheles*), the neck vertebrae show a tendency towards this condition.

Thus, as outlined above, the fossil specimen showed in many of its elements distinct differences from species in the other related families, leaving no doubt, however, that it did belong to the family Aegothelidae. Since it differed from the genera *Aegotheles* and *Megaegotheles* in that family, it represented an undescribed genus.

In the Order Caprimulgiformes the families other than the Owlet-nightjars (Aegothelidae) have highly developed wings and, therefore, well-developed wing-bones, by comparison with their small hindlimb bones. In the Owlet-nightjars the hindlimb bones are relatively much more developed. In the fossil *Quipollornis koniberi*, the comparison between wing and leg bones was not a marked one and shows that in this species, although it was an Owlet-nightjar on the basis of other characteristics, the longer hindlimbs of some present day Owlet-nightjars had not begun to develop.

The name given was *Quipollornis koniberi*, establishing a new genus and species. The generic name *Quipollornis* is based on the Aboriginal word *quipolly* for a 'water-hole containing fish' and the Greek word *ornis* for 'bird', while *koniberi* is the Aboriginal name of the tribe once inhabiting the area from which *Quipollornis* was recovered. The holotype specimen is preserved at the Australian Museum.

Based on the detailed studies of the Bugaldie skeleton, much has been learned about the history of the Owlet-nightjars. We know that the group had appeared in Australia by the Miocene. We also know that the first record of the Aegothelidae, *Quipollornis*, although it probably looked very similar to living forms, was not so specialized for a terrestrial mode of life as are its living relatives. Its wings were long relative to the legs, not vice versa, and like so many of the other caprimulgiforms it was probably an aerial insect feeder. It was a primitive member of a modern family, so typical of many birds of Miocene age. By its presence in Australia in the Miocene, it suggests that the uniqueness of Australia's avifauna extends back at least to this time and probably further.

In the accompanying picture the artist has endeavoured to show the foreground bird caught by the last rays of the sun as it pursues a moth in the dusk. The wings are shown to be fairly long and the markings of the bird are somewhat undistinctive by design. A myrtaceous twig, from a tree related to *Eucalyptus*, appears in the immediate foreground. The perching bird is seen against a distant background setting of more typical rainforest flora.

Ektopodon serratus

Stirton, Tedford and Woodburne, 1967

The 'Strange-toothed' Marsupial

CLASS: Mammalia
INFRACLASS: Marsupialia
ORDER: Diprotodonta
FAMILY: Ektopodontidae
GENUS: *Ektopodon*
SPECIES: *Ektopodon serratus*
AGE: Miocene

When the first specimens of *Ektopodon* were found in the Wipajiri Formation, a stream channel deposit at Lake Ngapakaldi in the Lake Eyre sub-basin of northern South Australia, by the late R.A. Stirton and his colleagues in 1962, they immediately provided a great enigma. In fact, its name means 'strange tooth', derived from the Greek words *ektopos* and *odous*. Only isolated teeth were found and they were so unusual, that when they were described as a new species, *Ektopodon serratus*, their discoverers (Stirton, Tedford and Woodburne, 1967) excluded them from the marsupials, which were the only mammal group then known from the Miocene in Australia.

With some doubt they were allied to the platypus, a monotreme that as a juvenile possesses a few very strange teeth with triangular cusp patterns and a large number of tooth roots. The cheek teeth (or molars) of *E. serratus* were relatively wider than those of any other known marsupial and were most unusual in having two transverse ridges (or lophs) with up to 9 cusps on each ridge. The cusps increase in size and slope towards one end of the ridges forming a rasp-like surface. In addition, the first upper molar has, at the front of the tooth, a shorter third row of several cusps. The teeth are so unusual that a new terminology had to be devised in order to describe them properly.

Further collecting during the early 1970s, at Lake Ngapakaldi, produced a few more isolated teeth and the suspicion that *Ektopodon* was not really a monotreme or a multituberculate, but in fact a marsupial, possibly related to the possums (Archer, 1976, 1981).

In 1976, a small mass of broken bone and tooth fragments was found near a small hill known as Mammalon Hill, at Lake Palankarinna, another dry salt lake east of Lake Eyre. Back in the laboratory, with the use of a binocular microscope, I was able to reconstruct the numerous fragments and chips of bones and teeth to produce an incomplete right lower jaw, with four teeth in it: a perfect lower premolar (P_3) and

Ektopodon had good binocular vision and was almost certainly arboreal, with a life style like that of a phalanger or possum. Its diet is uncertain as its unusual teeth can be interpreted in several ways, such as for crushing insects and grubs or grinding seeds. *Ektopodon* is shown here about to catch an insect. It is sitting in a *Casuarina* tree, evidence for which is found in the fossil pollen grains that occur with the fossil teeth at one locality.

F. KNIGHT 82

first lower molar (M_2) and somewhat damaged or incomplete lower molars M_3 and M_4. The sockets for the last lower molar (M_5) were empty, as were those of the incisor and possibly a tiny canine. The specimen was immediately recognizable as *Ektopodon* and was far better than any previously known specimen.

A few weeks after finding the Mammalon Hill jaw, work by Tom Rich, at Lake Tarkarooloo near Lake Frome, in northeastern South Australia, produced numerous bone fragments and isolated teeth of marsupials in channel deposits perhaps slightly older than the bones and teeth from Mammalon Hill. This site has since yielded several dozen complete or fragmentary *Ektopodon* teeth, as well as two toothless jaws and one jaw with only a fragment of tooth — but enough to identify it as *Ektopodon*. The specimens from Mammalon Hill and Lake Tarkarooloo are considered to be very closely related and probably the same species (Pledge, in press).

Ideas about *Ektopodon*'s relationships within the marsupials received a boost in the early 1970s when some isolated teeth showing characters intermediate between *Ektopodon* and the phalangers (such as the cuscus, *Phalanger*) were discovered at two new sites; Lake Palankarinna and Lake Pinpa, (Woodburne and Tedford, 1975). However, it was not until the discovery of a jaw at Mammalon Hill, that the marsupial relationship was established, since marsupials are characterized by, amongst other things, having four adult molars behind the premolars in each jaw, and *Ektopodon* had just this number.

Because of the information given by the Mammalon Hill *Ektopodon* jaw and a new upper jaw fragment (maxilla) of a related species, it was now possible to make a model of the complete molar dentition. But the anterior teeth remained a problem. Only the premolar of the lower jaw was known, a rather short-bladed tooth. Presumably therefore, the upper premolar would be rather similar, just as it is in many other marsupials. Sure enough, two fragments from Lake Tarkarooloo fitted the expected premolar pattern, although both were incomplete, and even together did not illustrate the whole tooth. When they are positioned according to the sockets of the upper jaw fragment the upper premolars line up with the anterior ridge of the first upper molar, to produce a longer cutting edge; all fitting into a pattern like pieces of a jigsaw puzzle.

The incisors are the least known of all the teeth, since none are preserved in the jaws. One isolated lower incisor from Lake Tarkarooloo is considered to be a good candidate because it is the right size and has a root of the right shape to fit the socket of the lower jaw found there. The upper incisors, however, are still a puzzle. They may have been adapted not so much for cutting as for plucking i.e. pulling leaves, fruit or seeds from their stems, whereas the premolars appear best suited for cutting, such as biting through twigs or possibly even seed husks. The molars are most distinctive: their structure can be likened to a rasp, that by sideways movement of the jaws would grind hard food particles to pulp. Microscopic examination of the worn occlusal surfaces (the areas where the teeth come in contact with each other) of some teeth does indeed show fine transverse wear striations. The transverse ridges of the teeth, meshing together, effectively prevent any back-and-forth movement, yet the longitudinally bladed premolars also functioned as a cutting mechanism, and could only do this if the molars were 'disengaged' or not in contact.

The fossil record of the Ektopodontidae is very poor, with five species, each known from one or, at best, two localities (Pledge, in press). Only one of the species has been named and thoroughly described. The fossils consist of only isolated teeth and a few fragmentary jaws.

Despite this poor fossil record some evolutionary patterns and trends are notice-

Excavation at Mammalon Hill is a slow, painstaking process because of the nature of the sediment in which the fossils occur.

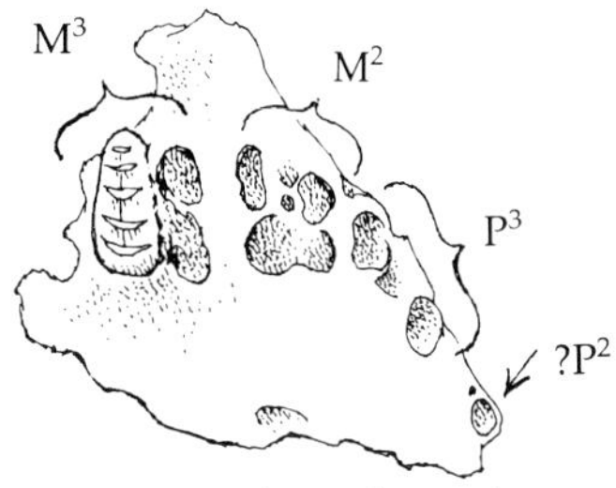

The upper jaw fragment of an ektopodont

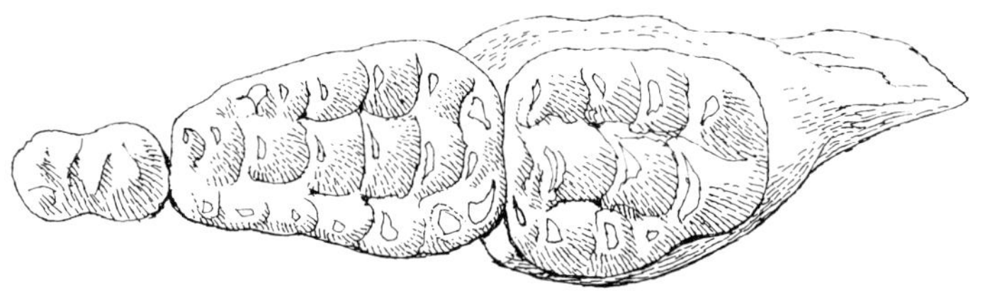

Right upper jaw of the extinct multituberculate *Meniscoessus robustus*, a Late Cretaceous form from North America. The rows of tubercles are oriented lengthwise on these teeth, whereas they are transverse on *Ektopodon* teeth. Multituberculates, which lived from the Jurassic until sometime in the Oligocene, were once thought to be related to *Ektopodon*.

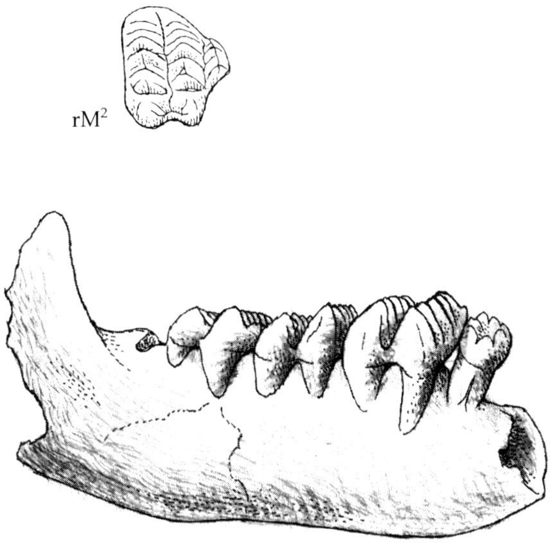

The right upper second molar and right jaw of *Ektopodon* sp.

able. The oldest species has the least number of cusps on the lophs, and its molars are relatively long. The Mammalon Hill teeth are broader and have more cusps on the lophs, and this trend reaches its extreme in *Ektopodon serratus*.

By comparison of the earliest and probably least specialized species with other marsupials, some idea of their relationships to other marsupials has been gained. No living species has teeth of this kind, but there are minor crenulations on the molars of some species of *Phalanger*, which resemble, somewhat, the cusp ridges of ektopodont teeth. *Ektopodon* and *Phalanger* have other minor features in common, so most palaeontologists working on this group regard the Ektopodontidae as having been derived from an early phalangerid (Woodburne and Clemens, in press).

In the reconstruction presented here, it is assumed that the Ektopodontidae are most closely related to the possums and their relatives (Phalangeridae), and in particular, to the cuscuses. Thus, a general body form can be visualized and a probable lifestyle inferred. For more detail, the fossils themselves were studied, particularly the mandibles (lower jaw), and the maxilla (upper jaw). The full palate of *Ektopodon* was reconstructed by mirror-imaging a drawing of the almost toothless maxilla, and adding the molar teeth of *Ektopodon*. Because the maxilla bone incorporates the base of the cheekbone and part of the eye-socket, it was also possible to reconstruct the facial part of the skull. This showed a pair of large, forward-directed eyes in a broad, blunt-faced skull. The massive lower jaws fit this image well.

The habits and place where *Ektopodon serratus* lived can now be inferred, based on characteristics of this hypothetical skull. Large eyes imply nocturnal or at least twilight activity. The forward-directed eyes would give stereoscopic vision with good depth perception, as in primates and hunting animals, and therefore suggest that *Ektopodon* was either arboreal or a hunter, or both. In either case, a considerable degree of manual dexterity would be expected. The form of the teeth, however, is nothing like that of a normal carnivore, although some, such as bears and sea otters, have evolved crushing teeth. However, as cuscuses and other possums enjoy an occasional insect, grub or young bird (Troughton, 1973), *Ektopodon* may have done likewise, possibly even concentrating on this food source rather than leaves and fruit. Another food possibility suggested by the teeth is that of seeds and nuts; that is, *Ektopodon* occupied a squirrel-like niche. There is even a possibility that *Ektopodon* was aquatic, living on worms, crustaceans, and molluscs in the abundant waterways of Mid-Miocene central Australia, and crushing them with its strange teeth. Thus, the rasping, crushing teeth of *Ektopodon* could effectively handle such foods as leaves, nuts, fruits, insects including caterpillars, worms, or even small shellfish.

Colouration of fossil animals is always, except in rare cases of preservation in tar or ice, a matter of conjecture, based on comparison with living relatives or with ecological equivalents. In the case of *Ektopodon* there are two possibilities, depending on the ecological niche chosen. For the omnivorous/herbivorous arboreal interpretation, *Ektopodon* probably would have had a coat colour like the cuscus: dark brownish grey above and pale underneath, with large pale blotches as camouflage (Troughton, 1973). The aquatic interpretation might require a rich or greyish brown coat, lighter beneath, rather like an otter, although a grey colour could also be expected because of the predominantly grey colour of the sediments and muddy streams at that time.

All in all, *Ektopodon* is still quite an enigma, and much more will be learned when more complete skull and skeletal material is found.

Wynyardia bassiana

Spencer, 1901

The Wynyard Marsupial

CLASS: Mammalia
INFRACLASS: Marsupialia
ORDER: Diprotodonta
FAMILY: Wynyardiidae
GENUS: *Wynyardia*
SPECIES: *Wynyardia bassina*
AGE: Miocene

No compendium of Australia's fossil vertebrates would be complete without an account of *Wynyardia bassiana*, Australia's first described Miocene land mammal, and until recently our oldest fossil marsupial. Also one of Australia's most elusive fossil marsupials, *Wynyardia* is currently known from but a single specimen, a relatively complete but badly damaged skeleton. This specimen was collected sometime prior to 1862 at the site of Fossil Bluff, a rugged coastal promontory near Wynyard on the north coast of Tasmania (Anon., 1862 cited in Mahoney and Ride, 1975).

Fossil Bluff was in many ways a most unlikely source for a specimen of this kind. In the first place, the sediments at Fossil Bluff are almost entirely of marine origin and although rich in fossil shells and corals, represent an unusual context for an undoubtedly land-dwelling mammal. Secondly, early studies of the marine component of the fauna indicated that the site might be of very great antiquity indeed, possibly dating to the earliest period of the Tertiary (50-60 million years ago). This great antiquity contrasted markedly with the very recent age of all prior fossil marsupial localities in Australia, none of which appeared to be much older than 1 or 2 million years.

Possibly as a consequence of the unusual nature of its context, the Fossil Bluff skeleton received little immediate attention. Indeed for almost forty years it remained not only unstudied but also unprepared from its ancient sandstone jacket. According to one commentator of the day, the eminent marine palaeontologist R. Tate, these early misgivings about the fossil were further compounded, during the late nineteenth century, by a growing belief that the skeleton was of a modern species and thus of likely recent origin (Tate, 1894). Although Tate failed to identify the source of this misconception, it may well have arisen out of an earlier reference to the Fossil Bluff specimen as that of 'a species of *Halmatarus*' (Johnston, 1888), a contemporary name for the larger kangaroos and wallabies. Whatever its origin, Tate was far from satisified with this determination, and set about a more detailed investigation of the fossil and its context. As a result of these studies, Tate soon became the specimen's most ardent advocate, convinced not only of its authenticity, but also of its great antiquity and correspondingly profound significance. On the question of the fossil's true identity, however, Tate wisely withheld judgement, referring the matter instead to his learned colleague, the notable natural scientist Sir Baldwin Spencer.

Spencer's findings on the fossil, for which he proposed the name *Wynyardia bassiana*, were presented to the Royal Society of Tasmania in 1900, and published in its 'Proceedings' the following year. In keeping with Tate's views as to its great age, Spencer claimed for *Wynyardia* a unique combination of primitive and advanced features. These alternately indicated a relationship with the primitively polyprotodont marsupicarnivores (marsupials which possess many equal-sized lower incisor teeth), and with the more specialized diprotodontan marsupials, those, such as the possums and kangaroos, which have greatly enlarged central lower incisors. Overall, *Wynyardia* was presented as an anatomically plausible and temporally well-situated 'missing link' uniting these two major groups of living marsupials. At that time both groups were thought to occur in the Americas as well as in Australia, and to have originated somewhere in the northern hemisphere. The discovery in Australia, of such an ancient and seemingly well-suited intermediate form, appeared to suggest otherwise however — that the Diprotodonta had evolved within Australia during the Early Tertiary, and only at a later date, dispersed to the Americas.

Not surprisingly, Spencer's conclusions met with a highly varied response from his international colleagues. The great majority, however, considered Spencer's claims regarding *Wynyardia* to be wholly extravagant, both with respect to his morphological observations, and the supposed great antiquity of the fossil. Beyond this, there was little consensus over the fossil except that many considered it to be much younger than either Tate or Spencer had envisaged, perhaps even as recent as Pliocene in age. As to its relationships, one popular viewpoint held *Wynyardia* to be a close relative of the modern phalangerid possums, the brush-tails and cuscuses (Bensley, 1903; Wood, 1924). Perhaps the most influential proponent of this view was Fredrick Wood Jones (1931); indeed as a result of his highly persuasive presentation, Jones lead many of his colleagues into the erroneous belief that *Wynyardia* was not only similar to the modern phalangerids, but was actually congeneric with the modern brush-tailed possums, *Trichosurus* spp. This misconception lead in turn to a marked lessening of interest in the fossil, a regrettable state of affairs that was destined to last for almost three decades.

Fortunately, however, this decline in interest in *Wynyardia* did not extend to the study of Tertiary palaeontology in general. Here the wheels of progress continued to turn, driven in a large part by the continual refinement of dating methods. These refinements were particularly dramatic in two areas of geological science; the use of marine microfossils for pan-global correlation, and the direct radiometric dating of volcanic rocks. Applying these techniques to the complex marine sequences of Tasmania and Victoria, geologists working during the 1930s and 1940s were, for the first time, able to construct a detailed and reliable regional chronology spanning the whole of the Tertiary period (Chapman and Crespin, 1935). Within this sequence, the Fossil Bluff Beds were placed roughly midway through the series, at, or just subsequent to, the Oligocene-Miocene boundary (Quilty, 1966). More recently this transition has been dated to around 21.5 million years ago (Abele and Page, 1974).

To the palaeontologists of the day, this growing awareness of the relatively great antiquity of the Fossil Bluff Beds, and thus of *Wynyardia*, posed a number of problems. In the first place, was it biologically plausible for the genus *Trichosurus* to have survived virtually unchanged since the Oligocene period? If not, then was *Wynyardia* less closely related to the modern possum than was commonly understood? Alternatively, could the fossil be somehow younger than the ancient marine sediments, perhaps even intrusive into the beds as was previously thought? These and similar uncertainties prompted several investigators to look anew at *Wynyardia* during the late 1950s.

The initial, nagging problem of the specimens' authenticity was tackled by Edmund Gill, then Curator of Fossils at the National Museum of Victoria. Using a

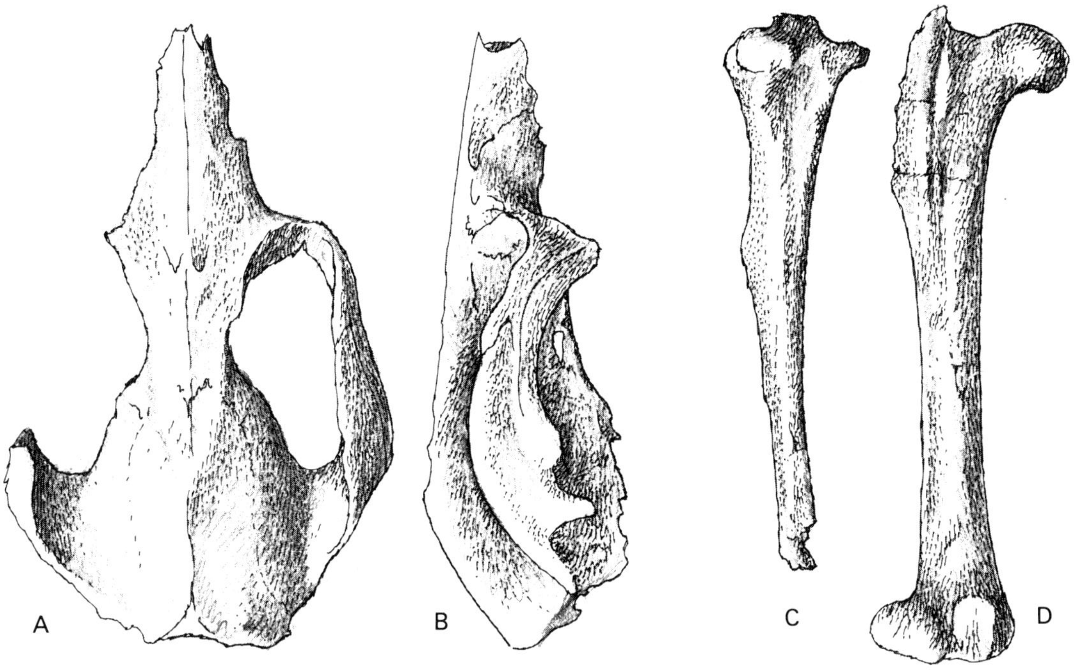

A selection of bones known from the type specimen of *Wynyardia bassiana,* preserved in the Tasmanian Museum in Hobart. (**a, b**) Skull from top and side; (**c**) tibia from back; and (**d**) femur from front. All redrawn from the original illustrations in Sir Baldwin Spencer's account of this animal in 1900.

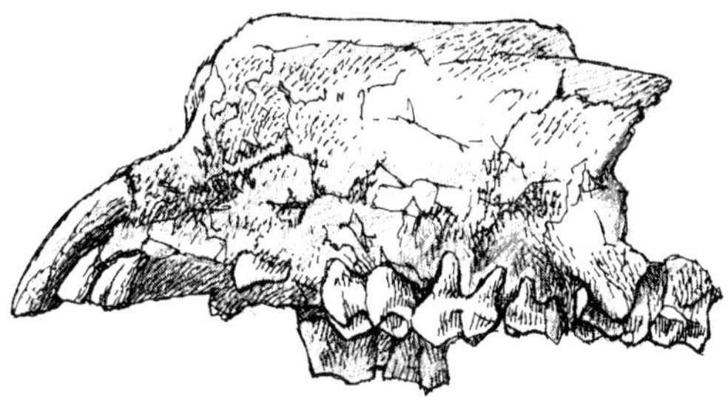

Side view of the snout of *Namilamadeta snideri,* a possible wynyardiid from the Miocene flood plain sediments of northeastern South Australia. This specimen is about 80 mm in length.

variety of chemical tests on the bones themselves, Gill set about to test whether the skeleton of *Wynyardia* could be distinguished from a number of undoubted Fossil Bluff fossils. His results, reported in 1954b, fell unequivocably in favour of *Wynyardia* being a genuine element of the Fossil Bluff fauna. *Wynyardia* was thus a truly ancient beast. But what sort of animal was it?

Shortly after Gill's pronouncement, David Ride presented his new findings on the anatomy and relationships of *Wynyardia* (Ride, 1964). In concordance with Spencer's analysis, Ride identified a number of decidedly primitive, non-possum like features in the fossil skull and skeleton. These were particularly marked in the previously unstudied ear region of the skull. Unlike Spencer, however, Ride was confident that the Miocene species had possessed the characteristic 'diprotodont' condition of the lower incisors, and could thus be included within the marsupial order Diprotodonta (used by Ride in a restricted sense to include only Australian forms). Combining these two lines of evidence, Ride concluded *Wynyardia* to represent a separate, very primitive branch within the Diprotodonta.

Over the last two decades, Ride's views regarding *Wynyardia* have achieved general currency among palaeontologists and zoologists alike. However, several recent developments emphasize the likelihood of further changes to the status of *Wynyardia*. The first of these arises from a recent study of the brain of *Wynyardia* (preserved as an impression on the inside walls of the braincase). On the basis of comparisons with many recent marsupials, the authors of this study Haight and Murray (1981), concluded *Wynyardia* to be a relatively generalized diprotodontan, similar in many respects to the modern phalangerid possums. Although other lines of evidence, notably that from the ear region, clearly precludes *Wynyardia* from placement within this family, the possibility that it lies close to the ancestry of the phalangerids certainly warrants further consideration.

The second, and even more dramatic development concerns the recent discovery of a variety of decidedly *Wynyardia*-like marsupials in Early Miocene sediments of the Lake Frome Basin (Tedford *et al.*, 1977). Unlike the Tasmanian fossil, this new material includes well-preserved teeth at all stages of wear. Interestingly, the worn cheek-teeth of these animals display a lophodont pattern (characterized by transverse shearing crests) such as is found in the modern kangaroos. Overall, however, the dental evidence is far from conclusive, and could variously be taken to indicate a relationship with the kangaroos, the extinct diprotodontids (both of which also possess lophodont cheek-teeth), the wombats or even with the koala. To date only one of these central Australian wynyardiids has been formally described; this is *Namilamadeta snideri*, a larger and more robust animal than *Wynyardia* (Rich and Archer, 1979).

Judging from this central Australian evidence, wynyardiids appear to have been quite abundant during the Early Miocene. By contrast, their remains are altogether absent in any more recent sediments. This apparent disappearance of the group at the start of the Miocene period correlates well with the initial stages of the kangaroo radiation (Flannery, Archer and Plane, 1983), and may well have come about through direct competition with this highly successful group.

The long history of investigation of *Wynyardia* and its geological setting provide a sound basis for reconstruction of the animal and its environment. Twenty million years ago, during *Wynyardia*'s day, Tasmania was an island refugia of similar size and shape to that of the present day (Deighton *et al.*, 1976; Gill, 1962). Coastal lowlands fronting onto a warm Bassian sea received high annual rainfall without marked seasonality (Dorman, 1966). These areas supported a lush broadleaf rainforest in which *Drimys* and the gymnosperm *Podocarpus* were common canopy trees (Harris, 1971). Upland areas, shrouded for much of the time in heavy cloud or swirling mists, probably supported less diverse forests dominated by species of *Nothofagus*, the Southern Beech (Kemp, 1978).

The lush lowland forests provided suitable habitats for a great diversity and abundance of vertebrates. This would most likely have included a diverse suite of marsupials including, amongst others, primitive kinds of kangaroos, early bandicoots and the extinct diprotodontids, a rich diversity of insectivorous bats, and a great variety of reptiles, frogs and birds, the latter including the sometimes gigantic mihirung birds or dromornithids.

Wynyardia bassiana was probably one of the larger marsupials of its day, perhaps

---

A *Wynyardia bassiana* picks its way carefully through a landscape covered with active basalt flows near Wynyard, Tasmania during Early Miocene time. In normal circumstances, it probably preferred to live in trees much like the modern Brush-tailed Possum, which it closely resembled in size and body form. Many marsupial groups, such as kangaroos and diprotodontoids, may have arisen from ancestors that were similar to *W. bassiana*. However, as these groups were well-established by the time the known specimen of *W. bassiana* lived, it could not have been the actual ancestor.

F. KNIGHT
83

comparable in size to the modern-day Spotted Cuscus of northern Australia and New Guinea. It was also quite likely similar to this species in overall build and habits, moving through its forest domain on all fours with a slow, but deliberate action. Although illustrated on the forest floor, *Wynyardia* was almost certainly at least partly arboreal and may, in fact, have spent much of its time aloft in the forest canopy.

The well-preserved cranium of *Wynyardia* suggests a relatively massive head, with well-developed external ears and somewhat forward-looking eyes, the latter similar to those of the living koala. Prominent muscular scars and ridges appearing on the skull and jaws indicate a powerful chewing action. Combined with the evidence of diprotodonty in *Wynyardia,* these features point to a primarily herbivorous lifestyle for the species. However, this would by no means exclude *Wynyardia* from having also indulged in opportunistic carnivory of the kind practised by many of the larger modern possums.

Despite the relatively complete knowledge of *Wynyardia,* many aspects of its anatomy, such as the nature of its tail, feet and teeth as well as its external appearance, remain speculative. In recreating these unknown, and in some cases, unknowable details we have borrowed extensively from the anatomy of the koala, a relatively primitive diprotodont, which might thus be expected to show some overall resemblance to *Wynyardia.*

Thylacoleo carnifex

Owen, 1859

A Marsupial Lion

CLASS: Mammalia
INFRACLASS: Marsupialia
ORDER: Diprotodonta
FAMILY: Thylacoleonidae
GENUS: *Thylacoleo*
SPECIES: *Thylacoleo carnifex*
AGE: Pleistocene

Few extinct mammals have aroused so much curiosity, so much conjecture as to their way of life and their relationship to other forms as the enigmatic Marsupial Lion, an animal described in 1859 by Sir Richard Owen, the noted British vertebrate palaeontologist as 'the fellest and most destructive of predatory beasts'.

This interpretation was soon challenged in 1871 when Owen wrote in response.

'. . . .several eminent and experienced investigators of Fossil Remains have endeavoured to determine to which of the groups specified the *Thylacoleo* was closely or immediately allied. Some have been led to the belief of its having been a Kangaroo, some have deemed it a Potoroo or Rat-Kangaroo, others would rank it with the arboreal Phalangers or Koalas; but all concur in repudiating its carnivorous character. . . . and have sought with more or less ingenuity, to invalidate the conclusions which I had been led to deduce from the parts of the fossilised remains. . . .'

Thus, Owen began the defence of his hypothesis; 124 years and more than 40 scientific papers later, consensus now favours Owen's original interpretation.

To understand the controversy we must examine the teeth of this remarkable animal. It has prominent front teeth, the first incisors, features indicative of marsupials belonging to the order Diprotodonta, a group which contains the possums, koalas, wombats and kangaroos. Most of these animals have a largely herbivorous diet. The cheek-teeth are dominated by huge, blade-like third premolars, which function in a similar manner to the flesh shearing carnassial teeth of carnivores. The molar teeth are small and reduced in number (one upper and two lower on each side). The canines (the large, stabbing teeth of the carnivores) are but small pegs. Owen (1871) argued that in the living carnivores the relative enlargement of the blade-like carnassials was a good measure of the ferocity of the species. Clearly *Thylacoleo* had to be particularly fierce! Furthermore, the reduction of the canine teeth to pegs in *Thylacoleo* was more than compensated for by the development of the first incisors, which were 'adapted for piercing, holding and lacerating, like the canine of a carnivore' (Owen, 1871). Indeed as Broom (1898) pointed out, much of the opposition to Owen's interpretation was based on the belief that 'diprotodonts can't be carnivors', because all others are herbivores. This opinion in turn had lead to suggestions of more fanciful diets: ripe fruit or eggs (Cope, 1882, 1884)! But Broom did not succeed in putting the matter to rest, for in 1929 Anderson drew attention to a paper by Charles De Vis (1883) of the Queensland Museum claiming that *Thylacoleo* was a marsupial 'hyaena' rather than a 'lion'. Anderson noted that hyaenas crushed and ate bone and that such animals 'have broad strong teeth with several cusps, and a well-developed cingulum (a low shelf around the base of the tooth) on the molars and premolars, the purpose of which is to protect the gums from injury by splinters of bone. The teeth of *Thylacoleo* are not of this type; nor is a cingulum to be found on them'.

After showing that *Thylacoleo* was not an 'hyaena' Anderson then took an illogical step; dismissing the carnivore hypothesis altogether he returned to the idea of a specialist plant-eater feeding on native cucurbits or cycad nuts, even though he was aware that 'there are difficulties in accepting the view that *Thylacoleo* could obtain adequate food from these sources'.

Further difficulties with the carnivore hypothesis arose from an analysis of the large incisor teeth set together at the front of the jaw like the beak of a parrot (Gregory, 1951). In this position they may have been able to pierce and lacerate but 'a living animal caught in such incisors could spin about as if on a pivot' (Gill, 1954a) and would be 'particularly unsuitable for catching and holding anything alive and struggling' (Marett Tims and Hopewell Smith, 1914).

During the 100 years after Sir Richard's description many skull and jaw fragments of *Thylacoleo* were collected from cave and swamp deposits throughout Australia, but it was not until 1956 that any idea of the portions of an individual skeleton was possible, when, during quarrying at Naracoorte in South Australia, an associated skeleton was found. During the next few years further partial skeletons were discovered in South Australian caves (Daily, 1960) and in the sand dunes of Lake Menindee (Tedford, 1967). Then in 1966 in a loam pit near Moree in New South Wales workmen unearthed an almost complete skeleton of this unusual animal, with only the hind feet and the tip of the tail missing. When the matrix was removed from the skeleton, it was found to contain the remnants of a pouch young (Finch, 1971). In 1969 the discovery of the Fossil Chamber in Victoria Cave, Naracoorte, lead to the unearthing of the remains of a further 14 animals, including well-preserved and articulated fore and hind feet. (Wells, 1974; Wells and Nichol, 1977). A picture of this unusual marsupial is now emerging. Finch (1971) has depicted *Thylacoleo* 'as an animal the size of a leopard with comparatively long limbs for its size and a large, heavy head supported on a thick muscular neck... The paws were strong, heavily clawed, and probably used in striking prey and for tearing it'.

Wells' and Nichol's (1977) examination of the forepaws demonstrated that the thumb is robust, capable of considerable movement and bears an exceedingly large, hooded claw. The remaining digits or fingers are long and slender, have small claws and only limited palmar flexion. When standing, the weight of the animal would have been borne by the ends of the toes in a similar manner to that of a cat. Notwithstanding, *Thylacoleo* would have had a powerful grasp through opposition of the thumb, not to the remaining fingers as in primates, but to a flattened bone in the wrist (the pisiform).

The structure of the hind foot suggests that the weight of the animal was distributed along the side of the foot (a plantigrade stance). Indeed the hind foot is remarkably similar in form and structure to that of the Brush-tailed Possum, even to syndactyly of digits II and III (that is, the two toes lie very close to one another, enclosed in a sheath of skin), a clear indication of its diprotodont ancestry. Wells and Nichol (1977) could find few anatomical differences in the hands of terrestrial and arboreal carnivores, and suggested that a hand well-adapted to climbing could equally well be adapted to holding prey.

Besides the information now available on the skeleton of *Thylacoleo*, studies of recently discovered skulls and teeth have allowed further insights into feeding habits,

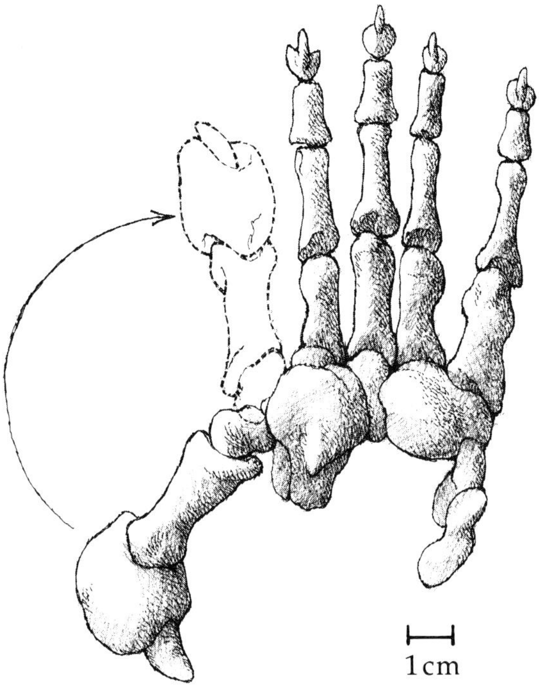

The hand (manus) of *Thylacoleo* seen from above. The sheathing bases for the retractible claws are clearly visible. The range of movement of the large pseudo-opposable thumb is indicated.

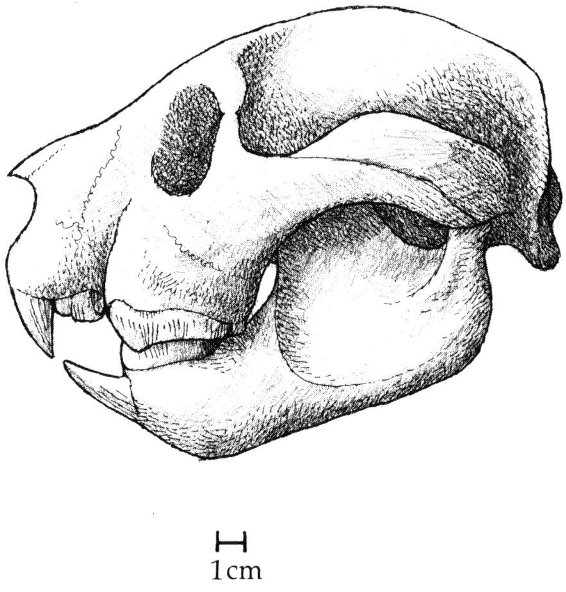

Side view of the skull and lower jaw of *Thylacoleo*.

clarifying the old controversies that have raged around this animal.

Wells, Horton and Rogers, (1982) have studied the structure and development of wear patterns on the teeth of one species, *Thylacoleo carnifex*. They examined the striations produced by the passage of food across the tooth enamel of known carnivores and herbivores and compared their findings with similar marks on the teeth of *Thylacoleo*. The patterns produced by food passing down the enamel surface of the teeth were similar to those on the teeth of known carnivores but differed from patterns preserved on herbivore teeth. They concluded that the lower teeth retained a stabbing or piercing function, while the upper teeth acted both as an anvil against which food is held and a stone against which the lowers are sharpened.

The large premolars have a crescent-shaped cutting edge. As these crescents pass one another during a bite, all the shearing forces are concentrated at two converging points. This process is further emphasized by the fine serrate enamel margins of the teeth, which produce an effect not unlike the cutting edge of a steak knife. Furthermore, wear on the upper incisors produced a blunted, semi-circular wear surface; in contrast, wear on the lower incisors maintained a sharp triangular outline from which fine, serrate ridges descended. Analysis of the jaw mechanics of the complete *Thylacoleo* material suggested a possum-like jaw musculature geared for stabbing with the lower incisors and cutting with the large premolars. All evidence pointed to a flesh-eating marsupial of diprotodontid ancestry, not a plant-eater.

The limb adaptations were for climbing and grasping rather than running, while the absence of interlocking canines as a means of holding prey was compensated for by the large clawed thumb. *Thylacoleo* may well have held its prey, killing it by strangulation or suffocation as do lions and cheetahs (Schaller, 1972); combined use of incisors and premolars would soon strip the skeleton of flesh and hide.

Convincing evidence of predation by thylacoleonids has come from studies of incised bones first reported from Queensland (De Vis, 1883) and Victoria (Spencer and

Walcott, 1912) and most recently from Lancefield in Victoria (Horton *et al.*, 1979; Horton and Wright, 1981). Horton and Wright (1981) described deep V-shaped cuts on ribs and limb bones of kangaroos and noted that 'The Lancefield bones suggest that the species (*Thylacoleo*) was a meat-eater and give no indication that it was a bone crusher'. Their suggestion gains support from the absence of completely severed bones in coprolites (scats), characteristic of the bone-eating and crushing Tasmanian Devil.

What then was the prey of *Thylacoleo*? Wells, Horton and Rogers (1982) point out that its size alone makes it an unlikely pursuer of arboreal species, which can move with ease through the highest branches. *Thylacoleo*'s remains are often found in association with extinct leaf-eating kangaroos of the genus *Sthenurus*. Perhaps *Thylacoleo* preyed on these and other kangaroo species and hauled its prey into trees, as does the leopard, thereby removing it from the reach of scavenging forms such as the Tasmanian Devil. This hypothesis accords well with all the available evidence and provides a niche for a carnivore, which avoids competition with the Thylacine (Tasmanian Tiger), *Thylacinus*, the only other large Australian marsupial with predominantly flesh-shearing cheek teeth.

Recent discoveries have extended the geographic range of *Thylacoleo* to the Pleistocene deposits of the Lake Eyre Basin; indeed it is present in most Late Pleistocene faunas throughout Australia. With the exception of the Victoria Cave deposit (Wells, *et al.*,1984), its numbers are usually low, which is consistent with an animal filling the top carnivore niche. *Thylacoleo* appears to have become extinct along with many large marsupial herbivores during a period of great aridity, approximately 18,000 years ago, a time when the great ice sheets of the northern hemisphere reached their zenith.

Until recently *T. carnifex* was the sole representative of the family Thylacoleonidae, but discoveries in mid-Tertiary sediments of Queensland, the Northern Territory and South Australia (Archer and Wade, 1976; Archer and Rich, 1982; Clemens and Plane, 1974; Pledge, 1977) attest to an older diversification of these unusual carnivores (Archer and Dawson, 1982). To date only teeth, skull and jaw fragments are known. While very much smaller in size than *T. carnifex*, Tertiary-aged species, such as *Wakaleo alcootensis, Wakaleo oldfieldi, Wakaleo vanderleuri* (mid-Miocene) and *T. hilli* (? Pliocene), all show the same characteristic shearing premolars that characterize their larger Pleistocene relative.

In the reconstruction two thylacoleonids are depicted dragging a freshly killed Grey Kangaroo into a tree. The grasping hand is clearly visible as is the possum-like hind foot; the long reach of the forearm is also evident. Tail length has been estimated from the taper of the proximal caudal or tail vertebrae, all that are known thus far from *Thylacoleo*.

Two *Thylacoleo*'s prepare to drag the carcass of a freshly killed Grey Kangaroo into a tree out of the reach of other scavengers.

F. KNIGHT 81

Ngapakaldia

Stirton, 1967

A Small Diprotodontoid

CLASS: Mammalia
INFRACLASS: Marsupialia
ORDER: Diprotodonta
SUPERFAMILY: Diprotodontoidea
GENUS: *Ngapakaldia*
SPECIES: *Ngapakaldia tedfordi* Stirton, 1967
Ngapakaldia bonythoni Stirton, 1967
AGE: Middle Miocene

Probably the best known Australian fossil mammal is *Diprotodon*. Prominantly displayed in most museums, this large cumbersome, rhinoceros-sized, quadrupedal herbivore has made an impact on museum goers and the interested layman since it was described by the distinguished English anatomist Sir Richard Owen (1838). It gave its name, which means 'two front teeth', to a whole group of animals that roamed the continent, Tasmania and New Guinea since early Tertiary time. Members of this group (the Diprotodontidae) became extinct in the Pleistocene and have no living descendants (Archer, 1977). The smallest and earliest known undoubted diprotondontoid is *Ngapakaldia*, the first tantalizing remains of which, a small toe bone, were found as scree on the western shore of Lake Palankarinna, east of Lake Eyre in 1953. It was immediately realized that this bone did not belong with any of the animals that were being excavated, and its affinities remained obscure until 1957 when Richard H. Tedford, now of the American Museum of Natural History, found further foot bones associated with teeth at Lake Pitikanta, a small salt lake to the north of Palankarinna. Each expedition to the area made further discoveries; in 1962 the University of California party drove their back hoe up the interdune valleys, north of Cooper Creek, to Lake Ngapakaldi, deep in the heart of the Tirari Desert where they discovered almost complete skeltons of this small diprotodontoid (Stirton, 1954, 1963).

The palaeontolgists working at Lakes Pitikanta and Ngapakaldi found, to their delight, that parts of the postcranial skeleton were included with almost every find. The animals were obviously gregarious herbivores which came, in considerable numbers, to the water's edge to drink, became bogged and were preserved in the greygreen muds. Individuals have been found, from one to five metres apart, some with their hind feet clasped together, an indication that the animal had died on its back. Others have been found with their hind limbs preserved but the front halves of their bodies cut off at the bedding plane. Amongst other individuals were those with only the front limbs preserved. Overall the most commonly found skeletal elements were the fore and hind foot and toe bones, the caudal or tail vertebrae, which taper to the smallest vertebra not much larger than a match head, plus the skulls, lower jaws and teeth. Two individuals were found entangled and buried together as they had struggled to scramble free of the mud.

As the late R.A. Stirton, of the University of California, and his students worked

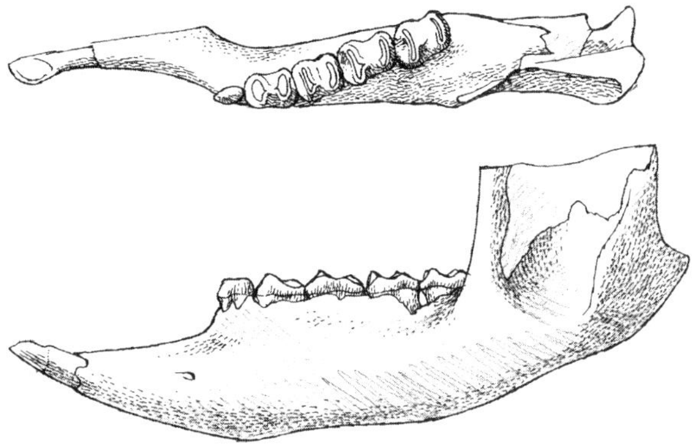

The lower jaw or mandible of *Ngapakaldia;* top and side view (after Stirton, 1967)

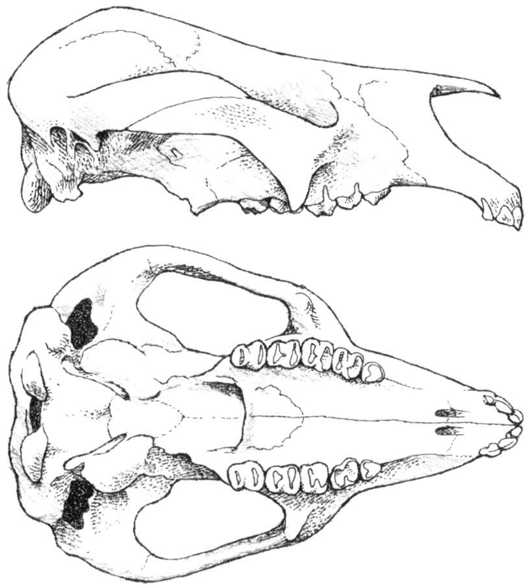

Side and bottom views of the skull of *Ngapakaldia* (after Stirton, 1967), about 250mm long.

at these sites it became obvious that in some parts of the Tirari Desert the small diprotodontoids were as common and numerous during the time in which the grey-green mudstones and dolomites of the Etadunna Formation were being deposited (Stirton, Woodburne and Plane, 1967), as was the huge *Diprotodon* at Lake Callabonna during the Pleistocene.

One of the great difficulties in collecting fossils in the Tirari Desert is surface weathering. The zone of grey-green claystone and white calcareous mudstone contains many specimens, indeed most of them, and consequently they are badly fractured and fragmented by the growth of gypsum crystals. At the surface these bones quickly become broken down by the elements and disintegrate. In some localities if one can dig beneath the surface layer to a depth of 40 cm or more, the strata are not affected, and it was in just such a situation that many of the finds in the Ngapakaldi Quarry were made. Fully articulated fore and hind feet were found and painstakingly removed,for here were the foundations on which these diprotodontoids stood, and it is in the feet that many animals show specializations for their particular life-style. Meticulous care in the removal of the foot bones from the quarry, keeping each bone in position, has enabled full reconstructions of the feet of *Ngapakaldia* to be made.

The hind feet of *Ngapakaldia* are syndactylous, that is, the second and third toes are thinner than the fourth and fifth and were enclosed in a single sheath of skin. In many kangaroos the syndactylous toes are reduced to match-stick thin splints, however in *Ngapakaldia*, while being reduced, they remain robust, a condition also found in the Brush-tailed Possum. The structure of the foot suggests a plantigrade condition, that is, the weight was distributed along the outside of the foot, a mode also seen, but in a more exaggerated way, in *Diprotodon*. Structures on the underside of the toe bones indicate the presence of powerful ligaments, and the laterally compressed and strongly curved last toe bones attest to the animal's powerful claws. Equally powerful claws and similar ligamental attachments are found on the fore foot. From this evidence it seems entirely conceivable that this small diprotodontoid could climb and spend at least part of its life in the trees. However, this is speculation, and we await a full description and evaluation of all the skeletal elements and their comparison with *Diprotodon* and *Zygomaturus*, the only diprotodontoids with complete skeletons, before more definitive statements can be made on its skeletal adaptations.

The skull, lower jaw, and dentition of the animal have been fully described by Stirton, who observed that *Ngapakaldia* was one of the oldest known diprotodontoids, represented by many specimens and already well advanced in its evolution. The molar teeth have two transverse lophs, the bilophodont condition that characterizes the family throughout the rest of its history. Further, the molars show no evidence of their derivation from animals which had the three-cusped triangular pattern characteristic of primitive marsupials. In well-preserved skulls the small sockets for the tiny, simple canine teeth are preserved. Canines are found in several living marsupials, including some kangaroos, but are missing in all younger diprotodontids. The expansion of the squamosal bone to exclude the alisphenoid bone, both skull bones, from the ear region is regarded as evidence that the nearest familial relationship of the Diprotodontidae is with the wombats.

Stirton further observed that *Ngapakaldia*, in contrast to all later members of the family, was more generalized in the rather normal profile of the skull; in the absence of two specialized processes on the lower jaw, and in having the canine teeth already mentioned. Specialized features in the skull led him to believe that the animal belonged in the family Palorchestidae, members of which are found in the mid-Tertiary of central Australia and the Pleistocene faunas of the eastern part of the continent. The incisor, premolar and molar dentition is, however, not so specialized as to place *Ngapakaldia* in direct ancestry to the later palorchestines.

We are left with a fascinating picture. Approximately fifteen million years ago, in what is now one of the driest, hottest and most inhospitable parts of the continent, mobs of large dog-sized *Ngapakaldia* moved in and out of forests, which surrounded lakes and lined the river courses, drinking at the water's edge and feeding both in the trees and perhaps out on the drier plains between the stream courses. As the climate deteriorated and the first major drying occurred, with sporadic, then later prolonged phases, the forest attenuated, the grassland expanded, and *Ngapakaldia*'s habitat diminished. When the lakes dried up for longer and longer periods, the animals in search of water, had further and further to go across the treacherous mud flats and many met their end on them.

Ngapakaldia poses many questions, perhaps the most fascinating of which is what were its ancestors and where shall we find them.

The reconstruction depicts a pair of *Ngapakaldia* during good times drinking at the water's, edge while in the background forest a small mob feeds quietly.

F. KNIGHT 82

Palorchestes

Owen, 1874

Large and Small Palorchestids

CLASS: Mammalia
INFRACLASS: Marsupialia
ORDER: Diprotodonta
FAMILY: Palorchestidae
GENUS: *Palorchestes*
SPECIES: *Palorchestes azael* Owen, 1874
Palorchestes parvus De Vis, 1895
AGE: *P. azael*, Pleistocene.
P. parvus Pliocene-Pleistocene.

In 1874 (1874b) Sir Richard Owen described the first-known specimen of *Palorchestes azael* as '. . . the finest fossil evidence which has yet reached me of an extinct kangaroo . . .'. The specimen, consisting of the anterior part of a skull, '. . . was discovered in the year 1851 by Dr Ludwig Becker in a bed of yellowish sand and clay mixed with very small shells, in the Province of Victoria, Australia . . .' (Owen, 1874b). Owen named the specimen *Palorchestes azael* and assumed, because of certain kangaroo-like features in the teeth, that it was the largest species of kangaroo then known.

Owen's assumption that *Palorchestes azael* was a kangaroo was only the first of many misconceptions about the nature of the species of *Palorchestes.* The very name *Palorchestes,* which means 'ancient leaper', is particularly inappropriate for this cumbersome quadruped. In the 100 years following Richard Owen's description several reconstructions have been attempted. Two of these graced the fossil gallery of the Australian Museum (Fletcher, 1945), and both took the form of a gigantic Grey Kangaroo (one of which was over 3 metres high in the resting position). These reconstructions probably did more to inspire graziers to give thanks for the end of the Pleistocene period than to give an idea of what *Palorchestes azael* really looked like.

In 1958, Jack Woods of the Queensland Museum published a paper suggesting that the species of *Palorchestes* were not kangaroos, partly because they lacked a masseteric foramen, a hole in the lower jaw that is a distinctive feature of all kangaroos. He suggested instead that the species of *Palorchestes* should be regarded as diprotodontids, a group of large extinct marsupials that includes the species of *Zygomaturus* and even *Diprotodon*, the largest marsupial that ever lived. He noted that the lower jaw, the zygomatic arch (a cheek bone) and some aspects of the teeth of *Palorchestes azael* more closely resemble those of diprotodontids than kangaroos. He further suggested that the kangaroo-like features (the tall cheek-teeth and well-developed longitudinal enamel bars of the molars), which led Owen to conclude that *Palorchestes azael* was a kangaroo, must have evolved independently in both kangaroos and palorchestids. Such specializations are often seen in animals that eat abrasive plants. More recently, Archer and Bartholomai (1978) have suggested that the species of *Palorchestes* and their relatives belong in their own family, the Palorchestidae.

Specimens of *Palorchestes* are rare and until 1980 the group was known only from teeth and jaws. Owen's description in 1876 of supposed limb bones of *Palorchestes azael* was, in fact, based on the remains of a large species of kangaroo.

During the 1970s, very unusual limb elements, including a humerus and a hindfoot were collected from a cave in the Wee Jasper area by Geoff Hope and his caving friends. The foot was prepared by Australian Museum staff members, and lent to us for study. Although we were not certain at the time, it seemed probable that it represented the first known limb-bones of a species of *Palorchestes* because it exhibited strikingly specialized deep and markedly compressed claws. Large claws of this distinctive kind were sometimes found as isolated bones in deposits containing the teeth of *Palorchestes*. Confirmation that these Wee Jasper bones did in fact belong to *Palorchestes* came in 1980 with the recognition by us that a partial skeleton of an enormous individual of *Palorchestes azael*, complete with lower jaws and teeth, had lain for many years in the collections of the Museum of Victoria. It possessed a humerus (upper arm bone) that was very similar to (although much larger than) that of the Wee Jasper specimen. It lacks locality data or other documentation, but is nonetheless a most important specimen of *Palorchestes azael* because its association of teeth and skeletal remains has now allowed other limb bones of *Palorchestes* species to be recognized. Most important among these are some 30 bones of *Palorchestes azael* (representing at least 3 individuals) from Foul Air Cave at Buchan in eastern Victoria. Several of these bones had previously been identified as belonging to a giant species of wombat or wombat-like animal (Scott, 1915). These remains, plus those from the unknown locality and the holotype skull fragment from Victoria, have been used as the basis for the present reconstruction of *Palorchestes azael*.

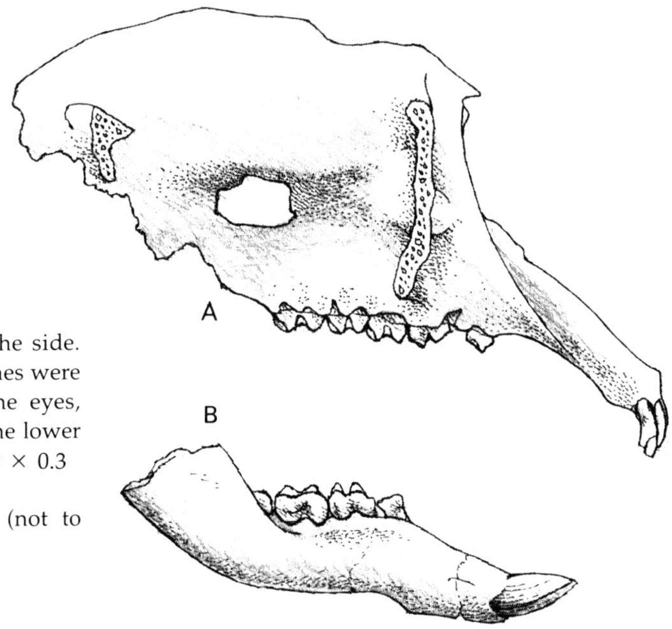

A. The skull of *Palorchestes painei* from the side. This skull shows clearly that the nasal bones were retracted backwards, almost to above the eyes, suggesting the presence of a trunk. On the lower jaw the shovel-like incisors can be seen; × 0.3

B. The lower jaw of *Palorchestes azael* (not to scale) ; × 0.3

A second, smaller species, *Palorchestes parvus*, was described in 1895 by Charles De Vis of the Queensland Museum. Until 1966, this species was thought to be restricted to the Chinchilla Sand of southeastern Queensland, which is Early to Middle Pliocene in age (3-5 million years old). However, since that time it has been found very rarely in a number of widespread localities, including sites of Pleistocene age (2 million-10,000 years ago). The Wee Jasper bones are now thought to belong to *Palorchestes parvus*, and this material, as well as skull remains from Queensland, has served as the basis for reconstructing this second species.

The remains of *Palorchestes azael* are rare also, a factor that has always made it difficult to study this species. Of the sixteen places at which its remains have been found, nine have yielded only a single specimen (a tooth or jaw fragment), and three others less than a handful of teeth. *Palorchestes azael* appears to have co-existed with *Palorchestes parvus* in at least four fossil sites: Strathdownie in southwestern Victoria; Buchan Caves in eastern Victoria; Wellington Caves in New South Wales; and Gore in southeastern Queensland. All of the sites yielding specimens of *P. azael* are thought to be Pleistocene in age, although only two have been dated using Carbon 14. The Pulbeena Swamp specimen (from Tasmania) is about 55,000 years old and the Spring Creek specimens (from Victoria) are about 20,000 years old.

The species of *Palorchestes* are at present placed in the family Palorchestidae (Archer and Bartholomai, 1978) along with the older species of *Ngapakaldia* and *Pitikantia* from Miocene (15 million years old) sediments in central Australia. These latter animals are large wombat-sized marsupials that have similarities in the base of the skull to the species of *Palorchestes* (Stirton, Woodburne and Plane, 1967), but have little else in common with them.

At present, three species of *Palorchestes* are recognized. The oldest is *P. painei* from 6-8 million year old rocks near Alice Springs. It is very similar to *P. parvus*, and to a lesser degree, *P. azael*. It illustrates that the genus was well differentiated by Late Miocene times. Even older palorchestid remains which probably represent a species of *Palorchestes*, are known from the mid-Miocene of Riversleigh in north-western Queensland. These remains clearly indicate that *Palorchestes* species have a pedigree at least as lengthy as that of most other marsupial families.

The distribution and abundance of *Palorchestes* bones can give some clues which are useful in reconstructing the ecology and habitat of these animals. Their remains are rare everywhere, indicating that perhaps they were solitary animals. This apparent rarity could have several causes. If one assumes that their rarity as fossils reflects their rarity as live animals, it is probable that they had low rates of reproduction and occupied relatively large territories or home ranges. Although their remains are often found in caves, there is no evidence that they inhabited them in the way that the ground sloths did in North and South America. Pliocene and Pleistocene remains of *Palorchestes* species appear to be almost totally confined to eastern coastal Australia. Today, as in the Pliocene and Pleistocene, this area supports extensive wet and dry sclerophyll forests. It may be that *Palorchestes* species were at least partly dependent on forests.

Although the fossil remains of *Palorchestes* have yet to be studied in detail, some features are clear. Bartholomai (1978), in reconstructing the head of *Palorchestes painei*, equipped it with a short trunk. This seems justified, first because of the highly retracted nasals bones (nose) shown by the skulls, a feature seen in other trunked mammals, and second, because it exhibits enormously enlarged infraorbital foramina (holes through which the arteries and nerves of the nose pass); an indication that a very large nerve and blood supply served the front of the rostrum. This last feature suggests some unusual structure requiring more service than an ordinary kangaroo-like nose. The more advanced Pleistocene species may have had even larger trunks than the primitive *P. painei*. A deep hole in the region where the jaws join (symphysial

---

Palorchestes azael and *Palorchestes parvus*, a large and a small *Palorchestes*, were surely among the most unusual mammals ever to have lived. The *Palorchestes*, seen in the foreground was the size of a large bull and may have been Australia's first tree-ripper. Its exceptionally powerful forearms, massive claws and bizarre head would surely have been enough to have inspired the legend of the bunyip — or at least a few nightmares among Australia's first Aboriginal inhabitants.

F. KNIGHT '82

sulcus) seems to indicate that the tongue was elongate and narrow, as is seen in giraffes. This adaptation, along with the trunk, suggests that *Palorchestes* species might have had to reach into trees and bushes for their food. The molars of *Palorchestes* species are very tall and complex, suggesting that a fibrous or abrasive diet was eaten.

The skeletons of the species of *Palorchestes* are highly unusual. Perhaps their most striking feature is the structure of the front legs. Relative to other marsupials, the area of the pectoralis muscle attachment on the humerus (upper arm bone) is greatly enlarged, forming a high, hooked process. Although this process is similar to that seen in wombats it is much better-developed in *Palorchestes*. The humerus of *Palorchestes azael* shows an overall similarity to that of wombats, except that it lacks the crest where the deltoid muscle attaches that forms such a prominent feature of the wombat humerus. The nature of the articulation of the upper and lower arm bones in *Palorchestes azael* is very unusual, and appears to indicate that this joint was immobile, the front legs being permanently locked in a partly flexed position. The only possible use that we can imagine for this adaptation would be to strengthen the already massive forearm. In *Palorchestes parvus*, a slight amount of flexibility of the elbow joint seems to have been retained. The ulna (a lower arm bone) in both species is straight and almost solid, the marrow cavity being tiny. The finger bones possess large, semi-circular articulations and appear to have been highly mobile, but only in a forwards-backwards direction. Each finger (digit) of the hand bore a massive, sharp, laterally-compressed claw, similar to that of a koala but far larger. These claws would not be useful for digging, but they would be ideal for ripping, tearing or climbing.

In comparison with the forelimb, the hindlimb was delicate. Toes four and five were equipped with the same kind of massive claws seen on the fingers of the hands, but toes two and three were reduced in size and bound closely together. This specialization is called syndactyly and is seen in all diprotodont marsupials and bandicoots. Syndactylous digits are used for grooming in many living marsupials. The species of *Palorchestes* may also have possessed a clawless opposable great toe similar to that seen in possums.

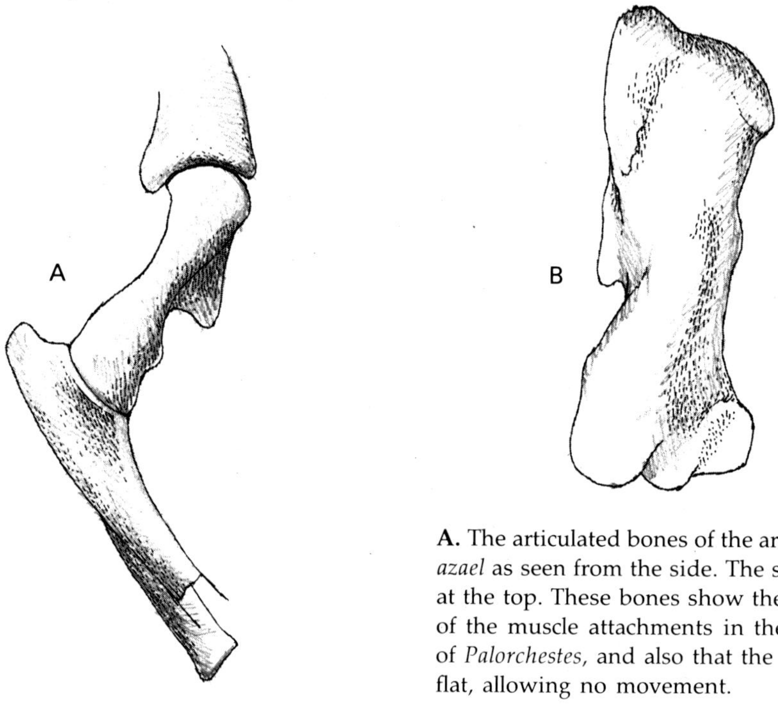

A. The articulated bones of the arm of *Palorchestes azael* as seen from the side. The shoulder blade is at the top. These bones show the massive nature of the muscle attachments in the arm of species of *Palorchestes*, and also that the elbow joint was flat, allowing no movement.

B. A rear view of the upper arm bone (humerus).

The massive claws must have presented a problem to the species of *Palorchestes*, assuming that they walked on the ground. To avoid damage to these claws they could have positioned the foot in several ways. They might have been knuckle walkers, like some great apes, or walked on the side of their feet, or had retractable claws such as those seen in cats. This last option, however, seems the most likely in view of the highly mobile phalanges. In spite of their large size it is also possible that these marsupials used their koala-like claws to climb, perhaps vertically or upside-down in a sloth-like manner.

The massive forearm equipped with grasping and ripping claws, the high crowned molars, the long tongue and the trunk would combine to suggest the possibility that the species of *Palorchestes* were 'tree rippers' and leaf and/or bark feeders. The deltoid crest seen on the humerus of wombats and *Palorchestes parvus*, and the slight mobility in the elbow joint of the latter species suggests that *P. parvus* may have used its forelimb in a significantly different way from *P. azael*. Perhaps, rather than tearing down tree branches, *P. parvus* uprooted shrubs and bushes in search of roots or tubers. Overall the features of the skeleton of species of *Palorchestes* suggest that they were filling a niche similar to that of elephants or (perhaps a better analogy) to that of some of the extinct ground sloths of the Americas. Elephants (and evidently some ground sloths) feed on leaves, twigs and branches torn from bushes and trees. Unfortunately, no living Australian marsupial represents an ecological, let alone a morphological analogue to these bizarre palorchestids. For this reason it is interesting to speculate about a possible 'vacant niche space' left in Australian ecosystems when the last of the palorchestids died.

Diprotodon

Owen, 1838

Diprotodon

CLASS: Mammalia
INFRACLASS: Marsupialia
ORDER: Diprotodonta
FAMILY: Diprotodontidae
GENUS: *Diprotodon*
AGE: Pleistocene

Diprotodon, a name familiar to most Australian primary school children, was the first fossil mammal from Australia given a scientific name. This was done in 1838 by the famous English comparative anatomist and palaeontologist, Sir Richard Owen. The name literally means 'two forward teeth' and refers to the two prominent incisors in the lower jaw that point straight forward from the jaw.

When Sir Richard first named *Diprotodon*, he had only a few scraps of teeth and a rather battered piece of jaw. During the next 40 years, as more bits and pieces of the skeleton became available to him, he gradually built up an idea of the structure of most of the skeleton as one would add bits to a jigsaw puzzle. He dutifully described the structure as the information gradually came to hand and finally, in his summary work of 1877 entitled *Researches on the Fossil Remains of the Extinct Mammals of Australia*, published a reconstruction of the entire skeleton as it was known to him. The parts of the animal then still completely unknown were the feet. In order to fill in this blank, he had his artist draw *Diprotodon* standing in the grass so as to hide those unknown quantities!

In Owen's lifetime, the mystery of the structure of the feet of *Diprotodon* was never solved. But in 1892, the year of his death, the locality which eventually was to provide the evidence to solve the puzzle was discovered at Lake Callabonna, east of the Flinders Ranges in South Australia. In 1953, Mr. Hector Brooks of Adelaide provided this account to R.A. Stirton (1954) about the discovery, which was one of the most exciting fossil vertebrate sites ever found in Australia.

> It seems the aboriginal who first found the bones, having heard the people around the Callabonna Station talking about a circus and its elephants, told them he knew where there were elephants bones, and that the elephants 'had been here in great numbers.' He then led Mr. F.B. Ragless to the site. Instead of elephant bones, Ragless guessed, the black man had found *Diprotodon* remains — and Ragless had heard there was a standing offer of £1,000 for the discovery of a skeleton of *Diprotodon*. The gentleman tried to keep the news in wraps, but somehow the station cook heard about it and quietly quit his job, hitch-hiked to the railroad, and went to Adelaide to collect the prize. The next thing Mr. Ragless knew, there were big headlines in the paper — his cook had found *Diprotodon* and was claiming the reward. Mr. Ragless rushed to Adelaide to expose the cook's claims as false. In the end, it seems there was so much confusion about the whole affair that no one got the reward — least of all the poor aboriginal.

On the basis of this information, the South Australian Museum did manage to appropriate £250 for a start at excavating the fossils. Beginning at the height of summer in January 1893, the first episode of digging went on for almost a year, and a narrative of that trip is provided by Stirling (1900). The work went on despite the natural discomforts of heat, flies, rabbit plagues and dust storms.

Six years after the fossils were excavated, a description of the feet of *Diprotodon* was published (Stirling and Zietz, 1899). Well might Owen have been intrigued had he lived to see these specimens, for few other mammals have feet remotely like them. The toes were reduced to little more than nubbins, the weight having been borne primarily by the bones of the wrist and ankle. A primitive feature retained in the hind foot was that the big toe opposed the others in such a fashion that if the toes of *Diprotodon* had been larger, they could have possibly grasped objects. This apparently was a relic from its arboreal ancestors, which would have found such an adaptation useful. Certainly an animal the size of *Diprotodon* would not have been climbing around in trees!

Despite the determined efforts of Owen and subsequently Sir Edward C. Stirling and A.H.C. Zietz, knowledge of the anatomy of *Diprotodon* was still not complete. On the penultimate page of his final paper describing the fossils from Lake Callabonna, Stirling (1913) made the comment that '. . . an undamaged and undistorted cranium [of *Diprotodon*] still remains of desideratum.' Although Owen had described a skull, much of the middle and top of the specimen available to him was missing. Despite the heroic efforts of restoration performed by Zietz on the material from Lake Callabonna, the specimens had suffered badly in the highly saline environment where they had been entombed.

In 1973, a 12 year-old school-girl, Kerry Hine, discovered a jaw of *Diprotodon* in a clay quarry owned by her family near Bacchus Marsh, west of Melbourne, Victoria. The discovery was reported to the Museum of Victoria and excavation at the site began immediately. Ultimately, one nearly complete skull and another dozen partial ones were recovered. Working with this material, the internal as well as external anatomy of the skull can now be accurately reconstructed.

Much of the skull has proven to be of incredibly delicate construction. Most of the skull roof is only a millimetre thick, much thinner than mammals not half as large, such as a Grey Kangaroo or sheep.

Although the head of *Diprotodon* is large, the animal is still a 'mental lightweight'. This trick is managed by having a skull which is in essence a box-within-a-box. The modest-sized brain was housed in the inner bone-box and between that and the outer box is a group of air chambers separated one from another by thin sheets of bone, which join the two bone-boxes to one another. Why should *Diprotodon* have had such a swollen head? If the outer bone-box did not exist and instead the muscles running from the skull to the jaws were attached to the inner box around the brain, the muscles would be oriented improperly to close the jaws. By enlarging the skull in this manner, the muscles were oriented advantageously and yet the head was not unduly massive. For the same reason, adult elephants have a similar arrangement in their skull. From the pattern of the air chambers in the two groups of animals, it is evident this adaptation evolved separately as their respective ancestors became large enough for this arrangement to become an advantage; an example of convergent evolution.

A less certain similarity between *Diprotodon* and elephants may have been in the structure of the nose. In the anterior part of the skull of *Diprotodon* is a prominent vertical plate of bone. As no living mammal has such a structure, its function is not immediately obvious. However, one possibility is that it served as a site for the insertion of large muscles. Such large muscles in the front of the skull might have been useful if the lips of *Diprotodon* were enlarged and highly mobile or if the animal had a trunk. If it did have a trunk, the muscles were arranged in a manner unlike

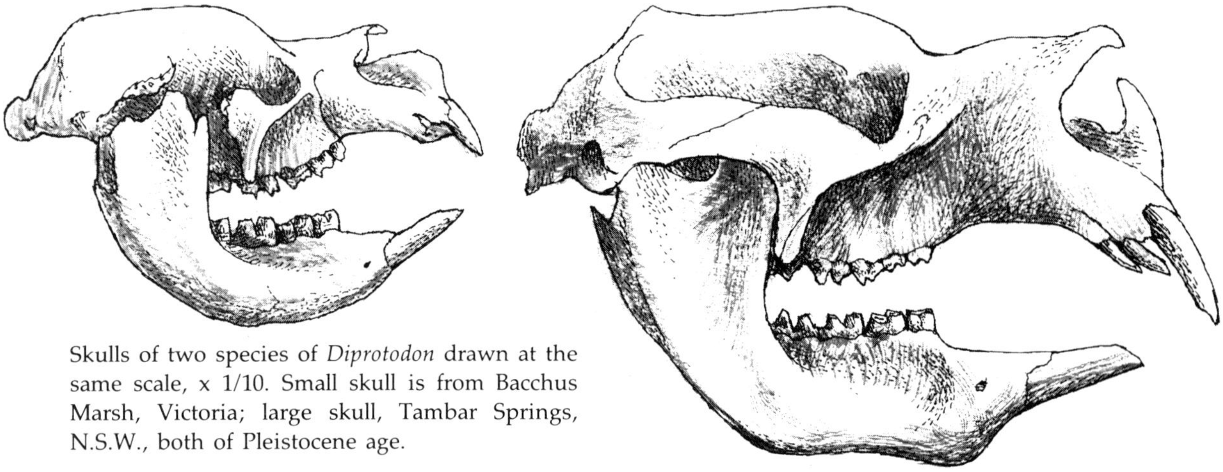

Skulls of two species of *Diprotodon* drawn at the same scale, x 1/10. Small skull is from Bacchus Marsh, Victoria; large skull, Tambar Springs, N.S.W., both of Pleistocene age.

those of a tapir or elephant, because neither of those animals has such a bony nasal septum. Unless a mummified carcass of a *Diprotodon* or an unambiguous rock drawing or carving is discovered someday, it will never be known for certain whether or not a trunk was present on these animals. Mummified carcasses of similar-sized ground sloths have been found in caves in extremely arid parts of North and South America, so it is not a complete flight of fancy to speculate that an intrepid spelunker might someday find such evidence.

Less mystery surrounds the diet of *Diprotodon*. The cheek-teeth are relatively simple structures with two cross lophs, suggesting a browsing habit. Gut contents collected at Lake Callabonna support this idea (Tedford, 1973).

Diprotodon may have owed its success to the spread of arid conditions in central Australia during the Late Cainozoic. The oldest specimens are of Early Pleistocene age when conditions in central Australia were becoming severe. This resulted in areas of forest being replaced by savannahs, steppe and finally deserts. The distribution of sites where *Diprotodon* has been found is concordant with this idea. They were centred in eastern, inland Australia, and none are known from Tasmania. By contrast, a diprotodontid first known in the Miocene, which persisted until the Late Pleistocene, *Zygomaturus*, apparently favoured forested environments and is known only from Tasmania, New Guinea, and coastal regions of mainland Australia. The separation between the two genera was not a rigid one, for there are a number of sites where both occur, but the general pattern is evident (Hope, 1982).

Ultimately, harsh climatic conditions may have been responsible for the disappearance of *Diprotodon* along with a number of other large terrestrial vertebrates at the end of the Pleistocene in Australia. An ongoing debate amongst a number of workers, pits this idea against the one that humans may have been responsible for this extinction episode either directly by hunting or indirectly by altering the habitat. On the basis of carbon 14 radiometric dates, it appears that humans were in Australia at least 10-15,000 years before *Diprotodon* became extinct. As yet there is no unequivocal published evidence that humans engaged in hunting *Diprotodon*. This is certainly not to say, however, that such interaction did not occur.

A baby *Diprotodon optatum*, bogged in a swamp at Lancefield, Victoria 26,000 years ago, is watched helplessly by its mother. During episodes of drought conditions, thousands of Grey Kangaroos also died and were buried in this same swamp when it was one of the few remaining sources of water in the district.

Diprotodon spp. were the last members of a diverse group of four-footed herbivorous marsupials, the diprotodontids, that were common in Australia during the Late Cainozoic. They had evolved by the Late Oligocene and persisted until at least 19,000 years ago.

F. KNIGHT 83

Female diprotodons evidently had a well-developed pouch, for the remains of juveniles have been found just in front of the pelvic region of adults. The pouches could have opened anteriorly as in kangaroos or posteriorly as in wombats.

At Bacchus Marsh, it appears that the accumulation there was the result of an event which destroyed the entire group at one time (Long, 1980). There, the approximately fifteen individuals preserved are all young adults. There are no remains of pouch young nor older adults with their teeth in an advanced state of wear. Among euros, it has been observed that during times of extreme stress such as a severe drought, breeding ceases and the older individuals die off first. The final group to die are the young adults of prime breeding age. The Bacchus Marsh diprotodons may represent an accumulation of the remains of just such a group. Such behavioral information can be gathered only by careful collecting and mapping of the bones in a fossil site, which was done at Bacchus Marsh. Without such data taken at the time of the excavation, a great deal of information can be lost forever.

There are at least two species of *Diprotodon* known, and ultimately more may be recognized. The difficulty in deciding exactly how many there are is related to the fact that at least seven species have been proposed, but none were based on a complete skeleton or even a complete skull and jaw. Often it is impossible to compare the specimens which are the basis for the various species, because they represent different parts of the animal. The answer to the species question can tell us something about how *Diprotodon* adapted to its environment. Was it a generalist capable of coping with a wide variety of conditions; an animal which moved freely about the landscape? If so, one would expect to see a few species whose ranges overlapped with one another. Or did it become specialized to cope with local conditions? In that case one would expect to recognize several geographically isolated species. Were the two or more species contemporaneous or did one succeed another, adapting to a changing environment by an altered morphology? Answers to all these questions await further collection and analysis.

Diprotodon is often referred to as a 'giant wombat'. Although the largest marsupial that ever lived, and thus a 'giant', *Diprotodon* was *not* a wombat. There is a real giant wombat, *Phascolonus gigas*, which although not as large as *Diprotodon*, was certainly no midget, being twice as long and tall and eight times as massive as any living wombat! It is true, however, that among living marsupials, wombats are probably the group most closely related to *Diprotodon* and the other diprotodontids. Just when the group that was the common ancestor to both split apart is unknown. In some of the oldest fossil marsupial sites in Australia, forms occur which are clearly wombats that lived alongside other forms which are diprotodontids. So the split occurred more than 15 million years ago. Because the record of Australian marsupials prior to that time is virtually non-existent, how much earlier the split occurred is a question yet to be answered.

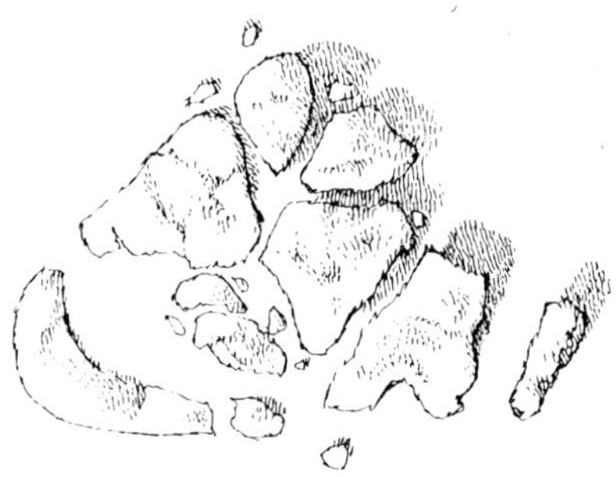

Footprint of a *Diprotodon* preserved on the lake floor of Lake Callabonna, South Australia. R.H. Tedford found entire trackways left by these animals as they lumbered across the lake more than 30,000 years ago; x 0.25.

Propleopus oscillans

(De Vis, 1888)

The Giant Rat Kangaroo

CLASS: Mammalia
INFRACLASS: Marsupialia
ORDER: Diprotodonta
FAMILY: Potoroidae
GENUS: *Propleopus*
SPECIES: *Propleopus oscillans*
AGE: Pleistocene

Among the tens of thousands of fossilized bones collected from the Pleistocene Darling Downs fossil deposits of Queensland during the last century, Charles De Vis (1888b) of the Queensland Museum recognized a very peculiar and then unique jaw bone. It belonged to a gigantic kind of rat kangaroo, which he named *Triclis* (later changed to *Propleopus*) *oscillans*. What excited the scientific community most about the find was that although the bone was closely similar to the jaw of the living Musky Rat Kangaroo (*Hypsiprymnodon moschatus*), which weighs about 500 grams, the fossil clearly came from an animal as large as the largest living kangaroos, that is about 70 kilograms in weight.

Until 1967 the jaw described by De Vis remained unique, but in that year a second specimen was described, this time from the Menindee Lakes area of New South Wales (Tedford, 1967). Since then, the remains (mostly isolated teeth) attributable to about 14 individuals have been found at widely scattered localities throughout southeastern Australia. Even with this number of fossils, *Propleopus oscillans* remains one of the rarest and least understood of Pleistocene marsupials.

Surprisingly, a second species of *Propleopus, P. chillagoensis*, from north Queensland, was described by Archer, Bartholomai, and Marshall in 1978, and subsequently, a further two species of giant rat kangaroo, still undescribed, have been found. All are extremely rare, and these four kinds represent quite an unexpected diversity for a group of animals that for nearly a century was known from a single jaw! Even now, with literally millions of fossil mammal bones found in Australia, the giant rat kangaroos are known from teeth and jaws attributable to less that 20 individuals. A single humerus (upper arm bone) thought to belong to *P. oscillans* is the only part of the skeleton other than jaws that is known, and even it is only tentatively identified, not having been found directly associated with dental remains.

When dealing with an extremely rare and little known animal such as *P. oscillans*, palaeontologists have to exercise a little more than the usual caution in reconstructing its lifestyle and environment. In this case, evidence can be drawn from three main areas: the distribution of the fossils and palaeoenvironment of the sites where they are found, the habits of the living rat kangaroos, and the morphology and nature of the fossils.

The remains of *P. oscillans* have been found at eight localities in southeastern Australia, from the Darling Downs of southeastern Queensland in the north to the northern Flinders Ranges of South Australia in the west. Remains attributable to one or two individuals have been found at each site. Age and palaeoenvironment are known in detail for only two of these localities; Lancefield in Victoria and Lake Menindee in New South Wales. The 26,000 year old Lancefield site, near Melbourne, is a swamp that has preserved a pollen record (Gillespie *et al.*, 1978). The pollen indicates that the area was a treeless savannah 26,000 years ago and that the climate was much colder and drier than at present. The 18,000 to 26,000 year old Menindee site, being further inland, was probably even more arid than Lancefield. The food resources available in such an environment were limited. The great majority of animals found at Lancefield (90%) were grazers, eating the most abundant food resource, grass. Based on the shape of the teeth, one concludes that *P. oscillans* was not a grazer, but could have eaten herbs, shrubs, meat or a combination of all three. The rarity of *P. oscillans* may yield clues as to dietary preference. While some Pleistocene herbivores are almost as rare as the giant rat kangaroos (e.g. *Palorchestes* species), all carnivores are rare relative to their prey.

The ten living species of rat kangaroos attain weights of between 500 grams and about 3 kilograms. They are widespread throughout Australia in a variety of habitats, from dense rainforest to desert. They include herbivores, omnivores and insectivores (*e.g.* the Musky Rat Kangaroo), and several species show a prediliction for meat when it is available. The Burrowing Bettong (*Bettongia lesueur*) has been known to scavenge the carcasses of sheep. The Rufous Rat Kangaroo (*Aepyprymnus rufescens*) has been reported to enter chicken pens and eat eggs. While most rat kangaroos have specializations of the stomach related to herbivory, the Musky Rat Kangaroo, the most primitive living type, and that most similar to the giant rat kangaroos, has a relatively unspecialized stomach.

Overall, the dentition of the species of *Propleopus* most closely resembles that of much smaller insectivores and omnivores, the Mountain Pigmy Possum (*Burramys parvus*)and the Musky Rat Kangaroo (*Hypsiprymnodon moschatus*). Where the species of *Propleopus* differ from these forms, the differences could be interpreted as adaptations to meat-eating. The lower incisors are very stout, sharp, and horizontally aligned. They lack the outside (buccal) band of enamel present in most other kinds of kangaroos, which is used to nip off herbage. While clearly not adapted to this usage, the form of the tooth and its wear allows for a constantly sharp enamel band to be maintained on the bottom edge and this, together with its stoutness, makes the tooth an ideal stabbing weapon. The premolars are the most striking part of the dentition. They form great 'buzz saw' shaped blades, which are strongly serrated and must have been capable of severing very tough matter. Their shape and the way in which they meet the opposing teeth (occlusal cycle) would have prevented food from slipping forward out of their reach. The microscopic wear patterns of the premolars are revealing. Previous studies (Wells *et al.*, 1982) have shown that there is a great difference between the wear on teeth of carnivores and herbivores. The coarse grooving pattern of microscopic wear present on the premolars of *P. oscillans* is more similar to that of carnivores such as dogs, the Thylacine and the Marsupial Lion (*Thylacoleo carnifex*) than it is to any of the herbivores, including other rat kangaroos. The molars show some unusual adaptations. In *Propleopus chillagoensis* the rear molars

---

Propleopus oscillans, killer kangaroo or strange, but peaceful, herbivore? These are some of the possibilities facing the scientist studying remains of the extinct Giant Rat Kangaroo. There is evidence supporting both interpretations. Here artist and scientist have taken a middle path, and portrayed *Propleopus* raiding an emu nest, a less spectacular, but more moderate view than that of the Giant Rat Kangaroo bringing down an adult emu!

KNIGHT '83

are greatly reduced in size, almost to the extent seen in the carnivorous marsupial lions of the genus *Wakaleo*. Also, the molars possess peculiar basal shelves (cingula), on the inside (lingual) of upper molars and the outside (buccal) of lower molars. Similar shelves are seen on the molars of carnivores such as the Tasmanian Devil (*Sarcophilus harrisi*), where they presumably prevent bone splinters from entering the gums. Such structures are not seen on the molars of any other kinds of kangaroo.

This figure shows the relative sizes of the extinct Giant Rat Kangaroo *Propleopus oscillans*, which may have weighed up to 70 kg, and the living Musky Rat Kangaroo, which weighs only 500 gm. These species are similar in tooth and jaw characteristics.

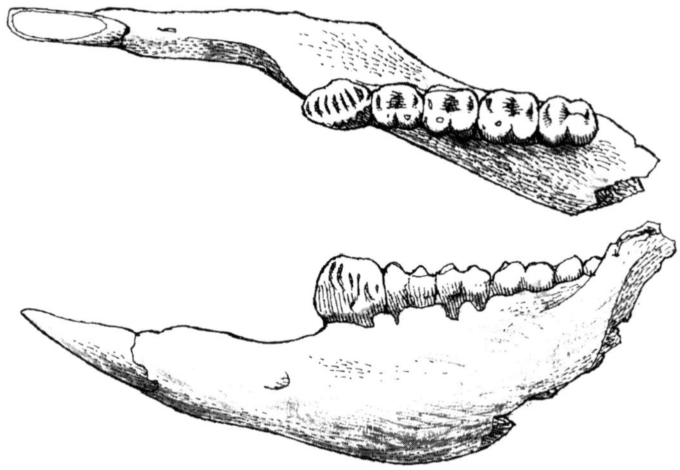

The lower jaw of the holotype of *Propleopus oscillans*. This specimen was found last century in southeastern Queensland. The stout incisor and blade-like premolar can be clearly seen ; viewed from above (top) and the side; × 0.7.

In general, *P. oscillans* possesses fewer of the adaptations to carnivory than does its northern relative *P. chillagoensis*. Although wear grooves (striae) and other adaptations indicate that it probably ate meat, the large molar area of *P. oscillans* also suggests that it may have taken a considerable amount of plant matter also. Perhaps it filled a similar niche to that of the African baboons, which are opportunistic feeders taking meat, eggs, insects, and a variety of plant matter.

While superficially the idea of a meat-eating kangaroo sounds rather outrageous, the weight of evidence is mounting. The idea that animals with herbivorous ancestors can adapt to carnivory has already been documented. The supreme mammalian carnivores of the Australian Pleistocene, the marsupial lions, arose from herbivorous, presumably possum-like ancestors. The giant rat kangaroos diverged from other kangaroo lineages at a very early stage, before specializations of stomach morphology occurred, which are designed to cope with a more strictly vegetarian diet. The scenario built up here can best be tested by the discovery of the skull and skeleton of a giant rat kangaroo. Until then, no definite conclusions can be drawn.

Perhaps only in Australia, where the lack of large mammalian carnivores could induce possums to turn into 'lions', is it possible that kangaroos could similarly have given rise to meat-eaters. While the fossil record of Australia doubtlessly has many surprises, the possibility that the species of *Propleopus* were 'killer kangaroos' is a most unexpected and intriguing realization.

Sthenurus

Owen, 1873

The Monodactyl Kangaroos

CLASS: Mammalia
INFRACLASS: Marsupialia
ORDER: Diprotodonta
FAMILY: Macropodidae
GENUS: *Sthenurus*
SPECIES: *Sthenurus tindalei* Tedford, 1966
S. n. sp.
AGE: Pleistocene

Among the bones of extinct vertebrates collected from the Wellington Caves of New South Wales by the colony's first Surveyor-General (and later Governor) Sir Thomas Mitchell, was part of the jaw of a kangaroo named *Macropus atlas* in 1838 by the renowned comparative anatomist of the British Museum Sir Richard Owen. As mentioned in Chapter 1 Owen described most of the fossil marsupials found in Australia in the nineteenth century sent to him from eastern and southeastern Australia as settlement fanned out across the continent. By 1873 sufficient material had accumulated to allow Owen to recognize *Macropus atlas* as a form distinct from the species of living kangaroos of the genus *Macropus*. He hypothesized that *atlas* represented a previously undescribed genus for which he supplied the name *Sthenurus*, meaning 'strong-tailed beast.' At the time Owen had only fragments of skulls and jaws of *Sthenurus*. Moreover he had confused the remains of species of *Sthenurus* with another genus of extinct kangaroos, *Protemnodon*, that he proposed in 1874 (1874b). Confirmation of the nature of the tail and the rest of the skeleton of *Sthenurus* did not come until a century later when articulated remains were found at Lake Callabonna and the Naracoorte area, both in South Australia. Owen was prescient indeed, for the new material indicated that the tail was stronger and shorter relative to body size than in living kangaroos!

Subsequent to Owen's discovery, remains of *Sthenurus* have been recognized in many other Pleistocene deposits (the last 2 million years) scattered across the southern coastal areas as well as inland to the Lake Eyre Basin. Although *Sthenurus* or its close relatives have been found in Tasmania and other continental islands of the south coast, such as Kangaroo Island, no remains attributed to this genus are known in northern Australia north of latitude 20°S or in New Guinea. These admittedly inadequate data suggest that *Sthenurus* may be a child of the temperate south of the continent, adapted to the developing open lands and seasonal climates that eventually came to characterize the environment of Australia.

Adaptation to open country and perfection of the browsing mode of feeding peculiar to kangaroos is evident in the skeletons of these creatures. The cheek teeth are composed of simple bi-ridged, intershearing crests without prominent longitudinal enamel linkages, and they maintain a fixed position in the jaw. These features are thought to be typical of living browsing kangaroos (Sanson, 1982). In *Sthenurus*, however, the jaws are short and thick and the jaw-closing muscles have a more favorable position to maximize power during chewing. The skeleton differs strikingly from that of the larger living kangaroos. The forelimbs show a greatly modified hand with very reduced thumb and little finger, but elongated central digits that bore long claws. The whole arm is less massive for the size of the animal than in living forms, with greater ability of rotation of the joints. Modifications of the shoulder blade in a surprisingly human fashion evidently enabled *Sthenurus* to reach high above its head to procure young leaves at the tips of branches. The short face of these animals gave them near stereoscopic vision, so important to manipulation of the arms and hands held in front or above the body.

Sthenurus, and its sister genera in the Sthenurinae, *Simosthenurus* and *Procoptodon*, were characterized by a unique monodactyl (one-toed) hind foot in which the central digit and its toe bones were flattened, the final toe bone hoof-like, while the lateral digits were all reduced to splints of bone encased in skin. Thus, the foot was superficially like that of the modern horse, whose ancestry also includes polydactyl forms, and like the horse, *Sthenurus* possessed a complex of elastic ligaments between toe bones that improved the spring flexion of the foot and thus aided the muscular system in providing power to the leap of these bulky animals.

No living kangaroo possesses these features, so we are left to speculate about how *Sthenurus* behaved and moved, from functional mechanical analysis of the skeleton. The impression given by the analysis is of a bulky-bodied, browsing kangaroo that moved about nearly exclusively in the bipedal mode. The modification of the hind foot seems to have been primarily to facilitate this type of locomotion in an animal whose forelimbs were more emancipated from locomotor function than in any living kangaroo. The arms were more mobile and had elongate hands capable of reaching to greater height relative to body size. The hand with its long claws was merely a hook, too modified to do much manipulation.

During Pliocene and Pleistocene times the species we place in the subfamily Sthenurinae were diverse in body size and form. Seventeen species have so far been described in the three genera in the tribe, and five of these belong to *Sthenurus* itself. In most later Pleistocene deposits we find that several sthenurine species coexisted. For example, at the site of Owen's first *Sthenurus*, Wellington Caves, there are five sthenurines present along with many other large marsupial herbivores. At Lake Callabonna just east of the northern tip of the Flinders Ranges in South Australia, three species of the genus *Sthenurus* have been found together with other large kangaroos and *Diprotodon*. Our analysis of these three forms suggests that differences in feeding mode and hence different browsing patterns may have allowed their coexistence and so reduced their competition with one another for food. This in turn suggests a richer environment in an area that is now nearly treeless blue-bush steppe and samphire-rimmed salt lakes. Pollen and plant macrofossils from the deposits yielding *Sthenurus* do indeed hint at a better vegetated environment, but still one of open country with a similar understory, yet with more trees, native pine, she-oak and mallee (Tedford, 1973).

Despite this diversity and adaptation to environments that still characterize much (but probably less today) of the Australia landscape, *Sthenurus*, like most other large marsupial herbivores, did not survive the shift to the modern distribution of climate at the close of the last ice age, about 11,000-12,000 years ago. The youngest dated sthenurines are represented by remains from southern Victoria (17,720 years BP,

Large leaf-eating sthenurine kangaroos. In the background *Sthenurus tindalei*, in the foreground the largest and as yet undescribed species from Pleistocene deposits at Lake Callabonna in South Australia.

F. KNIGHT

Flood, 1974), Kangaroo Island (16,000 years BP, Hope *et al.*, 1977), and Tasmania (14,450 years BP, Goede *et al.*, 1978).

All of these records are of small species of the genus *Simosthenurus*. Species of *Sthenurus* and *Procoptodon* seem to have become extinct at about the time of the last global glacial maximum (18,000 years BP). The pattern of extinction of the giant marsupial fauna suggests gradual decline in diversity from early in the last glacial period (40,000 years BP), when evidence of man begins to appear in Australia, to final extinction of all forms.

The only survivor of the sthenurine group may be a primitive sister-relative, the Banded Hare-wallaby (*Lagostrophus fasciatus*), that is extinct on the mainland today but survives on two islands off the Western Australia coast. Recent research (Flannery, 1983) has linked these small hare-wallabies to the sthenurines by a plexus of peculiar dental features that are uniquely shared by the tiny modern form and its proposed giant extinct relatives. The sthenurines may thus be an old division within the kangaroo family, although how much older than the rest of the family Macropodidae is uncertain, since the fossil record indicates only that rather modern-looking kangaroos, recognizable as relatives of living species, appear without clear antecedents at the beginning of Pliocene time (5 million years ago). Older kangaroos are known, but they represent very primitive groups not clearly related to the Pliocene or Recent species. Clarification of the origin of the sthenurines, the reason for their rise to diversity and rapid decline to extinction are problems in need of further evidence from the fossil record of Australia.

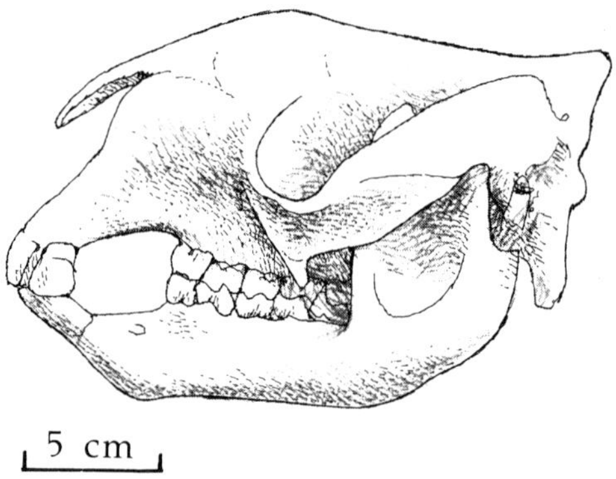

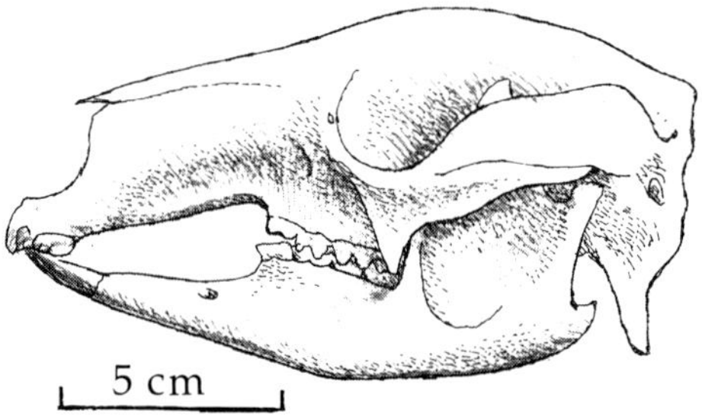

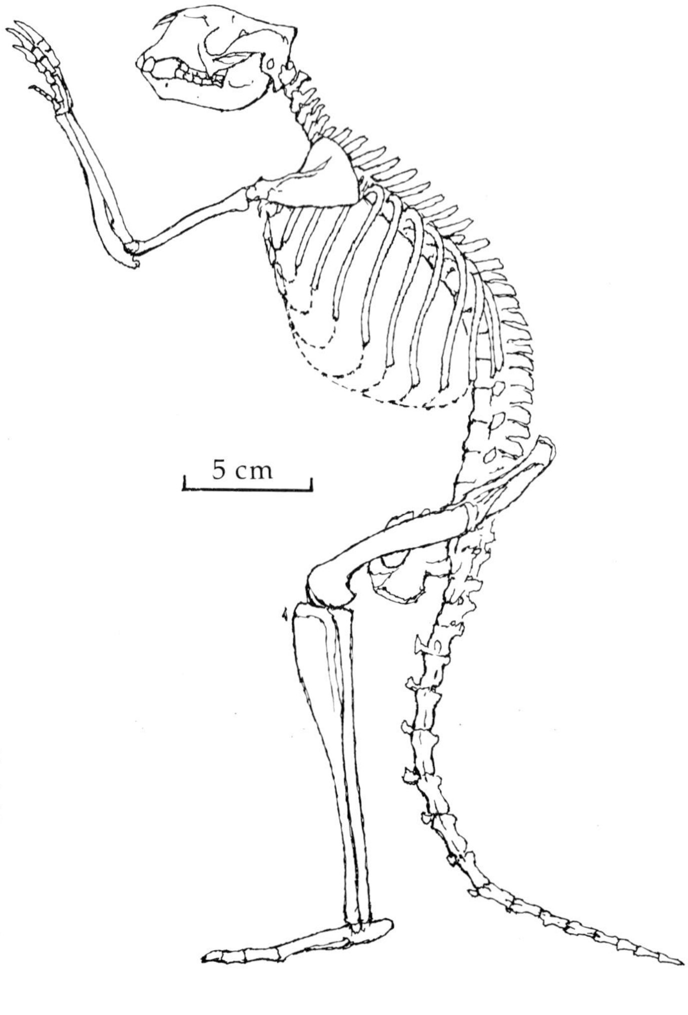

The short faced, deep skull of the large leaf-eating kangaroo is ideally adapted to maximize forces exerted by the jaw-closing muscles during chewing. It is in marked contrast with the long, slender skull of a modern grazing kangaroo. (below).

A large browsing kangaroo balances on its toes and tail as it reaches for a leafy bough.

Hipposideros (Brachipposideros) nooraleebus

Sigé, Hand and Archer, 1982

The Nooraleeba Bat

CLASS: Mammalia
INFRACLASS: Placentalia
ORDER: Chiroptera
SUBORDER: Microchiroptera
FAMILY: Hipposideridae
GENUS: *Hipposideros* *(Brachipposideros)*
SPECIES: *Hipposideros* (*Brachipposideros*) *nooraleebus*
AGE: Miocene

Freeing 15 million year old vertebrates from their limestone tombs is hard, rewarding work in the hot, sticky Gulf country of northwestern Queensland. The fossilized bones of lungfish, freshwater tortoises and crocodiles, dromornithid birds, huge diprotodontids, bizarre kangaroos, and other extinct creatures abound in the Carl Creek Limestone. These have made the outcrops of rocks on Riversleigh Station a palaeontologist's paradise since 1901 when the geologist W.E. Cameron (Cameron, 1901) first noticed the fossil bones near Riversleigh Station homestead. Over 80 years later, the Riversleigh faunal list has grown considerably and now gives significant new insight into the Middle Miocene creatures of northern Australia.

In the summer of 1976, near what is now referred to as 'Tedford's (1966) site D', Henk Godthelp from the Queensland Museum found something extremely rare about 15 metres south of the main limestone bluff, a locale now called Nooraleeba.

It was a graveyard for literally thousands of superbly preserved tiny insectivorous bats. Broken skulls, whole dentaries and complete postcranial elements packed the fine-grained limestone. Even the tiny ear (periotic) bones of the bats, those that contain the structures used in reception and transmission of the ultra-high-frequency sounds used by bats to detect their prey, were preserved in immaculate condition.

The curious habits of bats make them not only the objects of superstition and misunderstanding, but unfortunately also one of the most poorly represented mammal groups in the world's fossil record. Their small and fragile bones, so well-adapted for flight are usually crushed during fossilization. Their preferences for daytime roosting sites further hinder preservation, because their remains are normally destroyed by scavengers of the forest floors or often buried beneath the ground in caves.

Skeletal dissociation, either before or during fossilization, complicates the problems of delicacy and scarcity. In fossil deposits that do produce bats, usually more than one species of bat is represented. In these cases, it is difficult to determine, for example, which jaw fragment belongs to which femur fragment. As a result, traditionally most species determinations have been based on dental fragments only.

For all of these reasons, identifiable fossil bats are rare. In fact, throughout the world, only about 30 fossil bat genera have been described, these spanning a period of more than 50 million years. This contrasts dramatically with the 180 genera of living bats; bats being the world's second largest mammalian order (second only to rodents).

There are few Tertiary fossil bat sites in the world that provide either the kind of preservation or sample size for a single bat species that Riversleigh does. For these reasons alone, the Riversleigh bat locality is of worldwide interest.

In Australia, the ratio of fossil to modern bats is very low. Although bats comprise 25% of the modern non-marine mammal species in Australia, until the Riversleigh Station discovery, only three extinct Australian bats had been found. Isolated microchiropteran (insectivorous bat) lower molars were reported from the Pliocene sediments near Hamilton, western Victoria (Turnbull and Lundelius, 1970); and Archer (1978), described a structurally simple rhinolophid-like tooth from a Middle Miocene deposit in South Australia. In both cases the isolated teeth could not be certainly assigned to any particular bat family.

McKean (1975) described an extinct species of Long-eared Bat (*Nyctophilus howensis*) from a skull that was found by Jerry van Tets in a cave on Lord Howe Island, Tasman Sea. Its affinities with other Australian Long-eared Bats have not yet been refined.

The history of bat evolution and radiation in Australia, therefore, remained virtually unknown. Range expansions and contractions for living bat species were apparent from Pleistocene and sub-fossil records, but otherwise there was no fossil record to provide evidence as to when and from where bats first entered Australia.

Many zoogeographers (Hall, 1981) have proposed that bats could not have entered Australia until sometime in the Pleistocene (some 2 million years ago), when low sea levels would have made possible the crossing of water gaps from mainland Asia through the islands of the East Indian archipelago to New Guinea and Australia. Until recently the lack of an earlier fossil bat record could be seen as indirect support for this theory. Now, however, it would seem that Simpson (1961) was right — bats may have continuously filtered into Australia from Eurasia throughout at least the later Tertiary and the Quaternary.

The bulk of the Riversleigh bat material represents the remains of many individuals of a single bat species belonging to the family Hipposideridae. The remains show all of the characters for this family.

Most diagnostic at this level are characteristics of the highly modified hipposiderid pelvic and pectoral(shoulder) girdles. Hipposiderid bats are unique among all known mammals in that the pelvis possesses an opening (a pre-acetabular foramen) in the bone that lies in front of the articulation with the hind limb (Dobson, 1878). This is formed by a bridge of bone connecting the anterior tip of the pubic spine with the front end of the ilium (Walton and Walton, 1970). The characteristic hipposiderid pectoral (shoulder) girdle is a solid ring of fused bone made up of the seventh neck (cervical) vertebra, first and second body (thoracic) vertebrae, first and second ribs, and the entire presternum (Miller, 1907).

The superb preservation and representation of most skeletal elements of the Riversleigh species has enabled an almost complete reconstruction of this little hipposiderid bat. As Jepsen (1966) pointed out, living bats are so numerous compared to fossil bats and are so diverse in functional adaptation, that every bone in a fossil specimen may give information about the origin and evolution of bats. The preser-

---

The remains of small fish, crocodiles, tortoises and what appear to be algal encrustations, indicate that the depositional environment of the Nooraleeba Bat bones was a quiet, sunny, lime-enriched pool, like that depicted in the background. The bat bones were probably washed into the pool from an adjacent limestone cave or overhang (foreground). The two Nooraleeba bat species; the abundant *Hipposideros (Brachipposideros) nooraleebus* (left) and the less common megadermatid, a False Vampire Bat (right), are likely to have been cave-dwelling bats.

F. KNIGHT '82

Tertiary hipposiderids, (Leaf-nosed Bats) have been found in fossil deposits (dots) throughout western Europe, in Africa and now Australia. The fossils indicate that the geographic range of the family Hipposideridae was once much wider than it is today (hatched area).

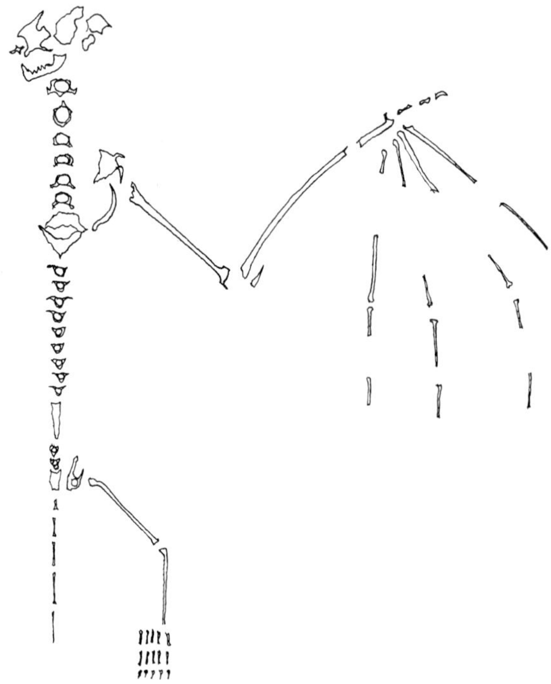

Partial reconstruction of the fossil hipposiderid bat from Riversleigh Station. The delicacy of bat bones and their habits have made them extremely rare in the fossil record.

vation of the Riversleigh bat presents one of the best opportunities so far known for study of the osteological adaptations for flying, echo locating, feeding, drinking and roosting in a Tertiary hipposiderid.

Modern hipposiderids are Old World cave-dwelling bats, generally restricted to the most humid parts of their tropical distribution. They are relatively well-represented in Tertiary fossil deposits throughout Europe, in northern and central Africa, and now Australia.

The Riversleigh Station bat is referred, on the basis of tooth characters, to the subgenus *Brachipposideros* (Sigé, 1968), which was previously known only from French fossil deposits. In France, at least six species of this group appear in the fossil record and extend from the Late Oligocene to Middle Miocene times (Sigé, Hand and Archer, 1982). Skeletal elements that can be referred to the French species are, unfortunately, rare. The Riversleigh specimens, unlike the French material, provide the first unambiguous information about the complete skeleton of any species of the subgenus *Brachipposideros.* Although a part of this group, a unique combination of dental characters makes the Riversleigh *Brachipposideros* (Sigé, Hand and Archer, 1982) distinct from all other known species. It most closely resembles the French Early Miocene bat *H. (B.) aguilari,* suggesting that the Nooraleeba hipposiderid had only recently arrived in Australia since its dental structure does not suggest a long period of evolution in isolation. Therefore, its arrival in Australia might have occurred roughly between 20 and 15 million years ago.

The Riversleigh *Brachipposideros* also exhibits a marked similarity (both in dental and skeletal characters) to Australia's only endemic hipposiderid genus, the Orange Horseshoe Bat, *Rhinonicteris aurantius* suggesting that it may have been closely related

Known geographic distribution of Tertiary species of the subgenus *Hipposideros* (*Brachipposideros*). The range probably once extended through areas of Asia.

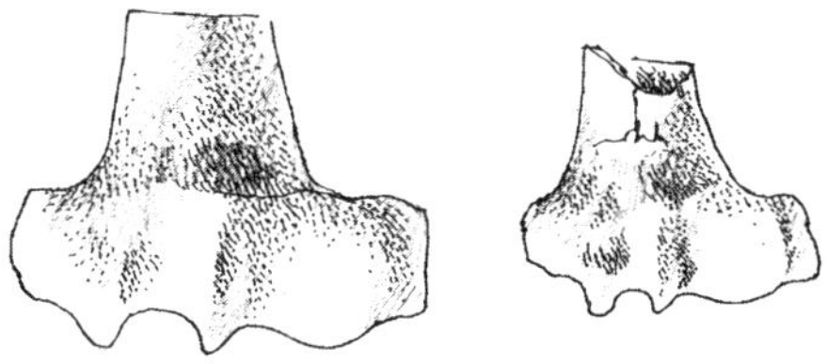

Like hipposiderids, the geographic range of species of the Old World False Vampires, the Megadermatidae, was once much wider than it is today (hatched area). Tertiary megadermatid sites are marked by dots.

Distal end of the left humerus of the Nooraleeba fossil False Vampire or megadermatid (left) shows great similarity to the same bone of Australia's only living megadermatid, the Ghost Bat (*Macroderma gigas*).

to the original Australian hipposiderid stock, which gave rise to the monotypic *Rhinonicteris*.

There is a second, much larger bat in the Nooraleeba fossil material represented by jaw fragments, an ear bone (periotic), shoulder bones (clavicle, scapula), vertebrae, and various incomplete limb bones (including a humerus, a femur and a tibia). At least two individuals produced these fossil fragments.

The material is referable again to an Old World family, the False Vampire Bats or Megadermatidae. Bats of this group are carnivores (not blood-feeders) and, compared to insect-feeders, exhibit conspicuous changes in the form as well as function of the teeth. This is most obvious in the cheek-teeth or molars, which have developed narrow, more blade-like crests for the efficient slicing of flesh. Characteristic dental features of megadermatids have been preserved in the fossilized jaw fragments of this second Nooraleeba bat. Another, even more diagnostic, feature may be seen in the humerus. Here, an outward displacement of the head of the bone (the capitulum) occurs such that the widest end of the humerus (the distal end) is nearly one-fourth the length of the whole bone (rather than the normal one-sixth ratio found in other bat groups) (Miller, 1907).

Megadermatids have been found in Europe, Africa and now Australia, in sediments deposited from the Oligocene to the Pliocene. The number of fossil species far outnumbers the five known living ones. Preliminary study of the Nooraleeba megadermatid bat material indicates that although it is only about half the size and at least specifically distinct from the only other known Australian megadermatid, *Macroderma gigas*, the Ghost Bat, these living and fossil Australian forms are closely related to each other. The Nooraleeba megadermatid is important not only because it provides information about the geographic origins and evolutionary relationships of Australia's megadermatids, but also because it may aid in refining knowledge of the time of immigration of megadermatids into Australia.

The Riversleigh fossil bat bones appear to have accumulated in the muddy bottom of a quiet, sunny, lime-enriched tropical pool containing fish, small freshwater tortoises and juvenile crocodiles. The bat bones are thought to have washed into the pool from an adjacent accumulation in a cave formed in the older Cambrian Thorntonian Limestone.

The precise age of the bat bone-filled pool is unknown, but because of the interpreted stage of evolution of the fossil marsupials preserved with the bats, it is presumed to be about Middle Miocene. This associated fauna so far is distinct from the typical vertebrate fossils found more widely at Riversleigh and thus has been named the Nooraleeba Local Fauna (Rich, T.H., *et al.*, 1982), local fauna referring to the collection of animals that represent forms living together at one time and place in the past.

The Nooraleeba bats make at least one further contribution to Australian palaeontology: they provide the first opportunity to use mammals to correlate non-Australian with Australian Tertiary mammal-bearing deposits.

Most inland Australian Tertiary mammal-bearing deposits have been correlated by interpreting the evolutionary stages of their marsupial species relative to that of marsupial species from better-dated faunas. As a consequence, there is a considerable risk of circular reasoning being incorporated into the dating concepts of these central Australian faunas. The presence of two cosmopolitan and highly vagile bat species in one of these otherwise tenuously dated Australian fossil faunas may provide an independent and much more reliable means for dating this, relative to the European sequences that are dated by marine sequences and radiometrically.

Among mammals, bats have traditionally been considered to be poor palaeozoogeographic indicators (Simpson, 1966; Butler, 1978) because of their ability to disperse more readily than more earthbound vertebrates. In this case, however, this same dispersal capacity may provide a very important link between the Tertiary bats lying beneath French vineyards and those, thousands of miles away, sleeping in stone with kangaroos.

The Riversleigh fossil *Brachipposideros* shows a marked similarity of the teeth and skeleton to Australia's only living hipposiderid, the Orange Horseshoe Bat, *Rhinonicteris aurantius.*

The Riversleigh fossil False Vampire is clearly related to Australia's *Macroderma gigas*. Like northern Australia's Ghost Bat, the fossil False Vampire probably fed on insects and small vertebrates, including birds.

Rhabdosteidae
The Lake Frome Freshwater Dolphin

CLASS: Mammalia
INFRACLASS: Placentalia
ORDER: Cetacea
SUBORDER: Odontoceti
FAMILY: Rhabdosteidae
GENUS AND SPECIES: Indeterminate at present
AGE: Middle Miocene

One of the most unusual occurrences of fossil mammals from Australia is that of dolphins from freshwater sediments in the Lake Frome area, South Australia. The first specimens were discovered by Pat Rich and Rod Wells in June and July 1971, during an expedition led by Dick Tedford of the American Museum of Natural History in New York. Other specimens were discovered later, between 1973 and 1980, by collectors from the Museum of Victoria, the South Australian Museum, and the American Museum of Natural History. It seemed at first that the dolphins were closely-related to a group of living 'river dolphins' (Family Platanistidae) (Tedford *et al.*, 1977), typified by the living, blind Ganges Dolphin *Platanista*. Later work suggests that the dolphins probably belong to an extinct family, Rhabdosteidae, unrelated to platanistids (Fordyce, 1983). The initial uncertainty about identification arose because none of the specimens is complete. Several specimens are represented by a distinctive bone, the periotic (an ear bone), which, in many groups of cetaceans, can be used accurately for identification to genus level.

The species is known from small, hard, erosion-resistant ear bones (periotic, ossicles, bulla) and less robust skull fragments, teeth, vertebrae and forelimb fragments. These represent perhaps six individuals. No articulated skeleton has been found, although one association of bones probably represents a single individual. Even in this case, most of the bones are broken and it is difficult to get an idea of the structure of the animal. For this reason, we must assume that body and skeleton proportions are similar to those of rhabdosteids from other localities, e.g. *Eurhinodelphis bossi* from the Middle Miocene of North America and *Argyrocetus patagonicus* from the Early Miocene of South America. Not enough bones are known to get any idea of variation resulting from growth or sex differences, and the fossils known do little more than suggest that only one species is represented. A few fragments for which measurements are known indicate that the skull was comparable in size to that of *Eurhinodelphis bossi*, perhaps over 1 metre long.

The freshwater dolphin fossils have been collected only from a limited area, about 20 kilometres across. They occur in outcrops of the Middle Miocene Namba Formation which is found in and around Lakes Namba, Pinpa and Yanda in the Lake Frome district, northeastern South Australia.

Periotics and incomplete skull remains leave little doubt that the fossils represent a species of the extinct family Rhabdosteidae. However, relationships with other species within the family are uncertain, either because comparable elements are not known or because the other species have not been described adequately. For these reasons, the Frome dolphins have not been placed in a new species or genus.

Although only a few bones of the dolphin are known, and the Family Rhabdosteidae is extinct, it is not difficult to reconstruct the skeleton and outline. Other rhabdosteids are relatively well known from Miocene rocks in Argentina, California, Virginia and Maryland (U.S.A.) and Belgium. These specimens represent small to medium-sized toothed whales from 2.5 to 4 metres long. Their skulls are superficially bird-like, with a narrow, medium to long 'beak' which can comprise over 90% of skull length. In many species the apical third of the beak is toothless and overhangs the shorter lower jaw, while the more posterior parts of both jaws carry scores of small conical teeth. Rhabdosteids appear to have had somewhat longer and more flexible necks than most living dolphins. Furthermore, their forelimbs were longer, but, like living dolphins, they lacked a flexible elbow joint. Other details of the skeleton of the Frome dolphins are not clear; for example, we do not know if the backbone comprised long or short vertebrae or their degree of flexibility.

All whales, dolphins and porpoises have a similar streamlined external body shape, which reflects hydrodynamic adaptation to rapid, energy-efficient movement through water. Horizontal tailflukes are always present, and most living species have a small dorsal fin in the middle of the back. The Frome dolphins probably exhibited all these features. Living toothed whales, dolphins and porpoises generally have a prominent forehead above the beak. This is caused by a mass of fatty tissue, the 'melon', which has been implicated in the transmission of high-frequency sound used in echo-location by many living species. Rhabdosteids, like all other odontocetes, possess bone features which suggest they echo-located and thus also possessed a melon.

The Frome dolphins were likely to have been similar externally to four living species of 'river dolphin'. The latter are unrelated to each other, but possess rather similar body forms presumably because of adaptation to similar freshwater habitats. Perhaps the Frome dolphins were light grey-brown, slightly darker above than below, with a long, low dorsal fin. The foreflippers most likely were rather broad and flexible, and showed traces of underlying digits.

Presumably the dolphins entered the freshwater systems of central Australia via an external drainage system, like the Murray River. Their closest relatives are found in marine rocks bordering North and South America, and, because the Frome dolphins are not radically different from these related species, it is unlikely that they had been isolated for long before the Middle Miocene. The known fossil record is too poor to tell if they were geologically long-term inhabitants of Australian freshwaters, but the number of fossils known from different localities suggests that they were relatively permanent elements, at least in ecological terms. The dolphins may have been the largest warm-blooded carnivores of central Australian freshwaters, and were exceeded in size perhaps only by some fish and crocodiles. If we can assume that the dolphins were similar in habit to living 'river dolphins', they would have eaten fish and large invertebrates. Acoustic ('sonar' or echo-locating) abilities were probably well-developed, and the dolphins must have been fast and efficient predators. Sophisticated social behaviour was likely, and the dolphins may have formed small schools.

The Lake Frome Freshwater Dolphin, a rhabdosteid, swimming in the shallow inland waters that in Miocene times occupied an area southeast of modern Lake Frome. Below are lungfish that also plied these waters.

F. KNIGHT

Why did the dolphins become extinct? All species of rhabdosteids disappeared worldwide by late in the Miocene, but the extinction of marine species could have had different causes to those of freshwater dolphins. It was probably Late Cainozoic reduction in extent of freshwater habitats, perhaps reflecting wider global climatic changes, which resulted in habitat reduction and, ultimately, extinction. We will never know exactly what caused the extinctions because the fossil record of the dolphins is too incomplete and our knowledge of habitat changes is too poor, so we can only speculate. Lungfish, which once lived with the Frome dolphins, survive today — why not the dolphins?

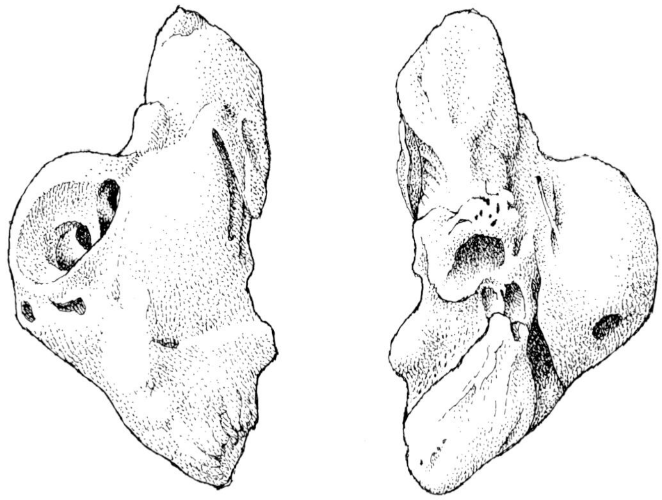

Right periotic (an ear bone) of a freshwater rhabdosteid dolphin from Lake Namba, Lake Frome area of South Australia. Dorsal view on left, ventral view on right. Length of the bone is about 37 mm.

How to care for a Vertebrate Fossil

Because the number of professional vertebrate palaeontologists in Australia is so small, amateur palaeontologists and geologists, mapping geologists, land-owners, or lucky passers-by will always play an important role in the discovery, collection and preservation of vertebrate fossils here. Ranken, the grazier in the Wellington Valley responsible for the early work on *Diprotodon* with Major Mitchell, is one of thousands of amateurs critical to the advancement of vertebrate palaeontology on this continent. Because Australian vertebrate fossils are quite rare, it is important therefore, for the lucky finder to know what to do once fossil bones, teeth, egg shells, or tracks are discovered. Much information can be lost if position in the rock sequence, orientation of the fossils and geographic location are not noted in detail. What is worse is the damage, often irreparable, that can be done to a fossil if it is not properly prepared before collection and transportation.

The behaviour pattern to remember, is that of contacting a professional palaeontologist as soon as possible. Several names, addresses and phone numbers of appropriate people and institutions are given in this book. Vertebrate palaeontologists use several approaches to ensure proper care of fossils both in the field and in the laboratory. A number of excellent books are available that discuss the techniques they use (Kummel and Raup, 1965; Macdonald, 1983; Rixon, 1976).

IN THE FIELD

The first and foremost rule for proper care of a vertebrate fossil is that it must be collected as complete as possible and with as much field data as possible. Once the specimen is collected, unless field data has been recorded it is probably lost forever — and so is much information about how, where, and when the animal lived and died and what happened to its remains as it settled into some sedimentary environment. Such a questionnaire as that modified from Ager (1963), and reprinted here, is well worth having on hand each time you go in the field, for it insures that most aspects of the field data will be examined.

Field questionnaire (modified from Ager, 1963).

The questions should be answered, as far as possible, for each separate rock unit.

Distribution

1. Are the fossils evenly distributed throughout the rock unit?
2. Are they in pockets, lenses, bands, or nodules?
3. Are they more abundant at any particular level in the unit?
4. Do they occur in reefs or shell banks?
5. Are the various kinds of fossils distributed in the same way?
6. Are the same fossils present as are seen in the same bed elsewhere?
7. Are any of the species present unusual in the studied area but more common elsewhere?
8. Approximately how many species are present?

Associations

1. What, very roughly, is the relative abundance of the different groups of fossils present?
2. Are there any obvious close associations, e.g., vertebrate bone with wood fragments?
3. Are there any obvious abstentees?
4. Are there any obvious derived fossils?
5. Are all growth stages present for each species? If not, which stages are present for which?

6. Are any of the other fossils attached to one another in any way?
7. Is there any evidence of the mixing of organisms from different environments, e.g., land plants and marine reptiles?

Preservation

1. Are there any unusual features about the preservation, e.g., colour patterns on mummified skin?
2. Are all the fossils preserved in the same way?
3. Are there any traces of 'soft' parts?
4. Are any delicate structures preserved, e.g., juvenile teeth or bird skulls?
5. Are the fossils worn or broken, and are some species more so than others?
6. Are the valves of bivalves (molluscs) separated? If so, are both valves present in equal numbers?
7. If the valves of bivalves are still joined together, are they tightly closed, partly open, or wide open?
8. If skeletons are present, are they apticulated or only partly so?
9. If higher-plant remains are present, which parts of the plants are they, (e.g., roots, stem, leaves, fruit)?

Relation to sediment

1. What is the nature of the enclosing sediment?
2. Are there any sedimentation structures, such as false bedding, slumps, ripple marks, and scour marks?
3. Is there any obvious relationship between the fossils and the nature and/or grade of the sediment (e.g., delicate bird bones in fine-grained sediment)?
4. Are any of the fossils obviously out of place in the sediment, e.g., terrestrial marsupials in marine limestone?
5. Are the fossils in nodules, if any, preserved in the same way as those in the surrounding sediment?
6. What is the nature of the infilling of any closed shells?
7. Are there any signs of a general disturbance of the sediment by organisms?
8. Are any of the fossils in position of life? If so, what percentage of each species?
9. Are any of the other fossils oriented in a particular way, e.g., long bones lying parallel to one another?

Form

1. Are there any noteworthy growth forms, e.g., delicately branched corals?
2. Are there any possible peculiarities of adaptation, e.g., only diving forms?
3. Are there any obviously pathological specimens or any that have been damaged during life?
4. Are there any borings, burrows, tracks, or trails?
5. Are there any other signs of organic activity, e.g., coprolites (fossil faeces)?
6. Are the fossils highly ornamented or smooth, or are there some of each?
7. Are there any signs of stunting or gigantism?
8. Are there any signs of unusual thickness of shell or excessive ornamentation?
9. Are there any signs of seasonal growth or of general change in growth rate or direction during life?

General

1. Are there any other phenomena of palaeoecological interest in the assemblage?
2. Give a title to the indigenous, or local, part (if any) of the assemblage in terms of its dominant members.

The techniques used by vertebrate palaeontologists vary depending upon the kind of fossils they find. As a first step, once bones or teeth have been found weathering out on the ground, a surface prospecting of the whole area is warranted, which will allow determination of where it is best to concentrate one's collecting efforts. After deciding where it is best to dig, systematic excavation begins, preferably on a grid system that allows both lateral and vertical mapping of fossils from some arbitrary datum. At this point a professional palaeontologist should be in attendance — for much information can be irretrievably lost if careful techniques are not used.

Many fossils, generally possum-sized or larger, will certainly appear as excavation proceeds, but numerous smaller ones will be missed unless the sediments in which they are buried can be sieved. Large quantities of such matrix need to be wet-washed through fly-wire screen. The water generally breaks down the sediments, such as sand, silt, or clay, and washes it off the fossils. The sediment generally passes through the screen, leaving behind a concentrate of stones, bones and teeth. This is dried and then sorted for the palaeontological 'goodies' that delight the determined bone-picker!

For fragile fossils another technique used by vertebrate palaeontologists is plaster-jacketing. First applied for preserving fossil bones during transport in the late nineteenth century, it has greatly increased the survival of bones from field to laboratory. Fossils that are to be plaster-jacketed are first exposed and a trench is dug around them. They are then preserved with dilute aquadere or some other soluble glue, many of which are suggested in Rixon (1976). Then a layer of wet toilet paper or newspaper is applied tightly to the bone, to act as a separator between the plaster and bone. Next, strips of hessian are cut into long lengths and dipped in a thin mixture of plaster and water and applied to the fossil bone in the same way as a doctor would apply a cast to a patient's broken leg. Once the plaster dries, and fast setting plaster is preferred, the cast is undercut, and the bone with adhering plaster is rolled over. The same procedure is carried out on the newly-exposed side, and again, once the plaster is dried, the bone is ready for transport back to the museum where it can be prepared and studied. Foam, or other kinds of padding such as newspaper, straw, or clothing can help minimize damage due to vibration during transportation. Both the screen-washing and the plaster-jacketing technique have revolutionized vertebrate palaeontology this century, allowing finer and finer detail to be gleaned from the fossil specimens.

BACK IN THE LABORATORY

Once the material has reached a safe repository, the often long procedure begins to prepare specimens; copy field data and identifications in permanent catalogues, including computerized data banks; prepare casts of specimens for exchange and display (Kelley, 1983); and finally store the fossils in some logical manner, which allows specimens to be retrieved for study at a moment's notice. To some extent, this aspect of palaeontology is the most difficult, for it is often difficult to gather sufficient funding and staff to allow rapid progress, once a collection arrives from the field. Whereas the excitement attached to expeditions in far away places with strange sounding names (like Lake Kittakittaooloo!) often draws numerous volunteers and sometimes funding, the day-to-day, sometimes monotonous, collection management work often appeals far less to both people and funding agencies. A recent upsurge in volunteer organizations such as the Friends of Museums attached to most natural history museums in Australia, are a heartening advance, providing responsible, volunteer staff for preparation and curation of material that might otherwise go unattended for years. Real problems still exist, however, in having sufficient, long term preparatory staff and in the purchase of equipment and materials for preparation, as well as storage.

In summary, keep in mind that any Australian fossil vertebrate, even though quite fragmentary, may be extremely important scientifically; a non-renewable resource. It is best to contact a palaeontologist immediately after vertebrate fossils have been found, so that an evaluation of the potential of the discovery can be made. It is better NOT TO DISTURB the material beyond first discovery, because non-recoverable and precious information can be lost forever.

List of names and addresses of vertebrate palaeontologists in Australia

In cases where a vertebrate palaeontologist needs to be contacted the following list should be helpful. Included are the names and addresses of professional vertebrate palaeontologists in Australasia, as well as the institutions that house the major fossil vertebrate collections in Australia. For each of the palaeontologists their area of speciality is indicated, although it does not always include all of the groups an individual has studied.

Institutions

The Australian Museum,
6-8 College Street.
Sydney, New South Wales, 2001
Tel: (02) 339 8111

The Australian National
Wildlife Collection,
CSIRO Division of Wildlife and
Rangelands Research
Gungahlin
Australian Capital Territory, 2602
Tel: (062) 41 1211

Bureau of Mineral Resources
Anzac Parade and Constitution Avenue
Canberra City
Australian Capital Territory 2601
Tel: (062) 49 9111

The Museum of Victoria
285-321 Russell Street
Melbourne, Victoria, 3000
Tel: (03) 669 9888

Museum and Art Gallery
of the Northern Territory
Darwin, Northern Territory, 5790
Tel: (089) 82 4211

The Queensland Museum
Gregory Terrace
Fortitude Valley
Queensland, 4006
Tel: (07) 52 2716

The Queen Victoria Museum
Wellington Street
Launceston, Tasmania, 7250
Tel: (003) 31 6777

The South Australian Museum
North Terrace
Adelaide, South Australia, 5000
Tel: (08) 223 8911

The Tasmanian Museum
5 Argyle Street
Hobart, Tasmania, 7000
Tel: (002) 23 1422

The Western Australian Museum
Francis Street
Perth, Western Australia, 6000
Tel: (09) 328 4411

Professional Vertebrate Palaeontologists

Mr. Ken Aplin
(mammals)
School of Zoology
University of New South Wales
P.O. Box 1
Kensington, N.S.W. 2033

Dr. Michael Archer
(mammals)
School of Zoology
University of New South Wales
P.O. Box 1
Kensington, N.S.W. 2033

Dr. Alan Bartholomai
(mammals, dinosaurs)
Queensland Museum
Gregory Terrace
Fortitude Valley,
Queensland, 4006

Mr. Robert Baird
(birds, especially songbirds)
Department of Earth Sciences
Monash University
Clayton, Victoria, 3168

Dr. Alexander Baynes
(mammals)
Western Australian Museum
Francis Street
Perth, Western Australia, 6000

Mr. Robert Bearlin
(whales and allies)
Department of Geology
University of Otago
Dunedin, New Zealand

Dr. Sandra Bowdler
(archaeological fauna)
Department of Prehistory
University of Western Australia
Nedlands, Western Australia, 6009

Dr. John Calaby
(mammals)
CSIRO Division of Wildlife
and Rangelands Research
P.O. Box 84 Lyneham,
Australian Capital Territory, 2602

Dr. Lyndall Dawson
(mammals)
School of Zoology
University of New South Wales
P.O. Box 1
Kensington, N.S.W. 2033

Dr. Eileen Finch
(mammals)
Department of Zoology
University of Western Australia
Nedlands, Western Australia, 6009

Dr. Timothy Flannery
(mammals)
Australian Museum
6-8 College Street
Sydney, New South Wales, 2001

Dr. Ewan Fordyce
(whales and allies)
Department of Geology
University of Otago
Dunedin, New Zealand

Ms. Suzanne Hand
(bats)
School of Zoology
University of New South Wales
P.O. Box 1
Kensington, N.S.W., 2033

Dr. Jeannette Hope
(archaeological fauna, mammals)
Department of Prehistory
Research School of Pacific Studies
Australian National University
Canberra City
Australian Capital Territory,
2601

Dr. David Horton
(archaeological fauna, especially mammals)
Australian Institute of Aboriginal Studies
Acton House Acton,
Australian Capital Territory, 2601

Dr. Richard Jenkins
(birds especially penguins)
Department of Geology and Mineralogy
University of Adelaide
Box 498, G.P.O.
Adelaide, South Australia, 5001

Ms. Karen Kefous
(fish, especially otolith analysis)
Department of Prehistory

School of General Studies
Australian National University
Canberra City
Australian Capital Territory, 2601

Dr. Anne Kemp
(fish, especially lungfish)
Queensland Museum
Gregory Terrace
Fortitude Valley, Queensland, 4006

Mr. Noel Kemp
(sharks)
Tasmanian Museum
5 Argyle Street
Hobart, Tasmania, 7000

Dr. John Long
(Palaeozoic fish)
Australian National University
Canberra City,
Australian Capital Territory, 2601

Dr. Jack Mahoney
(rodents)
Department of Geology
University of Sydney
Newtown, New South Wales, 2042

Mr. Greg McNamara
(mammals)
Department of Geology
James Cook University
Townsville, Queensland, 4810

Mr. Allan McEvey
(birds)
Museum of Victoria
285-321 Russell Street
Melbourne, Victoria, 3000

Mr. Charles Meredith
(birds, especially seabirds)
Department of Zoology
Monash University
Clayton, Victoria, 3168

Dr. Phil Millener
(birds)
National Museum of New Zealand
Private Bag,
Wellington, New Zealand

Dr. Ralph Molnar
(reptiles)
Queensland Museum
Gregory Terrace
Fortitude Valley,
Queensland, 4006

Dr. Peter Murray
(Pleistocene mammals)
Museum and Art Gallery
of the Northern Territory
Darwin, Northern Territory, 5790

Ms. Sue O'Connor
(archaeological fauna)
Department of Prehistory
University of Western Australia
Nedlands,
Western Australia, 6009

Mr. Mike Plane
(mammals)
Bureau of Mineral Resources
Anzac Parade and Constitution Avenue
Canberra City
Australian Capital Territory, 2601

Mr. Neville Pledge
(mammals)
South Australian Museum
North Terrace
Adelaide, South Australia, 5000

Dr. Pat Rich
(birds)
Departments of Earth Sciences
and Zoology
Monash University
Clayton, Victoria, 3168

Dr. Tom Rich
(mammals)
Museum of Victoria
285-321 Russell Street
Melbourne, Victoria, 3000

Dr. David Ride
(mammals)
School of Applied Science
Canberra College
of Advanced Education
P.O. Box 1
Belconnen,
Australian Capital Territory, 2616

Dr. Alexander Ritchie
(Palaeozoic fish)
Australian Museum
6-8 College Street
Sydney, New South Wales, 2001

Dr. Gordon Sanson
(mammals)
Department of Zoology
Monash University
Clayton, Victoria, 3168

Mr. Ron Scarlett
(birds)
Canterbury Museum
Rolleston Avenue
Christchurch, New Zealand

Dr. Meredith Smith
(mammals and snakes)
South Australian Museum
North Terrace
Adelaide, South Australia 5000

Dr. Anthony Thulborn
(reptiles)
Department of Zoology
University of Queensland
St. Lucia, Queensland, 4067

Dr. Susan Turner
(fish, especially thelodonts)
Queensland Museum
Gregory Terrace
Fortitude Valley,
Queensland, 4006

Dr. Jerry van Tets
(birds)
CSIRO Division of Wildlife
and Rangelands Research
P.O. Box 84 Lyneham,
Australian Capital Territory, 2602

Dr. Mary Wade
(reptiles)
Queensland Museum
Gregory Terrace
Fortitude Valley,
Queensland, 4006

Dr. Anne Warren
(amphibians)
Department of Zoology
LaTrobe University
Bundoora, Victoria, 3083

Dr. Jim Warren
(Palaeozoic fish)
Department of Zoology
Monash University
Clayton, Victoria, 3168

Dr. Rod Wells
(mammals)
School of Biological Sciences
Flinders University of South Australia
Bedford Park,
South Australia, 5042

Dr. Gavin Young
(Palaeozoic fish)
Bureau of Mineral Resources
Anzac Parade and Constitution Avenue
Canberra City
Australian Capital Territory, 2601

Select References

Abele, C., and Page, R.W. 1974. Stratigraphic and isotopic ages of Tertiary basalts at Maude and Aireys Inlet, Victoria, Australia. *Proc. Roy. Soc. Vic.* 6(2):143-150.

Agassiz, L., 1833-1843. Recherches sur les poissons fossiles. *Petitpierre, Neuchatel* 111:131-136.

Ager, D.V. 1963. *Principles of Paleoecology.* McGraw Hill, New York.

Anderson, C.A. 1925. Notes on the extinct chelonian *Meiolania*, with a record of a new occurrence. *Rec. Austral. Mus.*, 14:223-242.

——, 1929. Palaeontological notes No. 1. *Macropus titan* Owen and *Thylacoleo carnifex* Owen. *Rec. Austral. Mus.* 17:35-45.

——, 1930. Palaeontological notes, No. II. *Meiolania platyceps* Owen and *Varanus* (*Megalania*) *priscus* Owen. *Rec. Austral. Mus.* 17:309-316.

——, 1933. The fossil mammals of Australia. *Proc. Linn. Soc. N.S.W.* 58(1-2):9-25.

——, 1937. Palaeontological notes No. IV. Fossil marsupials from New Guinea. *Rec. Austral Mus.* 20(2): 73-78.

Andrews, C.W. 1906. *A Descriptive Catalogue of the Tertiary Vertebrates of the Fayum. Egypt.* British Museum (Nat. Hist.), London.

Anon. 1869. Discovery of the bones of an extinct animal, supposed to be the *Deprohadar* [sic.]. *The Sydney Mail*, 451(10), Feb. 20:5.

Appleby, R.M. 1979. The affinities of Liassic and later ichthyosaurs. *Palaeontology* 22:921-946.

Archer, M. 1972. *Phascolarctos* (Marsupialia, Vombatoidea) and an associated fossil fauna from Koala Cave Near Yanchep, Western Australia. *Helictite* 10(3):49-59.

——, 1976. Phascolarctid origins and the potential of the selenodont molar in the evolution of diprotodont marsupials. *Mem. Qd Mus.* 17:367-371.

——, 1977. Origins and subfamily relationships of *Diprotodon* (Diprotodontidae, Marsupialia). *Mem. Qd Mus.* 18(1):38-9.

——, 1978. Australia's oldest bat, a possible rhinolopid. *Proc. Roy. Soc. Qd.* 89:23.

——, 1981. A review of the origins and radiation of Australian mammals. *In*: A. Keast, ed., *Ecological Biogeography of Australia* Junk, The Hague: 1437-1488.

——, 1982a. A lion in possum's clothing. *Austral. Nat. Hist.* 20(11):373-379.

——, ed. 1982b. *Carnivorous Marsupials.* 2 vol, *Roy. Zool. Soc. N.S.W.*, Mosman.

——, and Bartholomai, A. 1978. Tertiary mammals of Australia: a synoptic review. *Alcheringa* 2:1-21.

——, ——, and Marshall, L.G. 1978. *Propleopus chillagoensis*, a new north Queensland species of extinct giant rat-kangaroo (Macropodidae: Potoroinae). *Mem. Nat. Mus. Vict.* 39:55-60.

——, and Clayton, G. eds., 1984. *Vertebrate Zoogeography and Evolution in Australasia.* Hesperian Press, Carlisle.

——, and Dawson, L. 1982. Revision of Marsupial Lions of the genus *Thylacoleo* Gervais (Thylacoleonidae, Marsupialia) and thylacoleonid evolution in the late Cainozoic. *In*: M. Archer, ed., *Carnivorous Marsupials, Roy. Zool. Soc. N.S.W.* 2:477-494.

——, and Rich, T.H. 1982. Results of the Ray E. Lemley Expeditions. *Wakaleo alcootaensis* n. sp. (Thylacoleonidae,Marsupialia),a new marsupial lion from the Miocene of the Northern Territory with a consideration of early radiation in the family. *In*: M. Archer, ed., *Carnivorous Marsupials. Roy. Zool. Soc. N.S.W.* 2:495-502.

——, and Wade, M. 1976. Results of the Ray E. Lemley Expeditions, Part 1. The Allingham Formation and a new Pliocene vertebrate fauna from northern Queensland. *Mem. Qd. Mus.* 17:379-397.

Auffenberg, W. 1972. Komodo dragons. *Nat. Hist.* 81(4):52-59.

Baird, R.F. (in press). Avian fossils from Quaternary deposits, "Green Waterhole Cave", southeastern South Australia. *Rec. Austral. Mus.*

——, Rich, P.V., and Walker, C.A. (in prep.). *Protocuculus*, the first cuckoo, from the Paleocene of Brazil.

Barnes, L.C. and Townsend, I.J., 1982, *Opal — South Australia's gemstone.* D.J. Woolman Govt. Printer. South Australia.

Barrett, C. 1946. *The Bunyip and Other Mythical Monsters and Legends.* Reed and Harris, Melbourne.

Bartholomai, A. 1978. The rostrum in *Palorchestes* Owen (Marsupialia, Diprotodontidae): Results of the Ray. E. Lemley Expeditions, Part 3. *Mem. Qd Mus.* 18:145-149.

Behrensmeyer, A.K. and Hill, A.P. 1980. *Fossils in the Making.* Univ. Chicago Press, Chicago.

Bensley, A.B. 1903. On the evolution of the Australian Marsupialia; with remarks on the relationships of the marsupials in general. *Trans. Linn. Soc. London* (*Zool.*), 9:83-217.

Berman, D.S. and McIntosh, J.S. 1978. Skull and relationships of the Upper Jurassic sauropod *Apatosaurus* (Reptilia, Saurischia). *Bull. Carnegie Mus. Nat. Hist.* 8:1-35.

Bowler, J.M. 1982. Aridity in the late Tertiary and Quaternary of Australia. *In*: W.R. Barker and P.J.M. Greenslade eds., *Evolution of the Flora and Fauna of Arid Australia*. Peacock Publications, Frewville: 35-45.

Bramwell, C.D. and Whitfield, G.R. 1974. Biomechanics of *Pteranodon*. *Phil. Trans. Roy. Soc. London*, B. 267:503-581.

Branagan, D.F. 1979. The idea of a museum. *In*: R. Strahan, *Rare and Curious Specimens. An Illustrated History of the Australian Museum*. Austral. Mus., Sydney: 1-6.

——, and Townley, K.A. 1976. The geological sciences in Australia — a brief historical review. *Earth Sci. Rev.* 12:323-346.

Broili, F. 1907. Ein neuer *Ichthyosaurus* aus der norddeutschen Kreide. *Palaeontographica* 12:139-162.

——, 1927. Ein *Rhamphorhynchus* mit Spuren von Haarbedeckung. *Sitz Ber. Bayer. Akad. Wiss.*:49-65.

Broom, R. 1898. On the affinities and habits of *Thylacoleo*. *Proc. Linn. Soc. N.S.W.* 23:57-74.

Brower, J.C. and Veinus, J. 1981. Allometry in pterosaurs. *Univ. Kansas, Paleo. Contrib.* 105:1-32.

Brown, H.Y.L. 1890. Report on journey from Warrina to Musgrave Ranges. *S. Austral. Parl. Pap.* 45.

——, 1892. Government geologist's report on country in neighborhood of Lake Eyre. *S. Austral. Parl. Pap.* 141.

——, 1894. Report of the government geologist for year ended June 30, 1894. *S. Austral. Parl. Pap.* 25.

Brunton, C.H.C., Miles, R.S., and Rolfe, W.D.I. 1969. Gogo Expedition 1967. *Proc. Geol. Soc. London* 1655:78-83.

Buffetaut, E. 1979. Revision der Crocodylia (Reptilia) aus den Gosau-Schichten (Ober-Kreide) von Österreich. *Beitr. Paleo. Österr.* 6:89-105.

Butler, P.M. 1978. Insectivora and Chiroptera. *In*: V.J. Maglio and H.B.S, Cooke, eds., *Evolution of African Mammals*. Harvard Univ. Press, Cambridge, London: 56-68.

Callan, R.A. and Tedford, R.H. 1976. New late Cainozoic rock units and depositional environments, Lake Frome area, South Australia. *Trans. Roy. Soc. S. Austral.* 100:125-167.

Cameron, W.E. 1901. Geological observations in northwestern Queensland. *Ann. Rep. Qd. Dep. Min.* 1901-1902.

Campbell, K.S.W. and Bell, M. 1977. A primitive amphibian from the Late Devonian of New South Wales. *Alcheringa*, 1:369-381.

Cas, R.A.F. 1983. A review of the facies patterns, palaeogeographic development and tectonic context of the Palaeozoic Lachlan Fold Belt of southeastern Australia. *Mem. Geol. Soc. Austral., Spec. Pub.* 10:1-104.

Chapman, F. 1914a. On a new species of *Ceratodus* from the Cretaceous of New South Wales. *Proc. Roy. Soc. Vict.* 27 (ns):25-27.

——, 1914b. *Australian Fossils, a Student's Manual of Palaeontology*. Melbourne and London.

——, 1934. *A Book of Fossils*. Shakespeare Head Press Ltd. Sydney.

——, and Crespin, I. 1935. The sequence and age of the Tertiaries of southern Australia. *Rept 22nd Meeting ANZAAS* 22:118-126.

Clark, B.D. and Bemis, W. 1979. Kinematics of swimming penguins at the Detroit Zoo. *Jour. Zool. (London)* 188(3):411-428.

Clarke, W.B. 1869. *Dinornis*, an Australian genus. *Geol. Mag.* 6:383-384.

Clemens, W.A. and Plane M. 1974. Mid-Tertiary Thylacoleonidae (Marsupialia, Mammalia). *Jour. Paleo.* 48:652-660.

Clift, W. 1831. Report by Mr. Clift, of the College of Surgeons, London, in regard to the fossil bones found in the caves and bone-breccia of New Holland. *Edinb. New Phil. Jour.* 10:394-395.

Colbert, E.H. 1946. *Sebecus*, representative of a peculiar suborder of fossil Crocodilia from Patagonia. *Bull. Amer. Mus. Nat. Hist.* 87:217-270.

——, 1967. A new interpretation of *Austropelor*, a supposed Jurassic labyrinthodont amphibian from Queensland. *Mem. Qd Mus.* 15(1):35-41.

Conway-Morris, S. and Whittington, H.B. 1979. The animals of the Burgess Shale. *Sci. Amer.* 241(1):122-133.

Cook, P.J. 1982. The Cambrian palaeogeography of Australia and opportunities for petroleum exploration. *APEA Jour.* 22:42-64.

Coombs, W.P., Jr. 1975. Sauropod habits and habitats. *Palaegeog. Palaeoclim. Palaeoecol.* 17:1-33.

——, 1978. The families of the ornithischian dinosaur order Ankylosauria. *Palaeontol.* 21:143-170.

Cope, E.D. 1882. The ancestry and habits of *Thylacoleo*. *Amer. Nat.* 16:520-522.

——, 1884. The Tertiary Marsupialia. *Amer. Nat.* 18:686-797.

Cott, H.B. 1961. Scientific results of an inquiry into the ecology and economic status of the Nile Crocodile (*Crocodilus niloticus*) in Uganda and northern Rhodesia. *Trans. Zool. Soc. London*, 29:1-211.

CSIRO 1979. Towards wise use of Antarctic krill. *Ecos*, 21:3-9.

Daily, B. 1960. *Thylacoleo*, the extinct marsupial lion. *Austral. Mus. Mag.* 13:163-166.

Deighton, I., Falvey, D.A. and Taylor, D.J. 1976. Depositional environments and geotectonic framework: southern Australian continental margin. *APEA Jour.* 1976:25-36.

De Jersey, N.J. 1963. Jurassic spores and pollen grains from the Marburg Sandstone. *Publ. Geol. Surv. Qd* 313:1-15.

Denison, R.H. 1941. The soft anatomy of *Bothriolepis*. *Jour. Paleo.* 15:553-561.
——, 1978. Placodermi. *In*: Hans-Peter Schultze, ed., *Handbook of Palaeoichthyology*, 2. Gustav Fischer Verlag, Stuttgart.
Dennis, K. and Miles, R.S. 1979a. A second eubrachythoracid arthrodire from Gogo, Western Australia. *Zool. Jour. Linn. Soc.* 67:1-29.
——, 1979b. Eubrachythoracid arthrodires with tubular rostral plates from Gogo, Western Australia. *Zool. Jour. Linn. Soc.* 67:297-328.
——, 1980. New durophagous arthrodires from Gogo, Western Australia. *Zool. Jour. Linn. Soc.*, 69:43-85.
——, 1981. A pachyosteomorph arthrodire from Gogo, Western Australia. *Zool. Jour. Linn. Soc.*, 73:213-258.
——, 1982. A eubrachythoracid arthrodire from Gogo, Western Australia. *Zool. Jour. Linn. Soc.* 75:153-166.
Desmond, A.J. 1975. *The Hot-Blooded Dinosaurs.* Blond and Briggs, London.
De Vis, C.W. 1883. On tooth-marked bones of extinct marsupials. *Proc. Linn. Soc. N.S.W.* 8:187-190.
——, 1888a. Australian ancestry of the crowned pigeon of New Guinea. *Proc. Roy. Soc. Qd.* 5:127-131.
——, 1888b. On an extinct genus of the marsupials allied to *Hypsiprymnodon*. *Proc. Linn. Soc. N.S.W.* (2nd ser.) 3:5-8.
——, 1888c. A glimpse of the post-Tertiary avifauna of Queensland. *Proc. Linn. Soc. N.S.W.* (2nd ser.) 3:1, 277-1, 292.
——, 1889. Additions to the list of fossil birds. *Proc. Roy. Soc. Qd.* 6:55-58.
——, 1891. Residue of the extinct birds of Queensland as yet detected. *Proc. Linn. Soc. N.S.W.* (2nd ser.) 6:437-456.
——, 1895. A review of the fossil jaws of the Macropodidae in the Queensland Museum. *Proc. Linn. Soc. N.S.W.* 10:75-133.
——, 1905. A contribution to the extinct avifauna of Australia. *Ann. Qd. Mus.* 6:3-25.
Dewey, J.F. 1972. Plate tectonics. *Sci. Amer.* 266:56-68.
Dickins, J.M. 1978. Climate of the Permian in Australia; the invertebrate faunas. *Palaeogeog. Palaeoclim. Palaeoecol.* 23:33-46.
Dobson, G.E. 1878. *Catalogue of the Chiroptera in the Collection of the British Museum (Nat. Hist.)* British Museum (Nat. Hist.), London.
Dorman, F.H. 1966. Australian Tertiary paleotemperatures. *Jour. Geol.* 74(1):49-61.
Douglas, J.G. and Williams, G.E. 1982. Southern polar forests: the Early Cretaceous floras of Victoria and their palaeoclimatic significance. *Palaeogeog. Palaeoclim. Palaeoecol.* 39:171-185.
Dugan, K.G. (1979), 1980. Darwin and *Diprotodon*: The Wellington Cave fossils and law of succession. *Proc. Linn. Soc. N.S.W.* 104(4):265-272.
Du Toit, A.L. 1937. *Our Wandering Continents: An Hypothesis on Continental Drifting.* Oliver and Boyd, Edinburgh and London.

Eaton, T.H. Jr. 1960. A new armored dinosaur from the Cretaceous of Kansas. *Univ. Kansas, Paleo. Contrib., Vert.* 8.
Eichwald, E.I. von 1840. Die Thier-und Pflanzenreste des alten roten Sandsteins und Bergkalks im Novgorodschen Governement. *Bull. Acad. (impériale) sci. Saint-Petersbourg*, 7:78-91.
Elliot, D.H., Rinaldi, C., Zinsmeister, W.J., Trautman, T.A., Bryant, W.A., and del Valle, R. 1975. Geological investigations on Seymour Island, Antarctic Peninsula. *Antarct. Jour. V.S.* 10(4):182-186.
Ettinghausen, C. Baron von, 1888. Contributions to the Tertiary flora of Australia. *Mem. Geol Surv. N.S.W., Dept. Mines.* Palaeo. 2:1-189.

Fergusson, C.L., Cas, R.A.F., Collins, W.J., Craig, G.Y., Crook, K.A.W., Powell, C.McA., Scott, P.A., and Young, G.C. 1979. The Late Devonian Boyd Volcanic Complex, Eden, N.S.W. *Jour. Geol. Soc. Austral.* 26:87-105.
Filatoff, J. 1975. Jurassic palynology of the Perth Basin, Western Australia. *Palaeontographica* (**B**) 154:1-113.
Finch, E. 1971. *Thylacoleo*, marsupial lion or marsupial sloth? *Austral. Nat. Hist.* 17(1):7-11.
Finlayson, H.H. 1938. On the occurrence of a fossil penguin in Miocene beds in South Australia. *Trans. Roy. Soc. S. Austral.* 62(1):14-17.
Flannery, T. F. 1983. Revision in the macropodid subfamily Sthenurinae (Marsupialia: Macropodoidea) and the relationships of the species of *Troposodon* and *Lagostrophus*. *Austral. Mamm.* 6:15-28.
——, Archer, M., and Plane, M. 1983. Middle Miocene kangaroos (Macropodoidea: Marsupialia) from three localities in northern Australia, with a description of two new subfamilies. *Bur. Min. Res. Jour. Austral. Geol. Geophys.* 7:287-302.
Fletcher, H.O. 1945. Palorchestes-Australia's extinct giant kangaroo. *Austral. Mus. Mag* 8(11):361-365.
Fletcher, J.J. ed. 1893. *The Macleay Memorial Volume. Linn. Soc. N.S.W.* Angus & Robertson and Turner and Henderson, Sydney.
Flood, J.M. 1974. Pleistocene man at Clogg's Cave: his tool kit and environment. *Mankind* 9:175-188.
Fordyce, R.E. 1983. Rhabdosteid dolphins (Mammalia: Cetacea) from the Middle Miocene, Lake Frome area, South Australia. *Alcheringa* 7:27-40.
Frakes, L.A. 1979. *Climates Throughout Geologic Time.* Elsevier, Amsterdam.
Frey, E. and Reiss, J. 1981. A new reconstruction of the pterosaur wing. *N. Jahrb. Geol. Pal., Abh.* 161:1-27.
Frith, H.J. 1962. *The Malleefowl.* Angus & Robertson, Sydney.

Gaffney, E.S. 1975. A phylogeny and classification of the higher categories of turtles. *Bull. Amer. Mus. Nat. Hist.* 155, (5):387-436.
Galton, P.M. 1980. Partial skeleton of *Dracopelta zbyszewskii* n. gen. and n. sp., an ankylosaurian dinosaur from the Upper Jurassic of Portugal. *Geobios* 13:451-457.
Gardiner, B.G. 1970. Osteichthyes. *In*: *McGraw Year Book of Science and Technology*: 284-286.
——, 1973. Interrelationships of teleostomes. *In*: P.H. Greenwood, R.S. Miles, and C. Patterson, eds., *Interrelationships of Fishes. Jour. Linn. Soc. (Zool.)* 52: suppl.:105-135.
——, and Bartram, A.W.H. 1977. The homologies of ventral cranial fissures in osteichthyans. *In*: S.M. Andrews, R.S. Miles and A.D. Walker eds., *Problems in Vertebrate Evolution*. Academic Press, London: 227-245.
——, and Miles, R.S. 1975. Devonian fishes of the Gogo Formation, Western Australia. *Colloques int. Centre natn. Rech. Scient.* 218:73-79.
Gervais, P. 1848-1852. Zoologie et paleontologie francaises (animaux vertebres) où nouvelles recherches sur les animaux vivants et fossiles de la France. Tome I. Tome II. Arthur Bertrand: Paris.
Gilbert-Tomlinson, J. 1968. A new record of *Bothriolepis* from the Northern Territory of Australia. *Bull. Bur. Min. Res. (Palaeont. Pap.)* 80:191-226.
Gill, E.D. 1954a, Ecology and distribution of the extinct giant marsupial, *Thylacoleo*. *Vic. Nat.* 71:18-35.
——, 1954b, The validity of the lower Tertiary possum, *Wynyardia bassiana*. *Austral. Jour. Sci.* 17(2):69.
——, 1962. Cainozoic. *Jour. Geol. Soc. Austral.* 9(2):233-253.
Gillespie, R., Horton, D.R., Ladd, P., Macumber, P.G., Rich, T.H., Thorne, R., and Wright, R.V.S. 1978. Lancefield Swamp and the extinction of the Australian megafauna. *Science*, 200: 1044-1048.
Glaessner, M.F. 1955. Pelagic fossils (*Aturia*, penguins, whales) from the Tertiary of South Australia. *Rec. S. Austral. Mus.* 11(4):353-372.
Goede, A., Murray, P., and Harmon, R. 1978. Pleistocene man and megafauna in Tasmania: dated evidence from cave sites. *Artefact* 3:139-149.
Goldfuss, G.A. 1831. [*Pterodactylus macronyx* at Banz]. *N. Jahrb. Min. Geol. Pal.* 1831:298.
Gregory, J.W. 1906. *The Dead Heart of Australia*. John Murray, London.
Gregory, W.K. 1951. *Evolution Emerging*. 2 vols. Macmillan, New York.

Haight, J.R. and Murray, P.F. 1981. The cranial endocast of the early Miocene marsupial *Wynyardia bassiana*: an assessment of taxonomic relationships based upon comparisons with recent forms. *Brain, Behavior, and Evol.* 19:17-36.
Hale, H.M. 1956. The first hundred years of the South Australian Museum 1856-1956. *Rec. S. Austral. Mus.* 12:1-225.
Hall, L.S. 1981. The biogeography of Australian bats. *In*: A. Keast, ed., *Ecological Biogeography of Australia*. Junk, The Hague: 1557-1583.
Harris, W.K. 1971. Tertiary stratigraphic palynology, Otway Basin, *In*: H. Wopfner, and J. G. Douglas, eds., *The Otway Basin of Southeastern Australia. Geol. Surv. S. Austral. Vict. Spec. Bull.*: 67-87.
Hecht, M. 1975. The morphology and relationships of the largest known terrestrial lizard, *Megalania prisca* Owen, from the Pleistocene of Australia. *Proc. Roy. Soc. Vic.* 87:239-250.
——, and Archer, M. 1977. Presence of xiphodont [sic.] crocodilians in the Tertiary and Pleistocene of Australia. *Alcheringa*, 1:383-385.
Heckel, P.H. and Witzke, B.J. 1977. Devonian world palaeogeography determined from distribution of carbonates and related lithic palaeoclimatic indicators. *Spec. Pap. Palaeontol.* 23:99-123.
Hill, D. 1981. The first fifty years of the Department of Geology of the University of Queensland. *Pap. Dept. Geol.* 10(1):1-68.
Hills, E.S. 1929. The geology and palaeontography of the Cathedral Range and Blue Hills in north western Gippsland. *Proc. Roy. Soc. Vict.* 41:176-201.
——, 1931. The Upper Devonian fishes of Victoria, Australia, and their bearing on the stratigraphy of the State. *Geol. Mag.* 68:206-231.
——, 1932. Upper Devonian fishes from New South Wales. *Jour. Geol. Soc. London* 88:850-858.
——, 1958. A brief review of Australian fossil vertebrates. *In*: T.S. Westoll, ed., *Studies on Fossil Vertebrates*. Athlone Press, London: 86-107.
Hochstetter, F. 1859. Notizen über einiger fossile Thierreste und deren Lagerstatten in New-Holland, gesammelt daselbst wahrend des Aufenthaltes Sr. Majestat Fregatte Novara im Monate December 1858. *Sber. K. Akad. Wiss. Wien.* 35:349-358.
Hoffstetter, R. 1959. Un dentaire de *Madtsoia* (Serpent géant du Paléocène de Patagonie). *Bull. Mus. Natn. Hist. Nat. Paris* (2e ser.) 31:379-386.
——, 1961a. Nouveaux restes d'un serpent boidé (*Madtsoia madagascariensis* nov. sp.) dans len Crétace Supérieur de Madagascar. *Bull. Mus. Natn. Hist. Nat. Paris* (2e ser.) 33:152-160.
——, 1961b. Nouvelles recoltes de serpents fossiles dans l'éocène supérieure du desert libyque. *Bull. Mus. Natn. Hist. Nat., Paris* (2e ser.) 33:326-331.
Hope, J. 1982. Late Cainozoic vertebrate faunas and the development of aridity in Australia. *In*: W.R. Barker, and P.J.M. Greenslade, eds., *Evolution of the Flora and Fauna of Arid Australia*. Peacock Publications, Frewville: 85-100.
——, Lampert, R.J., Edmondson, E., Smith, M.J., and van Tets, G.F. 1977. Late Pleistocene faunal remains

from Seton rock shelter, Kangaroo Island, South Australia. *Jour. Biogeogr.* 4:363-385.
Horton, D.R., Wells, R.T., and Wright, R.V.S. 1979. *Thylacoleo,* marsupial carnivore. Abs. *Aust. Mamm. Soc. Bull.* 6(1):25.
——, and Wright, R.V.S. 1981. Cuts in Lancefield Bones: Carnivorous *Thylacoleo,* not humans, the cause. *Archaeol. Oceania* 16:73-80.
Huxley, T.H. 1862. On the premolar teeth of *Diprotodon,* and on a new species of that genus. *Quart. Jour. Geol. Soc. London* 18:422-427.
——, 1887. Preliminary note on the fossil remains of a chelonian reptile, *Ceratochelys sthenurus,* from Lord Howe's Island, Australia. *Proc. Roy. Soc.* 42:232-238.

Jarvik, E. 1964. Specialisations in early vertebrates. *Ann. Soc. Roy. Zool. Belgique* 94:11-95.
Jell, P.A. 1983. An oniscomorph millipede from the Early Jurassic of Queensland — the first Mesozoic millipede. *Alcheringa* 7(3): 195-199.
Jenkin, P.M. 1957. The filter-feeding and food of flamingoes (Phoenicopteri). *Phil. Trans. Roy. Soc. London* 240(13):401-493.
Jenkins, R.J.F. 1974. A new giant penguin from the Eocene of Australia. *Palaeontology,* 17(2):291-310.
——, Jones, J.B., McGowran, B., Beecroft, A.S., and Fitzgerald, M.J. 1982. Lithostratigraphic subdivision of the Blanche Point Formation, Late Eocene, Willunga Embayment. *Q. Geol. Notes, Geol. Surv. S. Austral.* 84:2-7.
Jensen, R.A. 1975. Permo-Triassic stratigraphy and sedimentation in the Bowen Basin, Queensland. *Bur. Min. Res. Bull.* 154: 1-187.
Jepsen, G.L. 1966. Early Eocene bat from Wyoming. *Science* 154:1333-1339.
Johns, R.K. ed. 1976. *History and role of government geological surveys in Australia.* A.B. James, Govt. Printer, Adelaide.
Johnston, R.M. 1888. *Systematic Account of the Geology of Tasmania.* William Thomas Strutt, Gov. Printer, Hobart.
Jones, F.W. 1931. A re-examination of the skeletal characters of *Wynyardia bassiana,* an extinct Tasmanian marsupial. *Pap. Proc. Roy. Soc. Tas.* 1930:96-115.

Keast, A. 1981. The evolutionary biogeography of Australian birds. *In*: A. Keast, ed., *Ecological Biogeography of Australia.* Junk, The Hague: 1568-1635.
Kelley, K. 1983. A workshop manual 1983. The moulding and casting of dinosaurs. *Mus. Victoria,* Melbourne.
Kemp, A. and Molnar, R.E. 1981. *Neoceratodus forsteri* from the Lower Cretaceous of New South Wales. Australia. *Jour. Paleo.* 55:211-217.
Kemp, E.M. 1978. Tertiary climatic evolution and vegetation history in the southeast Indian Ocean region. *Palaegeog., Palaeoclim., Palaeoecol.* 24(4):169-208.
Kemp, T.S. 1982. *Mammal-like Reptiles and the Origin of Mammals.* Academic Press, London and New York.
Kennett, J.P. 1982. *Marine Geology.* Prentice-Hall, Englewood Cliffs.
Krefft, G. 1870. Description of a giant amphibian allied to the genus *Lepidosiren,* from the Wide Bay District, Queensland. *Proc. Zool. Soc. London,* 16:221-224.
Kuhn, O. 1938. Die Crocodilier aus dem mittleren Eozän des Geiseltales bei Halle. *Nova Acta Leopoldina,* (nf), 6:313-329.
Kummel, B. and Raup, D. 1965. *Handbook of Paleontological Techniques.* W.H. Freeman and Co.: San Francisco.

Lane, E.A. and Richards, A.M. 1963. The discovery, exploration and scientific investigation of the Wellington Caves, New South Wales. *Helictite* 2:1-55.
Langston, W., Jr. 1973. The crocodilian skull in historical perspective. *In*: C. Gans, and T.S. Parsons, eds., *Biology of the Reptilia,* 4, Morph. D., Academic Press, London.
——, 1975. Ziphodont crocodiles: *Pristichampsus vorax* (Troxell), new combination, from the Eocene of North America. *Field. Geol.* 33:291-314.
Long, J. 1980. The taphonomy of the Bacchus Marsh *Diprotodon* site. *Rec. Bur. Min. Res. Geol. Geophy.* 1980/67:49.
——, 1982. The history of fishes on the Australian Continent. *In*: P.V. Rich and E.M. Thompson, eds., *The Fossil Vertebrate Record of Australasia,* Monash Univ. Offset Printing Unit, Clayton: 53-86.
——, 1983. New bothriolepid fishes from the Late Devonian of Victoria. *Palaeontology,* 26:295-320.
——, Turner, S., and Kemp, A. 1982. Contributions to Australian fossil fish biostratigraphy. *In*: P.V. Rich and E.M. Thompson, eds., *The Fossil Vertebrate Record of Australasia,* Monash Univ. Offset Printing Unit, Clayton: 119-143.
Longman, H.A. 1922. An ichthyosaurian skull from Queensland. *Mem. Qd Mus.* 7:246-256.
——, 1924. Some Queensland fossil vertebrates. *Mem. Qd Mus.* 8:16-28.
——, 1926. A giant dinosaur from Durham Downs, Queensland. *Mem. Qd Mus.* 8:183-194.
——, 1927a. The giant dinosaur: *Rhoetosaurus brownei. Mem. Qd Mus* 9:1-18.
——, 1927b. Australia's largest fossil — the *Rhoetosaurus* dinosaur. *Aust. Mus. Mag.* 3:97-102.
——, 1929. Palaeontological notes. *Mem. Qd Mus.* 9:247-251.

——, 1930. *Kronosaurus queenslandicus.* A gigantic Cretaceous pliosaur. *Mem. Qd Mus.* 10:1-7.
——, 1941. A Queensland fossil amphibian. *Mem. Qd Mus.* 12:29-32.
——, 1945. Fossil vertebrates from Gore Quarries. *Mem. Qd Mus.* 12:164.
Lydekker, R. 1887. *Catalogue of the Fossil Mammalia in the British Museum (Nat. Hist.)* (5). British Museum, (Nat. Hist.), London.
——, 1891. *Catalogue of the Fossil Birds in the British Museum (Nat. Hist.).* Longmans, London.
——, 1896. *A handbook to the Marsupiala and Monotremata.* Edward Lloyd, London.

Mack, G. 1956. The Queensland Museum, 1855-1955. *Mem. Qd Mus.* 13(2):107-119.
Macdonald, J.R. 1983. *The Fossil Collector's Handbook. A Paleontology Field Guide.* Prentice-Hall, Englewood Cliffs.
Mahoney, J.A. and Ride, W.D.L. 1975. Index to the genera and species of fossil Mammalia described from Australia and New Guinea between 1838 and 1968. *W.A. Mus. Spec. Publ.* 6, Perth.
Marett Tims, H.W. and Hopewell Smith, A. 1914. *In: Tomes' Manual of Dental Anatomy.* 7th ed.:418.
Marples, B.J. 1952. Early Tertiary penguins of New Zealand. *Palaeont. Bull. N.Z.* 20:1-66.
——, 1953. Fossil penguins from the mid-Tertiary of Seymour Island. *Scient. Rep. Falkld. Isl. Depend. Surv.* 5:1-15.
McDougall, I., Embleton, B., and Stone, D. 1981. Origin and evolution of Lord Howe Island, southwest Pacific Ocean. *Jour. Geol. Soc. Austral.* 28:155-176.
McGowan, C. 1972. Evolutionary trends in longipinnate ichthyosaurs with particular reference to the skull and forefin. *Life Sci. Contr. Roy. Ont. Mus.* 83:1-48.
McGowran, B. 1978. Stratigraphic record of early Tertiary oceanic and continental events in the Indian Ocean region. *Mar. Geol.* 26(1/2):1-39.
McKean, J.L. 1975. The bats of Lord Howe Island with the description of a new nyctophiline bat. *Aust. Mamm.* 1:329-332.
McCoy, F. 1867. On the occurrence of *Ichthyosaurus* and *Plesiosaurus* in Australia. *Ann. Mag. Nat. Hist.* 19(3):355-356.
——, 1869. On the fossil eye and teeth of the *Ichthyosaurus australis,* (McCoy) from the Cretaceous formations of the source of the Flinders River, etc. *Trans. Proc. Roy. Soc. Vict.* 9(2):77-78.
Merrilees, D. 1968. Man the destroyer: late Quaternary changes in the Australian marsupial fauna. *Jour. Roy Soc. W. Austral.* 51:1-24.
Miles, R.S. 1970. New fossil fishes from Western Australia. *Hemisphere,* 14:9-12.
——, 1971a. Palaeozoic Fish. *In: McGraw-Hill Yearbook of Science and Technology*: 312-314.
——, 1971b. The Holonematidae (placoderm fishes), a review based on new specimens of *Holonema* from the Upper Devonian of Western Australia. *Phil. Trans. Roy. Soc. London(B)* 263:101-234.
——, 1977. Dipnoan (lungfish) skulls and the relationships of the group: a study based on new species from the Devonian of Australia. *Zool. Jour. Linn. Soc.* 61:1-328.
——, and Dennis, K. 1979. A primitive eubrachythoracid arthrodire from Gogo, Western Australia. *Zool. Jour. Linn. Soc.* 66:31-62.
——, and Young, G.C. 1977. Placoderm interrelationships reconsidered in the light of new ptyctodonts from Gogo, Western Australia. *In*: S.M. Andrews, R.S. Miles, and A.D. Walker, eds., *Problems in Vertebrate Evolution, Linn. Soc. Symp. Ser.* 4. Academic Press, London: 123-198.
Miller, A.H. 1963. The fossil flamingoes of Australia. *Condor* 65(4):289-299.
Miller, G.S. 1907. The families and genera of bats. *U.S. Nat. Bull.* 57:1-282.
Mitchell, T.L. 1838. *Three Expeditions into the Interior of Eastern Australia, with Descriptions of the Recently Explored Region of Australia Felix, and of the Present Colony of New South Wales.* 2 vol., T. and W. Moore, London.
Molnar, R.E. 1980. An ankylosaur (Ornithischia: Reptilia) from the Lower Cretaceous of southern Queensland. *Mem. Qd. Mus.* 20:77-87.
——, 1981. Pleistocene ziphodont crocodilians of Queensland. *Rec. Austral. Mus.* 33: 803-834.
——, and Pledge, N.S. 1980. A new therapod dinosaur from South Australia. *Alcheringa* 4:281-287.
——, and Thulborn, R.A. 1980. First pterosaur from Australia. *Nature,* 288:361-363.
Moyal, A.M. 1976. *Scientists in the Nineteenth Century Australia: a Documentary History.* Cassell Australia. Sydney.
Moy-Thomas, J.A. and Miles, R.S. 1971. *Palaeozoic Fishes.* Chapman and Hall, London.

Newsome, A.D., and Rochow, K.A. 1964. Vertebrate fossils from Tertiary sediments of Central Australia. *Austral. Jour. Sci.* 26(11):352.
Newton, A. 1893. Palaeontological discovery in Australia. *Nature,* 47(1226):606.

Oliver, W.R.B. 1930. *New Zealand Birds.* Fine Arts, Wellington.
Olson, S.L. 1973. Evolution of the Rails of the South Atlantic Islands. *Smith. Contr. Zool.* 152:1-53.
——, 1980. The significance of the distribution of the Megapodiidae. *Emu* 80:21-24.
——, 1983. Evidence for a polyphyletic origin of the Piciformes. *Auk* 100:126-133.
——, (in litt.). A selective synopsis of the prequaternary fossil record of birds. *In*: D. Farner and J. King eds., *Avian Biology.* Academic Press, New York.

——, and Steadman, D.W. 1979. The humerus of *Xenicibis,* the extinct flightless ibis of Jamaica. *Proc. Biol. Soc. Wash.* 92(1):23-27.
——, and Wetmore, A. 1976. Preliminary diagnoses of two extraordinary new genera of birds from the Pleistocene deposits in the Hawaiian Islands. *Proc. Biol. Soc. Wash.* 89(18):247-258.
Owen, R. 1838. Fossil remains from the Wellington Valley, Australia, *In*: T.L. Mitchell, *Three Expeditions into the Interior of Eastern Australia.* 2nd ed. Boone, London: 359-363.
——, 1859a. Description of some remains of a gigantic land lizard (*Megalania* [*Varanus*] *prisca,* Ow.). *Phil. Trans. Roy. Soc. London,* 149:43-48.
——, 1859b. On the fossil mammals of Australia, Part I. Description of a mutilated skull of a large marsupial carnivore (*Thylacoleo carnifex,* Owen) from a calcareous conglomerate stratum, eighty miles S.W. of Melbourne, Victoria. *Phil. Trans. Roy. Soc. London,* 149:309-322.
——, 1860. Description of some remains of a giant land-lizard (*Megalania prisca* Owen) from Australia. *Phil. Trans. Roy. Soc. London*: 45-46.
——, 1871. On the fossil mammals of Australia. 4. Dentition and mandible of *Thylacoleo carnifex* with remarks on the arguments for its herbivority. *Phil. Trans. Roy. Soc. London* 161:213-266.
——, 1872. Untitled. *Proc. Zool. Soc. London*: 682-683.
——, 1873. On the fossil mammals of Australia. Family Macropodidae. Genera *Macropus, Pachysiagon, Leptosiagon, Procoptodon,* and *Palorchestes.* Pt. IX. *Proc. Roy. Soc. London,* 21:386-387.
——, 1874a. On a *Dinornis* Pars 19: containing a description of a femur indicative of a new genus of large wingless birds (*Dromornis australis*) from a post-Tertiary deposit in Queensland. *Trans. Zool. Soc. London,* 8(6):381-384.
——, 1874b. On the fossil mammals of Australia. Part 9. *Phil. Trans. Roy. Soc. London,* 164:783-803.
——, 1876. On the fossil mammals of Australia. Part 10. Family Macropodidae: Mandibular dentition and parts of the skeleton of *Palorchestes;* additional evidences of *Macropus titan, Sthenurus* and *Procoptodon. Phil. Trans. Roy. Soc. London* 166:197-226.
——, 1877. Researches on the fossil remains of the extinct mammals of Australia with a notice of the extinct marsupials of England. 2 vols. J. Erxleben, London.
——, 1881. Description of some remains of the gigantic land-lizard (*Megalania prisca,* Owen) from Australia. Part II. *Phil. Trans Roy. Soc. London* (1880) 171:1037-1050.
——, 1882a. On an extinct chelonian reptile (*Notochelys costata,* Owen), from Australia. *Quart. Jour. Geol. Soc. London* 38:178-183.
——, 1882b. Description of some remains of the gigantic land-lizard (*Megalania prisca,* Owen) from Australia. Part III. *Phil. Trans. Roy. Soc. London* (1881), 172:547-556.
——, 1886. Description of fossil remains of two species of a megalanian genus (*Meiolania*) from 'Lord Howe's Island'. *Phil. Trans. Roy. Soc. London* 177:471-480.

Pan Kiang 1981. Devonian antiarch biostratigraphy of China. *Geol. Mag.,* 118:69-75.
Pentland, W. 1832. On the fossil bones of Wellington Valley, New Holland, New South Wales. *New Phil. Jour.* 12:301-308.
Pescott, R.T.M. 1954. *Collections of a Century. The History of the First Hundred Years of the National Museum of Victoria.* Nat. Mus. Vict. Melbourne.
Playford, P.E. 1970. The Devonian 'Great Barrier Reef' of Western Australia. *Austral. Nat. Hist*: 339-343.
——, 1976. Devonian reef complexes of the Canning Basin W.A. *25th Int. Geog. Congress Excursion Guide* No. 38A.
——, 1980. Devonian 'Great Barrier Reef' of Canning Basin of W.A. *Bull. Amer. Assn. Petroleum Geol.* 64(6):814-840.
——, and Lowry, D.C. 1967. Devonian reef complexes of the Canning Basin, Western Australia. *Bull. Geol. Surv. W. Austral.* 118:1-150.
Pledge, N. 1977. A new species of *Thylacoleo* (Marsupialia: Thylacoleonidae) with notes on the occurrences and distribution of Thylacoleonidae in South Australia. *Rec. S. Austral. Mus.* 17:277-283.
——, 1980. Macropod skeletons, including *Simosthenurus* Tedford, from an unusual 'drowned cave' deposit in the southeast of South Australia. *Rec. S. Austral. Mus.* 18(6):131-141.
——, 1981. The giant rat-kangaroo *Propleopus oscillans* De Vis, (Potoroidea: Marsupialia) in South Australia. *Trans. Roy. Soc. S. Austral.* 105(1):41-47.
——, (in press). A new species of *Ektopodon* from the Miocene of South Australia. *Univ. Calif. Publ. Geol. Sci.*
Poplin, F. 1980. *Sylviornis neocaledoniae* n.g., n.sp. Ratite éteint de la Nouvelle Calédonie. *C.N.R.S. Paris,* 290(D):691-694.
——, Mourer-Chauviré, C., and Evin, J. 1983. Position systématique et datation de *Sylviornis neocaledoniae,* mégapode géant de la Nouvelle Calédonie. *C.N.R.S. Paris* 297(3):99-102.
Press, M.M. 1979. *Julian Tenison Woods.* Catholic Theological Faculty, Sydney.

Quilty, P.G. 1966. The age of Tasmanian marine Tertiary rocks. *Austral. Jour. Sci.* 29(5):143-144.
Quirk, S. and Archer, M. 1983. *Prehistoric Animals of Australia.* Austral. Mus. Sydney.

Rage, J.-C. 1974. Squamates. *In*: F. deBroin, E. Buffetaut, J.-C. Koniguer, J.-C. Rage, D. Russell, P. Taquet,

C. Vergnaud-Grassini and S. Wenz. La fauna de Vertébrés continentaux du gisement d'In Becetr (Sénonien du Niger). *C.r. hebd. séanc. Acad. Sci. Paris* 279:470-471.

Ranken, C.G. 1916. *The Rankens of Bathurst.* S.D. Townsend, Sydney.

Repetski, J.E. 1978. A fish from the Upper Cambrian of North America. *Science* 200:529-531.

Retallack, G.J. 1981. Late Carboniferous to middle Triassic megafossil floras from the Sydney Basin. *In*: C. Herbert, and R. Welby. A Guide to the Sydney Basin. *Geol. Surv. N.S.W. Bull.* 26:383-430.

Rich, P.V. 1979. The Dromornithidae, a family of large, extinct ground birds endemic to Australia. *Bur. Min. Res. Bull.* 184:1-196.

——, 1981. Feathered leviathans. *Hemisphere* 25(5):298-302.

——, and Baird, R.F. (in prep.). The palaelodids of Australia.

——, and Balouet, C. 1984. The waifs and strays of the bird world or the ratite problem revisited, one more time. *In*: M. Archer and G. Clayton, eds. 1984. *Vertebrate Zoogeography and Evolution in Australasia.* Hesperian Press, Carlisle: 447-455.

——, and Gill, E. 1976. Possible dromornithid footprints from Pleistocene dune sands of southern Victoria, Australia. *Emu* 76(4):221-223.

——, and Green, R.H. 1974. Bird footprints at South Mt. Cameron, Tasmania. *Emu* 74:245-248.

——, Long, J., Warren, A., Rich, T., Molnar, R., and Molnar, B. 1982. An all too brief and superficial history of Australian vertebrate paleontology. *In*: P.V. Rich and E.M. Thompson, eds., *The Fossil Vertebrate Record of Australasia*, Monash Univ. Offset Printing Unit, Clayton: 2-26.

——, and McEvey, A.R. 1977. A new owlet-nightjar from the early to mid-Miocene of eastern New South Wales. *Mem. Nat. Mus. Vict.* 38:247-253.

——, ——, and Baird, R.F. (in press). The Scrub-birds (*Atrichornis*) and Lyrebirds (*Menura*) of Australia: An osteological comparison. *Mem. Austral. Mus.*

——, and Scarlett, R.J. 1977. Another look at *Megaegotheles*, a large owlet-nightjar from New Zealand. *Emu* 77(1):1-8.

——, and Thompson, E.M. eds 1982. *The Fossil Vertebrate Record of Australasia.* Monash Univ. Offset Printing Unit, Clayton.

——, and van Tets, G.F. 1982. Fossil birds of Australia and New Guinea: their biogeographic, phylogenetic, and biostratigraphic input. *In*: P.V. Rich and E.M. Thompson, eds., *The Fossil Vertebrate Record of Australasia.* Monash Univ. Offset Printing Unit, Clayton: 235-384.

Rich, T.H. and Archer, M. 1979. *Namilamadeta snideri*, a new diprotodontan (Marsupialia, Vombatoidea) from the medial Miocene of South Australia. *Alcheringa* 3:197-208.

——, ——, Plane, M.D., Flannery, T.F., Pledge, N.S., Hand, S.J., and Rich, P.V. 1982. Australian Tertiary mammal localities. *In*: P.V. Rich and E.M. Thompson, eds., *The Fossil Vertebrate Record of Australasia*, Monash Univ. Offset Printing Unit, Clayton: 525-572.

——, and Hall, B. 1979. Rebuilding a giant. *Austral. Nat. Hist.*, 19(9):310-313.

Ride, W.D.L. 1964. A review of Australian fossil marsupials. *Jour. Proc. Roy. Soc. W. Austral.* 47:97-131.

Ritchie, A. 1969. Ancient fish of Australia. *Austral. Nat. Hist.* 16:218-223.

——, 1974. From Greenland's icy Mountains...a Detective story in stone. *Austral. Nat. Hist.* 18:28-35.

——, 1975. *Groenlandaspis* in Antarctica, Australia and Europe. *Nature*, 254:569-573.

——, and Gilbert-Tomlinson, J. 1977. First Ordovician vertebrates from the Southern Hemisphere. *Alcheringa* 1:351-368.

Rixon, A.E. 1976. *Fossil Animal Remains. Their Preparation and Conservation.* Athelone Press, Univ. London.

Roberts, J. 1981. Control mechanisms of Carboniferous brachiopod zones in eastern Australia. *Lethaia* 14: 123-134.

Rolfe, W.D.I. 1966. Phyllocarid crustacean fauna of European aspect from the Devonian of Western Australia. *Nature* 209:192.

Romer, A.S. and Lewis, A.D. 1959. A mounted skeleton of the giant plesiosaur *Kronosaurus. Breviora* 112:1-15.

——, and Parsons, T.S. 1977. *The Vertebrate Body.* 5th ed. W.B. Saunders and Co. Philadelphia.

Rosen, D.E., Forey, P.L., Gardiner, B.G., and Patterson, C. 1981. Lungfishes, tetrapods, palaeontology and plesiomorphy. *Bull. Amer. Mus. Nat. Hist.* 167:163-264.

Runnegar, B. 1979. Ecology of *Eurydesma* and the *Eurydesma* fauna, Permian of eastern Australia. *Alcheringa* 3:261-285.

Sanson, G.D. 1982. Evolution of feeding adaptations in fossil and Recent macropodids. *In*: P.V. Rich and E. M. Thompson, eds., *The Fossil Vertebrate Record of Australasia*, Monash University, Offset Printing Unit, Clayton: 489-506.

Scarlett, R.J. 1968. An owlet-nightjar from New Zealand. *Notornis* 15:254-266.

Schaller, G.B. 1972. *The Serengeti Lion.* University of Chicago Press, Chicago.

Scott, H.H. 1915. A monograph of *Nototherium tasmanicum*. *Rec. Geol. Surv.* Tas. 4:1-47.

Sharov, A.G. 1971. Novie letaooschchie reptilii iz Mezozoia Kazakhstana i Kirgizii. *Akad. Nauk SSSR, Paleo. Inst. Trudi* 130:104-113.

Sigé, B. 1968. Les Chiropteres du Miocene inferieur de Bourziques. I.-Etude systematique. *Palaeovertebrata* 1(3):65-133.

——, Hand, S.J., and Archer, M. 1982. An Australian Miocene *Brachipposideros* (Mammalia, Chiroptera) related to Miocene representations from France. *Palaeovertebrata* 12(5):149-172.

Simpson, G.G. 1932. The supposed association of dinosaurs with mammals of Tertiary type in Patagonia. *Amer. Mus. Novitates* 566.

——, 1933. A new fossil snake from the *Notostylops* beds of Patagonia. *Bull. Amer. Mus. Nat. Hist.* 67:1-22.

——, 1938. *Crossochelys*, Eocene horned turtle from Patagonia. *Bull. Amer. Mus. Nat. Hist.* 74:221-254.

——, 1946. Fossil penguins. *Bull. Amer. Mus. Nat. Hist.* 87(1):1-99.

——, 1957. Australian fossil penguins with remarks on penguin evolution and distribution. *Rec. S. Austral. Mus.* 13(1):51-70.

——, 1961. Historical zoogeography of Australian mammals. *Evolution* 15:431-436.

——, 1966. Mammalian evolution on the southern continents. *N. Jb. Geol. Palaont. Abh.* 125:1-18.

——, 1971a. A review of the pre-Pliocene penguins of New Zealand. *Bull Amer. Mus. Nat. Hist.* 144(5):319-378.

——, 1971b. Review of fossil penguins from Seymour Island. *Proc. Roy. Soc. London*, (B), 178(1053):357-387.

——, 1975. Fossil penguins. *In*: B. Stonehouse, ed., *The Biology of Penguins*, Macmillan, London and New York: 19-41.

——, 1976. *Penguins Past and Present, Here and There.* Yale University Press, New Haven and London.

Smith, A.G., Hurley, A.M. and Briden, J.C. 1981. *Phanerozoic Paleocontinental World Maps.* Cambridge Earth Science Series, Cambridge University Press, Cambridge.

Smith, J.L.B. 1958. *Old Fourlegs, the Story of the Coelacanth*: Pan Books, London.

Smith, M.J. 1976. Small fossil vertebrates from Victoria Cave, Naracoorte, South Australia. IV. Reptiles. *Trans. Roy. Soc. S. Austral.* 100(1):39-51.

Spencer, B. 1901. A description of *Wynyardia bassiana*, a fossil marsupial from the Tertiary beds of Table Cape, Tasmania. *Proc. Zool. Soc. London*: 776-794.

——, and Walcott, R.H. 1912. The origin of cuts on bones of Australian extinct marsupials. *Proc. Roy. Soc. Vict.* 24:92-133.

Stein, R.S. 1975. Dynamic analysis of *Pteranodon ingens*: a reptilian adaptation to flight. *Jour. Paleo.* 49:534-548.

Stensiö, E.A. 1927. *The Downtonian and Devonian Vertebrates of Spitsbergen. Part I, Family Chephalaspidae.* Skr. Svalband Nordishavet Result. Norske Statsun. Spitsbergeneksped. Olso 12.

——, 1948. On the Placodermi of the Upper Devonian of East Greenland. 2. Antiarchi: subfamily Bothriolepinae. *Palaeozoologica Groenlandica* 2:1-622.

Stirling, E.C. 1896. The newly discovered extinct, gigantic bird of South Australia. *Ibis* 7(11):593.

——, 1900. Physical features of Lake Callabonna. *Mem. Roy. Soc. S. Austral.* 1:i-xv.

——, 1913. Fossil remains of Lake Callabonna. Part IV. 1. Description of some further remains of *Genyornis newtoni*, Stirling and Zietz. 2. On the identity of *Phascolomys* (*Phascolonus*) *gigas*, Owen, and *Sceparnodon ramsayi*, Owen, with a description of some remains. *Mem. Roy. Soc. S. Austral.* 1:111-178.

——, and Zietz, A.H.C. 1896. Preliminary notes on *Genyornis newtoni*; a new genus and species of fossil struthious bird found at Lake Callabonna, South Australia. *Trans. Roy. Soc. S. Austral.* 20:171-190.

——, and ——, 1899. Fossil remains of Lake Callabonna. Part 1. Description of the manus and pes of *Diprotodon australis*, Owen. *Mem. Roy. Soc. S. Austral.* 1:1-40.

——, and ——, 1900. Fossil remains of Lake Callabonna. 1. *Genyornis newtoni.* A new genus and species of fossil struthious bird. *Mem. Roy. Soc. S. Austral.* 1(2):41-80.

Stirton, R.A. 1954. Digging down under. *Pacific Discovery* 7(2):3-13.

——, 1963. Fossil marsupials and drifting sands. *Austral. Nat. Hist.* 14(6):180-184.

——, 1967. The Diprotodontidae from the Ngapakaldi fauna, South Australia. *Bur. Min. Res. Bull* 85:1-44.

——, Tedford, R.H. and Miller, A.H. 1961. Cenozoic stratigraphy and vertebrate palaeontology of the Tirari Desert, South Australia. *Rec. S. Austral. Mus.* 14:19-61.

——, ——, and Woodburne, M.O. 1967. A new Tertiary formation and fauna from the Tirari Desert, South Australia. *Rec. S. Austral. Mus.* 15:427-462.

——, Woodburne, M.O., and Plane, M. 1967. A phylogeny of the Tertiary Diprotodontidae and its significance in correlation. *Bur. Min. Res. Bull.* 85:149-160.

Stonehouse, B. 1968. *Penguins.* Arthur Barker Ltd., London, and Golden Press, New York.

Strahan, R. 1979. *Rare and Curious Specimens.* The Austral. Mus. Sydney.

Tate, R. 1894. Unrecorded genera of the Older Tertiary fauna of Australia, including diagnoses of some new genera and species. *Jour. Proc. Roy. Soc. N.S.W.* 27:167-197.

Tedford, R.H. 1966. Fossil mammals from the Carl Creek Limestone, northwestern Queensland. *Bur. Min. Res. Bull.* 92:217-237.

——, 1967. The fossil Macropodidae from Lake Menindee, New South Wales. *Univ. Calif. Publ. Geol. Sci.* 64:1-165.

——, 1973. The diprotodons of Lake Callabonna. *Austral. Nat. Hist.* 17(11):349-354.

——, Archer, M., Bartholomai, A., Plane, M., Pledge, N.S., Rich, T., Rich, P., and Wells, R.T. 1977. The discovery of Miocene vertebrates, Lake Frome area, South Australia. *Bur. Min. Res. Jour. Aust. Geol. Geophys.* 2:53-57.

——, Banks, M.R., Kemp, N.R., McDougall, I., and Sutherland, F.C. 1975. Recognition of the oldest known fossil marsupials from Australia. *Nature* 255:41-42.

Thompson, A.L., ed. 1964. *A New Dictionary of Birds*. McGraw-Hill Book Co., New York.

Thulborn, R.A. 1983. A mammal-like reptile from Australia. *Nature* 303:330-331.

——, and Warren, A.A. 1980. Early Jurassic plesiosaurs from Australia. *Nature* 285:224-225.

Tindale, N.B. 1951. Comments on supposed representations of giant bird tracks at Pimba. *Rec. S. Austral. Mus.* 9:381-382.

Toombs, H.A. and Rixon, A.E. 1959. The use of acids in the preparation of vertebrate fossils. *Curator* 11:304-312.

Traynor, M.G., ed. 1983. *National Conference of Preparators and Technicians*. Mus. Victoria, Melbourne.

Troughton, E.P. 1973 (revised). *Furred Animals of Australia*. Angus and Robertson, Sydney.

Truswell, E.M. and Harris, W.K. 1982. The Cainozoic palaeobotanical record in arid Australia: fossil evidence for the origins of an arid-adapted flora. *In*: W.R. Barker, and P.J.M. Greenslade. *Evolution of the Flora and Fauna of Arid Australia*. Peacock Publications, Frewville: 67-76.

Turnbull, W.D. and Lundelius, E.L. 1970. The Hamilton fauna. A late Pliocene mammalian fauna from the Grange Burn, Victoria, Australia. *Fieldiana, Geol.* 19:1-163.

Vallance, T.G. 1975. Origins of Australian geology. *Proc. Linn. Soc. N.S.W.* 100:13-43.

——, 1978. Pioneers and leaders — a record of Australian palaeontology in the nineteenth century. *Alcheringa* 2:243-250.

van Tets, G.F. 1974. A revision of the fossil Megapodiidae (Aves), including a description of a new species of *Progura* De Vis. *Trans. Roy. Soc. S. Austral.* 98(4):213-224.

——, and Rich, P.V. (in prep). A giant eagle from the Pleistocene of Australia.

——, and Smith, M.J. 1974. Small fossil vertebrates from Victoria Cave, Naracoorte, South Australia. III. Birds (Aves). *Trans. Roy. Soc. S. Austral.* 98(4):225-227.

Walcott, C.D. 1892. Preliminary notes on the discovery of a vertebrate fauna in Silurian (Ordovician) strata. *Geol. Soc. Amer. Bull.* 3:153-172.

Walton, D.W. and Walton, G.M. 1970. Post-cranial osteology of bats. *In*: B.H. Slaughter and D.W. Walton, eds., *About Bats — a Chiropteran Symposium*. Southern Methodist Univ. Press; Dallas: 93-122.

Warren, A.A. 1977. Jurassic labyrinthodont. *Nature* 265:436-437.

——, 1981. A horned member of the labyrinthodont Superfamily Brachyopoidea from the Early Triassic of Queensland. *Alcheringa* 5:273-288.

——, and Hutchinson, M.N. 1983. The last labyrinthodont? A new brachopoid (Amphibia, Temnospondyli) from the early Jurassic Evergreen Formation of Queensland, Australia. *Phil. Trans. Roy. Soc. London.* (B). Biol. Sci. 303(1113):1-62.

Wegener, A. 1924. *The Origin of Continents and Oceans*. Dover, London.

Wellnhofer, P. 1978. Pterosauria. *In*: O. Kuhn and P. Wellnhofer, eds., *Handbuch der Paläoherpetologie*, 19. Gustave Fischer Verlag, Stuttgart.

Wells, R.T. 1974. Reconstructing the past. Excavations in fossil caves. *Austral. Nat. Hist.* 18(6):208-211.

——, Horton, D.R., and Rogers, P. 1982. *Thylacoleo carnifex* Owen (Thylacoleonidae): marsupial carnivore? *In*: M. Archer, ed., *Carnivorous Marsupials*. *Roy. Zool. Soc. N.S.W.* 2:576-586.

——, Moriarty, K., and Williams, D.L.G. 1984. The fossil vertebrate deposits of Victoria Fossil Cave Naracoorte: An introduction to the geology and fauna. *Aust. Zool.* 21(4):305-333.

——, and Nichol, B. 1977. On the manus and pes of *Thylacoleo carnifex*. Owen (Marsupialia). *Trans. Roy. Soc. S. Austral.* 101(6):139-146.

Welsby, T. 1905. *Schnappering and Fishing*. Outridge Printing Co., Brisbane.

Wetmore, A. 1967. Re-creating Madagascar's giant extinct bird. *Nat. Geog.* 132(4):488-493.

White, E.I. 1925. Two new fossil species of *Epiceratodus* from South Australia. *Ann. Mag. Nat. Hist.* 9(16):139-146.

——, 1952. Australian arthrodires. *Bull. Brit. Mus. (Nat. Hist.) (Geol.)* 1:249-304.

White, M.E. 1981. Revision of the Talbragar Fish Bed flora (Jurassic) of New South Wales. *Rec. Austral. Mus.* 33:695-721.

White, T.E. 1935. On the skull of *Kronosaurus queenslandicus* Longman. *Occ. Pap. Boston Soc. Nat. Hist.* 8:219-228.

Whitehouse, F.W. 1955. The geology of the Queensland portion of the Great Australian Artesian Basin. *Qd. Parl. Pap.* 1955(2):653-676.

Wiman, C. 1905. Über di alttertiären Vertebraten der Seymourinsel. *Wiss. Ergebn. schwed. Sudpolarexped.* 3(1):1-37.

Williams, D. 1981. *Genyornis* eggshell (Dromornithidae: Aves) from the Late Pleistocene of South Australia. *Alcheringa* 5(2):133-140.

Williams, I.S., Tetley, N.W., Compston, W., and McDougall, I. 1982. A comparison of K-Ar and Rb-Sr ages of rapidly cooled igneous rocks: two points in the Palaeozoic time scale re-evaluated. *Jour. Geol. Soc. London* 139:557-568.

Wood, H.E. 1924. The position of the 'sparassodonts': with notes on the relationships and history of the Marsupialia. *Bull. Amer. Mus. Nat. Hist.* 51:77-101.

Woodburne, M.O. and Clemens, W.A. (in press). A new genus of Ektopodontidae and additional comments on *Ektopodon serratus*. *Univ. Calif. Publ. Geol. Sci.*
——, and Tedford, R.H. 1975. The first Tertiary monotreme from Australia. *Amer. Mus. Novitates.* 2588:1-11.
Wood Jones, F. 1931. A re-examination of the skeletal characters of *Wynyardia bassiana*, an extinct Tasmanian marsupial. *Pap. Proc. Roy. Soc. Tas.* 1930:96-115.
Woods, J.T. 1958. The extinct marsupial genus *Palorchestes* Owen. *Mem. Qd Mus.* 13:177-193.
Woodward, A.S. 1888. Note on the extinct reptilian genera *Megalania* Owen and *Meiolania*, Owen. *Ann. Mag. Nat. Hist.* 6(1):185-189.
——, 1890. The fossil fishes of the Hawkesbury series at Gosford, New South Wales. *Mem. Geol. Surv. N.S.W.* 4:1-55.
——, 1908. The fossil fishes of the Hawkesbury series at St. Peters. *Mem. Geol. Surv. N.S.W.* 10:v,1-32.

Yalden, D.W. and Morris, P.A. 1975. *The Lives of Bats.* David and Charles: Newton Abbot.
Young, G.C. 1981. Biogeography of Devonian vertebrates. *Alcheringa* 5:225-271.
——, 1984. Reconstruction of the jaws and braincase in the Devonian placoderm fish *Bothriolepis. Palaeontology,* 27:635-661.
——, and Gorter, J.C. 1981. A new fish fauna of Middle Devonian age from the Taemas/Wee Jasper region of New South Wales. *Bur. Min Res. Bull.* 209:85-128.

Zinsmeister, W.J. 1982. Review of the Upper Cretaceous — Lower Tertiary sequence on Seymour Island, Antarctica. *Jour. Geol. Soc. London.* 139(B):779-785.

Index